Land Snails

of

Great Smoky Mountains National Park and Southern Appalachians

Tennessee & North Carolina

Daniel C. Dourson

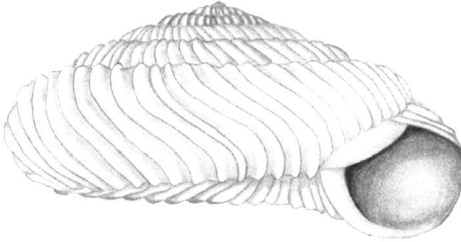

Important Contributions from
Keith Langdon & Judy Dourson

Illustrated by Rachael Grabowski

Snail Cartoons by A. B. Bobby J. Osborne

All photographs by Daniel Dourson

(unless otherwise stated)

Published by
Goatslug Publications
1128 Gritter Ridge Rd.
Stanton, KY 40380
Email: theroguebiologist@gmail.com

ISBN: 978-0-615-89696-0

Front Cover: The colorful shell of a mountain tigersnail is illuminated by the glow of a full moon.

DLiA
Discover Life In America

CUMBERLAND MOUNTAIN
CMRC
RESEARCH CENTER

LINCOLN MEMORIAL
U N I V E R S I T Y

Goatslug Publications
Bakersville, NC 28705

Contents

Plight of the Environment 1

Introduction 3

Map of study area 11

Map of the GSMNP 12

Prelude to the Land Snails 17

How to Collect Land Snails...31

Shell Morphology...33

Basic Anatomy of Land Snails 36

PART I

Land Snails of the Great Smoky Mountains National Park 38

Abundance Categories for Land Snails of the Park 39

Identifying land shells 41

Pictorial Key to the Land Snails 43

Species Accounts of Land Snails....................................49

    Shells Taller than Wide, Simple Lip...........................50

    Shells Taller than Wide, Reflected Lip.........................56

    Shells Wider than Tall, Simple Lip (with or without teeth) 76

    Shells Pill-shaped with Narrow Aperture 178

    Shells Wider than Tall, Reflected Lip 190

Native Slugs of the Great Smoky Mountains National Park 232

Beyond Identification of Gastropods......................................255

PART II

Land Snails of the Southern Appalachians (not recorded from the GSMNP) 258

Pictorial Key to Land Snails of the Southern Appalachians 259

Exotic Slugs..312

Native Aquatic Snails of the GSMNP 317

Non-toxic Invasive Land Snail Control....................................319

Glossary...320

Bibliography..325

Index of Scientific Names 330

Index of Common Names..333

Acknowledgments

This manuscript would not have been possible without the support and assistance of numerous individuals and organizations. I offer my sincere gratitude for the contributions that allowed me to move forward information about land snails, both historical and current.

A special thanks to Bobby Martin, of Martin Microscope and Cumberland Mountain Research Center at Lincoln Memorial University, whose generosity precipitated the use of the combined photographic set up of a Wild M420 Apozoom macroscope with a Jenoptik ProgRes C5 camera and i-Solution Lite software to create perfectly focused images of the micro-snails (under 5 mm). Many of the snails in this category had never been photographed and the only images available were simple line drawings, due in part to their diminutive size and the lack of equipment available to provide the proper depth of field; Jochen Gerber, Collections Manager and Rudiger Bieler, Curator of Invertebrates, Field Museum of Natural History, Chicago, Illinois for many specimen loans that were photographed and used in this book; John Slapcinsky and loans from the Florida Museum o f Natural History; Ronald S. Caldwell, Director of Cumberland Mountain Research Center at Lincoln Memorial University, for assistance in locating and obtaining pertinent research papers for the book. Acid deposition map by Jim Renfro, Air Resource Specialist with the GSMNP.

Collectors: Without the interest and diligence of the following individuals who have searched the Great Smoky Mountains landscapes for the obscure and often secretive land snails throughout the years, this book would not have been possible: Henry A. Pilsbry, Leslie Hubricht, John Slapcinsky, Ronald Caldwell, Mark Gumbert, Daniel Douglas and G. Thomas Watters. A special thanks to Amy and Wayne Van Devender for their numerous county records included in the NC maps. Their contributions greatly improved our knowledge of the land snails of the region. Thanks to Jeannie Hilton (past Executive Director of Discover Life in America) for organizing the Snail Blitz and the Karst Quest Citizen Science programs and assisting in snail survey work and Todd Witcher, current DLIA Director, for support of snail research through small grants and citizen science projects. Thanks to Paul Super and Susan Sachs of the Appalachian Highlands Science Learning Center for incorporating land snail collecting into summer programs and enthusiasm in assisting with my first documentation of the land snails of the park, an online publication, Land Snails of the Great Smoky Mountains National Park (Eastern Region). Thanks to the Smoky Mountains Institute at Tremont for including land snail surveys as part of the Citizen Science program. Thanks to Dr. Ernest Bernard and Penny Long of the University of Tennessee, for the specimens collected in a pitfall trap project in the park. Dr. Frank Anderson of Southern Illinois University provided records as well as Becky Keller, whose records of snails found at high elevations during thesis work are included; Dr. Harold Keller for documenting the presence of *Anguispira jessica* 23 m above the ground in the tops of a tree while surveying for slime molds. Finally, thanks to the following park personnel for their support and assistance for the past 12 years: Adrian Mayor, Becky Nichols, Chuck Parker, and Chuck Cooper.

Illustrations and Photo Contributions: Thanks to Rachael Grabowski for outstanding art work throughout the text, many redrawn from Pilsbry's four volume treatises and Burch's *How to Know the Eastern Land Snails* (1962); thanks to A. B. Bobby J. Osborne, Savannah School of Art and Design, for the cartoons throughout the book; thanks to Charles Wilder for the stunning images that capture the essence of the GSMNP photographed on numerous hikes and Steve Bohleber for the image of Mt. Le Conte in the fall; Jeffrey C. Nekola, Department of Biology, University of New Mexico, and Brian F. Coles, Mollusca Section, Department of Biodiversity, National Museum of Wales, United Kingdom for use of photos and illustrations from *Pupillid Land Snails of Eastern North America*; Jochen Gerber, photos of *Anguispira* species; Aleta Karstad, watercolor images of the slugs from *Identifying Land Snails and Slugs of Canada* by F. Wayne Grimm et al. (2010); Wayne Van Devender, Appalachian State University, for SEM images and photos; Jodi White-McLean for images of *Philomycus* love dart; Joris Koene for images of love darts; Rex Meredith, for excellent line drawing of the child and the salamander; George Boorujy for illustration of the great tit, and finally to Art Bogan, North Carolina Museum of Natural Sciences, Research Curator of Aquatic Invertebrates for his assistance in obtaining specimens of aquatic snails to photograph and his review and images of *Pleurocera canaliculatum* and *Leptoxis praerosa* on page 318.

Reviewers: I would like to express my deep gratitude to the following reviewers: Keith Langdon, who has spent a large part of his career working as Inventory and Monitoring Coordinator for the Great Smoky Mountains National Park. Keith's critique of the book and his knowledge of the park significantly improved the quality. Input from Keith on revisions to this new reprinting greatly improved the text.

Jochen Gerber, Field Museum of Natural History, Chicago for his willingness to devote extensive time and effort to the review of malacological content; John Slapcinsky, Florida Museum of Natural History, for devoting time to provide a candid review of the text; and finally, my wife, Judy Dourson, for field assistance, database, maps, literature searches, and editing.

Plight of the Environment

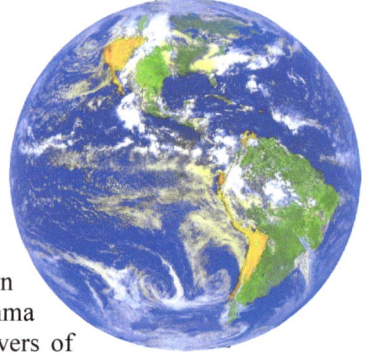

Given the state of the environment, I often ponder the future of the planet and the dilemma of our species. What were once pristine rivers of vitality are now gutters of spoil, oceans of prosperity turned to waters of poverty, and forests untainted by human activities are becoming but distant memories. Our world is not infinite but a place of boundaries. Clearly, our species has pushed many of these boundaries and beyond. Our ever-increasing demand for Earth's limited resources has led us down a dangerous pathway, tearing away the delicate strata of life on which we all depend. In an age of enlightenment, why are we still losing ground on which we stand? Many blame corporate and government entities but blame does not fall on the heads of a few.

In the last century, we left our agrarian lifestyle behind, trading our kinship to the land for a lifeless relationship with technology. It is a perplexing time for us all, living in a world of machines but still entirely dependent on the environment. Spending much of our time trapped behind walls of concrete and steel, we have become desensitized to a planet in peril. While we may be better aware of the plight, our species turns a blind eye and continues on a reckless course of consumerism, exhausting the planet's finite resources. The transfer of this uncertainty to our children has implications that are far worse. According to Richard Louv, author of "Last Child in the Woods," an exceptional book about our human disconnect to the natural world, the lack of connectivity to planet Earth has created a generation that may well be suffering a non-medical condition aptly labeled "nature-deficit disorder".

Environmental ethics is not a way to imagine, it's a way of life, a life that honors earth and its inhabitants—a life that considers no less than the whole. The decisions we make every day, singly or collectively, have far-reaching and often irreversible consequences to the land— land that our progeny and fellow creatures will inherit. If we are to protect the things around us, we must all think in terms of how our every action, from flushing our toilets to the foods we eat, affects our planet and the life force within it. Becoming better stewards of the land, air, and water makes sense, regardless of climate change. Caring for the earth, our only home, is everybody's responsibility. If we fail as competent caretakers of our native soil, all that follows including our children, will surely suffer the consequences.

Sustainable Development

The term "sustainable development" promotes the contentious idea that Earth's finite resources can be sustained indefinitely. But can long term development of limited resources really be sustainable? An excerpt from the document, *Global Diversity Strategies* has this to say; "Development has to be both people-centered and conservation based. Unless we protect the structure, function, and diversity of the world's natural systems—on which our species and all others depend—development will undermine itself and fail. Unless we use our Earth's resources sustainably and prudently, we deny people their future. Development must not come at the expense of other groups or later generations, nor threaten other species' survival (WRI 1992)".

The world as seen through the eyes of a child

~Introduction~

The Great Smoky Mountains National Park (GSMNP) is an outstanding illustration of undomesticated environment, unlike anything else in North America. The Park encompasses 211,246 hectares in area—over 2,100 square kilometers or around 800 square miles—safeguarding one of the finest examples of unbroken temperate forest left on planet Earth. It has been designated both a World Heritage Site and International Biosphere Reserve. The park is part of the Blue Ridge Province of the Southern Appalachians where elevations can range from a low 250 m in elevation in yawning valleys to magnificent jade-colored peaks reaching heights of more than 2000 meters.

The wide range of elevation coupled with diverse forest and soil types favor a wealth of biodiversity. The flora and fauna have remained largely intact in these primordial mountains and several taxa assemblages (e.g. salamanders and slime molds), reach their greatest diversity found anywhere. More than 100 species of trees thrive within park boundaries with an additional 1500 flowering plants. There are approximately 200 species of birds, 66 species of mammals, 50 species of fish, 39 species of reptiles and 43 species of amphibians, 30 of which are salamanders. Millipedes, mushrooms and bryophytes reach record diversity in the park as well. There are currently 146 native land snail species reported from the GSMNP. This represents roughly 28 percent of the total land snail fauna of eastern North America and more species than occur in the states of Ohio, Indiana and Illinois combined. Of the 146 species, 10 land snails and forms are restricted to the park.

Past work by malacologists Henry Pilsbry, John B. Burch and Leslie Hubricht established the fundamental foundation for the study of land snails in the eastern United States, including the Great Smoky Mountains National Park. Pilsbry compiled previous literature and described new taxa with his comprehensive four volume treatises. Burch provided keys and Hubricht contributed much-needed range maps. Although these fine works were comprehensive at the time, since then, numerous forms have been elevated to species and there have been a number of taxon revisions. Moreover, remote and inaccessible areas of the southern Appalachian Mountains (including the GSMNP) continue to harbor undiscovered snails, as evidenced by the recent descriptions of six new species from mountainous regions of Tennessee and North Carolina (Slapcinsky and Coles 2004; Dourson 2012).

Prior to the Discover Life in America's **All Taxa Biodiversity Inventory** (ATBI), precious little was known of the park's remarkable land snail diversity. Historical records preceding the (1998) ATBI project included around 91 species. Since then 55 species have been added, including two land snails unknown to science (Dourson 2012). While the species diversity of land snails in the park is well documented, there is a dearth of basic biological information for most species.

Land Snails of the Great Smoky Mountains National Park and Southern Appalachians is presented in two parts, **Part I** specifically covers 146 land snails; 27 species with restricted distributions surrounding the park. **Part II** expands its reaches beyond park boundaries to include an additional 44 species found in the Southern Appalachian Mountains from around Mount Rogers, Virginia, south to Chilhowee Mountain, Tennessee (see map on page 11). In total, 203 land snail species are covered in this text; an astonishing **54 species endemic to the region.** Also illustrated are 15 aquatic snails found around or in the GSMNP. The book is fully illustrated with hundreds of color photographs and drawings. A pictorial key for both parts is included and was designed for both the beginner and advanced malacologist alike. The book includes interesting facts about land snail ecology such as diet, reproduction, defense strategies, parasites, overall benefits and harmful effects on wildlife and humans. Also included is a comprehensive section devoted to exotic slugs.

Environment and Land Snail Distribution

Without question, calcium carbonate is an essential mineral to land snails for regulation of bodily processes, reproduction, but most importantly, shell-building (Burch 1962; Fournie et al. 1984; Hickman et al. 2003). Land snails obtain calcium in several ways including consuming soil particles from calcareous substrates, eating decaying leaf matter (Wareborn 1970; Burch and Pearce 1990; Nation 2005), almost certainly by ingesting Physarales slime molds (which precipitate amorphous calcium carbonate) and gleaning calcium from the shells and bones of deceased animals. *Triodopsis platysayoides* (an endemic West Virginia land snail) have been documented feeding on the vacant shells of land snails including its own kind (Dourson 2008).

Land snail abundance (number of shells) and land snail diversity (number of species) have long been associated with a variety of geological and ecological factors. Studies have shown for example, that terrestrial gastropods living around carbonate cliffs can exhibit large and diverse populations (Nekola 1999) but show significant declines in abundance in as little as 50 m from a calcareous source (Kalisz and Powell 2003) or limestone cliffline (Dourson 2007). Other research has demonstrated while limestone may impact abundance, it has little affect on diversity. Land snail scarcity has reported associations with low soil pH (Burch 1955), declining soil cations, specifically calcium (Petranka 1982), increasing coniferous presence (Jacot 1935; Karlin

1961), and increasing elevation (Petranka 1982). The influence of pH on land snails is thought to be indirect, the main effect of a low Ph being a lowering of the amounts of soil cations, principally calcium (Karlin 1961; Cameron 1970). But low abundance in non calcareous (acidic) areas may only give the illusion of low diversity. Douglas (2011) found that land snail diversity on acid soils covered by old growth forests at Lilley Cornett Woods in Kentucky were analogous to limestone soils at Floracliff Nature Sanctuary along the Kentucky River Palisades. In the Southern Appalachians, geological sources of calcium occur in limestone, dolomite, calcareous shale, sandstone and the igneous-derived amphibolite, all which have calcium and/or Mg in varying amounts, supplying the necessary calcium for shell building. But what about land snails thriving on non-calcareous substrate like heath balds? Where and how these land snails obtain sources of calcium are poorly understood. In regions lacking calcareous substrates, land snails may rely on abscissed leaves of deciduous trees or herbaceous plants (e.g. stinging nettle) as a primary source for calcium.

McHargue and Roy (1932) in a study of several species of deciduous forest trees found that the amount of calcium in leaves expressed as a percentage of dry weight ranged from 1.64 to 7.8%, with the higher values occurring towards the end of the growing season. Other studies have shown that unlike other macro-nutrients that are reabsorbed by trees prior to leaf abscission, foliar calcium concentrations in deciduous trees increase throughout the growing season and peak at senescence (Guha and Mitchell 1966; Potter et al. 1987). Calcium is a relatively immobile nutrient and the re-absorption of calcium may not be as high a priority to deciduous species as other nutrients. Further, Gosz et al. (1973) reported that dead birch leaves could form a significant pool of calcium on the forest floor, since the concentrations of calcium remained high in dead leaves 12 months after abscission. Vesterdal and Raulund-Rasmussen (1998) found that the nutrient content of the forest floor under a single species of tree varied with the soil type, but they also found variation between different species on the same soil type. The idea that certain deciduous species contain higher levels of foliar calcium than others was supported by Arthur et al. (1993), who reported that yellow birch had relatively high levels of foliar calcium. Gosz et al. (1972) found foliar calcium concentrations were higher in yellow birch than in maple or beech. Ricklefs and Matthews (1982) looked at the leaf chemistry of 34 species of broad leaved deciduous trees and found that yellow birch had higher than average calcium concentrations. Jenkins (2007) found that calcium levels in dogwood leaves growing in the GSMNP were the most significant source of calcium in acidic forests, although many have since died due to dog-wood anthracnose.

Additional factors such as gradient (slope), litter moisture, elevation and microhabitat (leaf litter, moss and logs) can significantly affect the presence or absence of land snails. Coney et al. (1982) found more species of land snails on steep slopes than on more moderate ones. Petranka (1982) found that 15 of the 56 land snail species found on Black Mountain, Kentucky showed some preference for slope, with 9 species showing an affinity for increasing slope. The importance of leaf litter moisture (thought to be a factor of slope) to land snails was emphasized by Boycott (1934), Getz (1974), Pollard (1975), and others. Although aspect was reported to markedly affect microclimate (Geiger 1965; Braun 1940), Petranka's (1982) study found no environmental variable to be significantly correlated with aspect. With respect to elevation, Petranka (1982) reported that pH, potassium, calcium, and magnesium levels would decrease with increasing elevation and that the number of snail species and the number of individuals found per site would also decrease with elevation. In a study by Coney et al. (1982) conducted in the Hiwassee River Basin of Tennessee, the most important environmental factors influencing the presence or absence of land snail species was microhabitat (leaf litter, moss and logs, $P<0.05$ for 27 species), followed in decreasing order of importance by slope ($P<0.05$ for 15 species), rock type ($P<0.05$ for 13 species), stages of forest succession ($P<0.05$ for 12 species), soil pH ($P<0.05$ for 8 species), elevation ($P<0.05$ for 7 species) and soil moisture ($P<0.05$ for 6 species).

Less well-known is how the convergence of large physiographic and geophysical landscape edges serve to bridge distinctive regions and their allied terrestrial gastropod communities (Dourson 2007; Dourson and Beverly 2009; Douglas 2011). Neighboring land masses and geology may, in effect, be the driving force behind the distribution and mixing of some snail faunas, acting as travel corridors for dispersal that results in remarkably high land snail diversity in comparatively small places (Dourson 2007).

In the early 1900's, Pilsbry considered the broad Valley and Ridge Provinces of eastern Tennessee to be a physiographic barrier to the intermingling of land snails from the Southern Appalachians (including the GSMNP) with snails found on Pine and Black Mountain, Kentucky. Collections since by Branson and Batch (1968, 1988) and Petranka (1982), have ameliorated the barrier concept, revealing some interesting biogeographical associations. The land snails of the southeastern mountains of Kentucky have more affinity with land snails occurring in the Great Smoky Mountain-Blue Ridge physiographic sections of the Cumberland Province, with some divisions of the snail fauna showing relationships with Virginia and West Virginia. This view was further supported by Hubricht et al. (1983), whose land snail collections in southeastern Kentucky showed further evidence of the region's faunal affinities to the Ridge and Valley and Blue Ridge Physiographic Provinces. A recent snail inventory at Bad Branch Nature Preserve in Letcher County and Breaks Interstates Park in Pike County Kentucky demonstrated additional support for these biogeographical affinities, adding one new species, *Paravitrea lamellidens,* previously thought to be endemic to the GSMNP area. Of the 64 species found at Bad Branch, approximately eleven were representative of southern mountain ranges, eight representative with West Virginia and Virginia and nine have affinities with more northerly faunas (Dourson and Beverly 2009). Breaks Interstate Park with its copious land snail fauna (81 species), demonstrates still another great gastropod mixing pot (Dourson and Beverly, unpublished data).

Both Bad Branch State Nature Preserve and Breaks Interstate Park are positioned within great physiographic and geophysical landscape edges, located along the Pine Mountain massif of southeastern Kentucky, bordered to the south by the Valley and Ridge Provinces and to the north by the Cumberland

Plateau. All three regions contain their own snail affiliations. Moreover, the amalgamation of these snail rich ecoregions provided a number of terrestrial gastropods an opportunity to coexist. This large landscape edge, together with the wide range of soil types, elevation and slope gradient has brought together a fascinating assemblage of species not frequently occurring in the state of Kentucky. The Central Knobstone Escarpment in Powell County, Kentucky forms another large physiographic and geophysical landscape edge on Furnace Mountain where the Cumberland Plateau, the Knobs, and the Outer Bluegrass regions of Kentucky converge. The merging of these distinct regions was also shown to harbor an exceptional number of land snail taxa, a reported 61 species co-existing within a 5 acre mesic hillside (Dourson 2007). Interestingly, a number of the snails documented at Furnace Mountain were established well beyond their eastern and western limits in the state (Branson 1973; Hubricht 1985; Branson and Batch 1968). While the majority of land snails found at Furnace Mountain study were by and large common and wide ranging species, the snails of Pine Mountain showed a much tighter affiliation with a particular region (e.g. Great Smoky Mountains National Park) and were generally rarer land snail species.

High Elevation Forests in the GSMNP and Land Snails
A study on land snails comparing five high elevation forest types (northern hardwood, spruce/fir, beech gaps, heath balds and grassy balds) in the GSMNP (figure 1, page 8) found that northern hardwood forests were the most speciose (Dourson and Langdon 2012). Yellow birch, a common component of northern hardwood forests may be an important factor, providing land snails the necessary calcium from abscised leaves. Aged leaves of yellow and sweet birch represented a significant portion of the diet of *Triodopsis platysayoides* (Dourson 2008), a land snail restricted to acidic sandstone substrate in the Cheat River Gorge of West Virginia. *Discus macclintocki,* a snail of algific talus slopes in Iowa, has a diet of yellow birch, maple, and dogwood leaves (Nekola 1999).

Surprisingly, land snails were found in high numbers in heath balds but the source of calcium in these acidic (primarily rhododendron), environments remains unknown. One heath bald harbored 10 species, rivaling northern hardwood sites in terms of species richness, and exceeding northern hardwoods sites in terms of numbers of individuals. It was previously speculated that heath balds were poor snail habitat, a result of highly acidic soils, but this theory is now in question. One explanation might exist in the composition of species found. More than half the shells were that of *V. latissimus*, a semi-slug with a largely proteinaceous shell which requires less calcium carbonate for shell building, allowing it to thrive in more acidic environments. This would explain its relative abundance in heath balds, but there were respectable numbers of calcareous shelled snails as well, including *Stenotrema altispira* and *Mesomphix subplanus*. There are other groups of snails (e.g. *Glyphyalinia*) reported to thrive in calcium poor habitats (Nekola 2008), but the source of calcium within acidic environments such as heath balds, specifically the foods in which land snails sequester these calcium sources from, remains unknown; this awaiting further investigation.

Spruce/fir and beech gap forests were nearly equal in terms of snail diversity and abundance (figure 1), although studies have shown that coniferous forests are generally snail poor when compared to deciduous forests (Jacot 1935; Karlin 1961). Beech gaps, however, tend to be beech monocultures and calcium content of beech tree leaves is low when compared to yellow birch or flowering dogwood (Lutz & Chandler 1946, Nation 2005). Coney et al. (1982) found that as beech increased in importance, snail abundance decreased. Other research has shown that leaf litter of beech had the lowest calcium content of hardwood species (Lutz & Chandler 1946) and only species of pines are reported to contain less calcium than beech. The least productive site sampled was a grassy bald which yielded only 3 species. Grassy balds are thought to contain few species because of acidic soils. Another limiting factor may be the thickly-woven grass and sedges characteristic of grassy balds which may act as a deterrent to large diameter species, restricting their movement.

Within park boundaries the total land snail fauna documented above 1400 meters stands at forty-nine taxa (Dourson and Langdon 2012), several species endemic and restricted to increasing attitude. High elevation forests act like islands, trapping these gastropods in cooler environs that are separated from each other by warmer river valleys. As a result these snails have remained anchored to high Appalachian peaks, unable to move and exchange their unique genetic material. These ecosystems have more than likely been the most important forces in engineering new species in the southern mountains. One ill-fated disadvantage of being tied to altitude however is that these gastropods will likely be at most risk from climate change and high elevation acid deposition (see map on page 9).

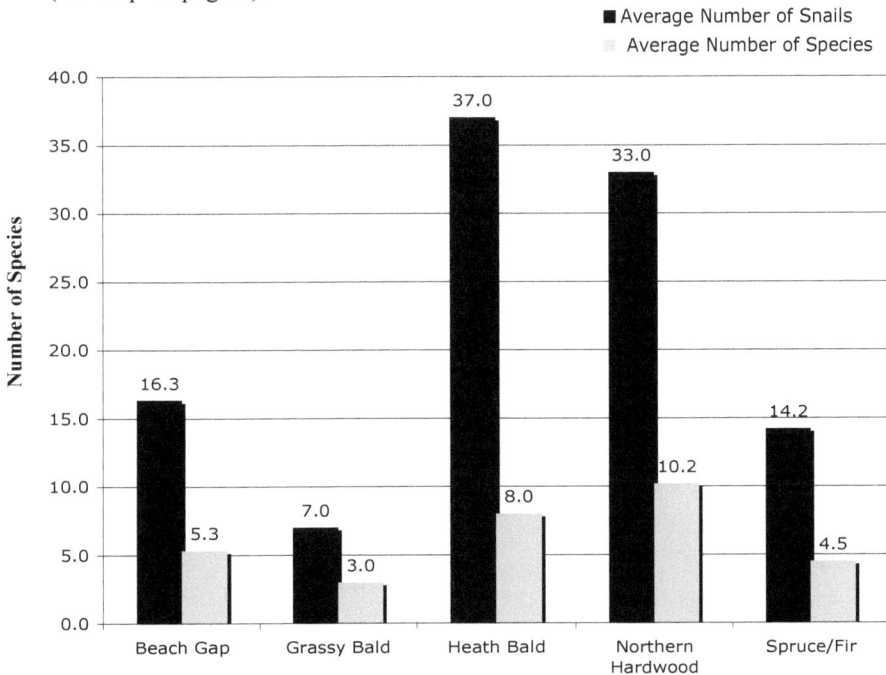

Figure 1. **Average number of snails and species in 24, 30 m^2 plots, at elevations greater than 1400 (4500 feet) meters in the GSMNP (Dourson and Langdon 2012).**

Mid to Lower Elevation Forests in the GSMNP and Land Snails

Mid to lower elevation cove hardwood forests in the GSMNP usually harbor greater species diversity and abundance than do the higher climes, especially if limestone is added to the mix. Cades Cove, for example, is rich in calcareous-rock formations, resulting in greater diversity that exceed all other sampled habitats of comparable size found within the park. Fifty-eight species were documented in White Oak Sinks, in an area of around two hectares (Douglas et. al 2013). Another exceptional site is located in a talus slide above Forney Creek where thirty-eight species, including the newly described *Carychium arboreum* (Dourson 2012) were documented in a quarter acre (slightly calcareous) rocky hillside (Dourson and Gumbert 2004). Most sites encompassing a few acres in the Southern Appalachian Mountains however will typically contain around twelve species, with richer sites holding twenty to twenty-five species. The poorest land snail positions in the park are clearly the pine stands. These areas offer little for snails and few species other than *Ventridens, Glyphyalinia* and *Zonitoides* live in these highly acidic soils. Finding a half dozen or so species in habitats dominated by pine, is indeed a gratifying number. A few gastropods have been found in park caves, the majority of which are washed in through torrent events. Surprisingly, no troglobitic (cave) snails have been discovered during collection trips into these under-worlds.

Modeled Acid Deposition (Nitrogen and Sulfur) -- 1999-2003 Average

Great Smoky Mountains National Park (NC/TN)

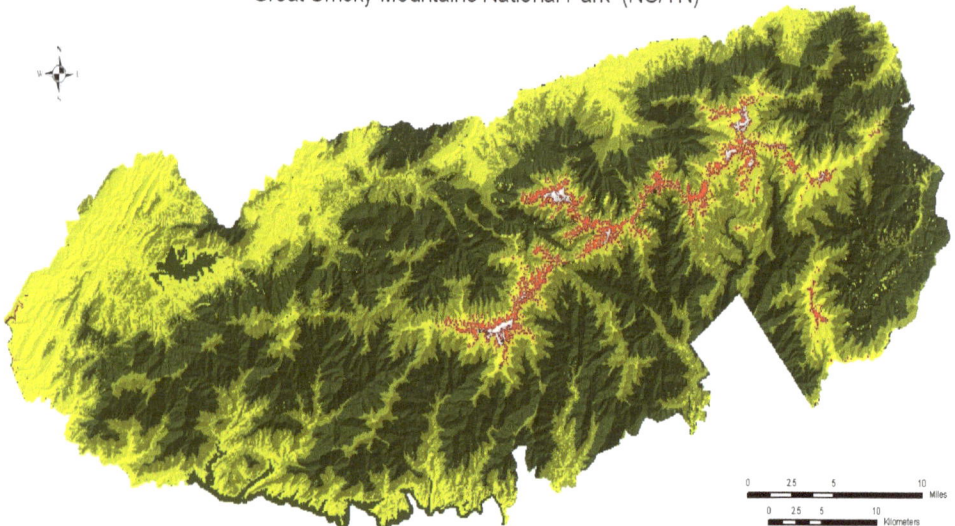

Annual Total Deposition (kg/ha/yr)

Sulfur		Nitrogen
0 - 13.3		0 - 12.0
13.3 - 16.6		12.0 - 15.0
16.6 - 24.9		15.0 - 22.5
24.9 - 35.3		22.5 - 32.0
35.3 - 49.7		32.0 - 45.0

Twin Creeks Natural
Resources Center GIS
Great Smoky Mts. National Park
Richard Schulz, June 30, 2006

9

Specific Land Snail Habitat

There is scarcely a place where one cannot find at least a few examples of land snails, even in your own backyard. The most successful searches begin by turning over leaf litter; the more you uncover, the greater number of gastropods you'll find, especially around outcrops of north-facing limestone (a). Using a three or four prong garden hand-rake works best and saves your finger tips from the sharp con-veyers of pain (thorns and bro-ken glass) and the occasional leaf-covered copperhead. Tree crotches (b) that contain abun-dant litter-fill will often support large numbers and diverse populations of small species, upwards to fourteen taxa. The base of large diameter trees such as black walnut, butternut, bass-wood, buckeye and birch also appear particularly rich (pers. obs.). Prime real estate for na-tive slugs is found in natural forests under exfoliating bark of rotting hardwood logs (c) in advanced stages of decay, par-ticularly those logs that form log-bridges over small ravines and streams.

Map of Area

The scope of this book covers the Great Smoky Mountains National Park entirely and portions of the Southern Appalachians of Tennessee and North Carolina occurring within the shaded-dashed circle.

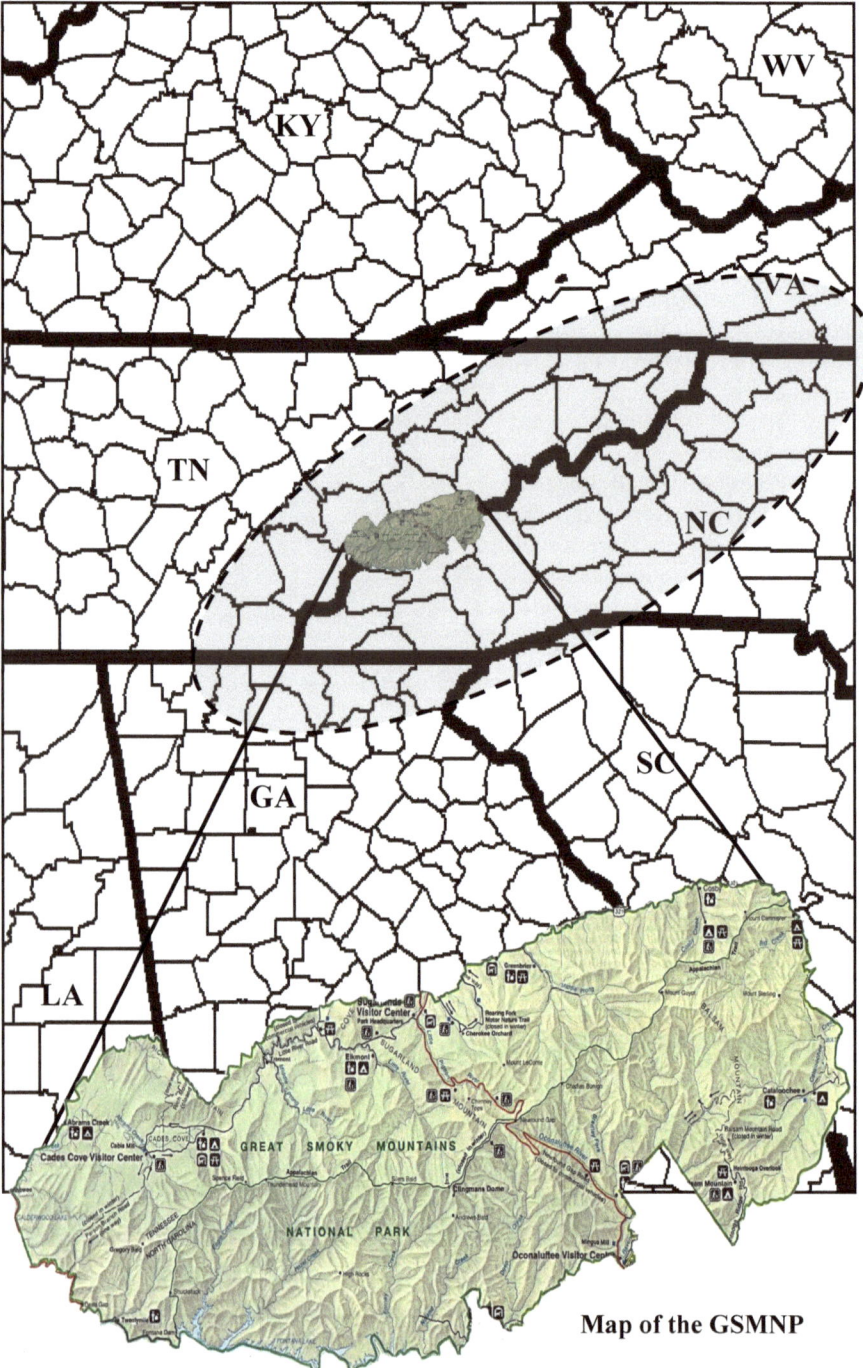

Map of the GSMNP

Map of Great Smoky Mountains National Park

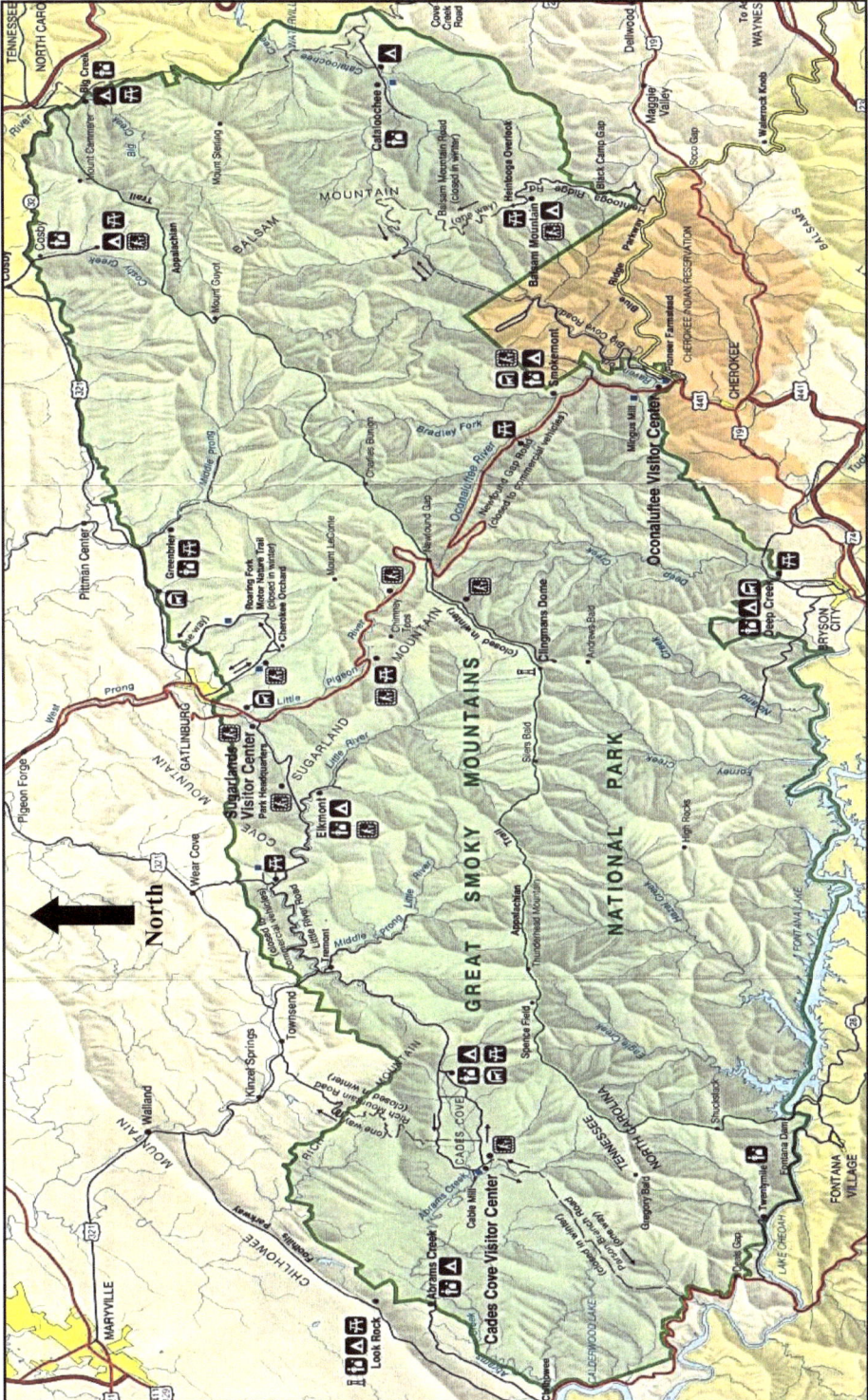

North

Map of the GSMNP, courtesy of the Great Smoky Mountains National Park

Essence of the Great Smoky Mountains National Park
Images by Charles Wilder

Early spring in the Smoky Mountains

Mouse Creek Falls

The Chimneys

The Smoky Mountains

14

Henry Whitehead Homestead

Black bellied Salamander

Prologue to the
Land Snails

Land snails come in every color, form and size imaginable; many beyond belief. Take, for example, terrestrial gastropods in the genus *Opisthostoma* (three species illustrated above). These eye-catching gems have taken calcium carbonate sculpturing to extremes. No less complicated than a Roman Cathedral, their architecture represents the most spectacular achievements in convoluted evolution ever seen in a land snail. The genus is known only from limestone outcrops and caves of Borneo. (Illustrations by Jaap Vermeulen).

Land snails are part of a large and diverse group of organisms known as Gastropods, belonging to the enormously diverse Phylum Mollusca. Containing nearly 100,000 described species world-wide, mollusks include aquatic snails (both marine and freshwater), land snails, terrestrial slugs, sea-slugs, sea-hairs, limpets, bivalves (clams, oysters, and mussels), squids, octopuses, and the famous nautiluses. Octopuses are considered the most highly evolved of the mollusks having feet divided into a number of prehensile and skillful tentacles capable of twisting the lid off a glass jar to access the food within. Although snails and slugs are slow moving, squids are among the fastest animals known, exceeding underwater speeds of more than 70 mph in short bursts, a result of their water-propelled jet propulsion. At 18 meters long and weighing in at around two tons, the giant squid is the largest living invertebrate and have the biggest eyes of any animal, around 15 inches. Mollusks are also the longest lived multi-cell animals known, one species of bivalve, the ocean quahog, *Arctica islandica* living more than 500 years.

There is scarcely a place on the dry surface of the world, outside the polar regions, where one cannot find at least a few examples of gastropods. (Abbott 1989). From hot, nearly waterless deserts to cold mountain tops, land snails are thriving. Despite their biological and physiological limitations, land snails have developed efficient mechanisms for coping with freezing, starvation, and desiccation. For example, when conditions become increasingly dry, snails cover the aperture of their shells with an epiphragm, a mucous sheet that hardens, sealing in critical moisture and slowing desiccation. Some snails can remain dormant for years, resuming activity during wet weather. Land snails in North America are typically nocturnal and are most active in wet weather when temperatures are between 50-75 degrees Fahrenheit.

Even though mollusks rank as one of the most numerous and speciose groups of organisms on Earth, they remain largely unstudied. As a result, little is known of their importance's in many ecosystems and land snails, like most invertebrates, suffer from being a conservation "blind-spot". As snail research moves forward, however, our understanding of the value of these organisms is increasing. Research has shown, for example, that land snails play an important role in micronutrient cycling in terrestrial ecosystems (Dallinger et al. 2000), disperse plant seeds and fungal spores (Richter 1980) and have been shown to predict vertebrate conservation priorities (Moritz et al. 2001). Further, live snails and their vacant shells provide a food and calcium carbonate source to many systematic groups. These include but are not limited to ants, firefly larva, snail-killing flies (Foote 1959); *Cychrine* beetles, which feed chiefly on land snails (Symondson 2004), Arachnids including harvestmen, carnivorous snails, numerous species of salamanders (Petranka 1998), turtles and frogs (Burch 1962), a variety of small mammals including shrews, mice, and moles (Reid 2006), snakes (Lee 1994, Dourson 2012), a variety of passerine birds (Graveland et al. 1994; Graveland 1996), thrushes, ruffed grouse and wild turkey (Martin et al. 1951), bats, (Bonato et al. 2004; Thabah et al. 2007) and primates including humans.

While a building body of evidence suggests the importance of mollusks in present-day ecosystems, their historical value is less well known; namely their contributions made to existing plant communities, animals and, in particular, caves. In the past, the colossal accumulation of deceased mollusks, corals, and tiny creatures known as Foraminifera (that have calcareous skeletons), provided the necessary building material to create limestone, where caves are essentially formed. Many species uniquely adapted to caves roost or otherwise live in these vast underworlds, a number of these species occurring nowhere else. These ancient shells have also provided the necessary limestone in cement to form the foundations of our cities and homes.

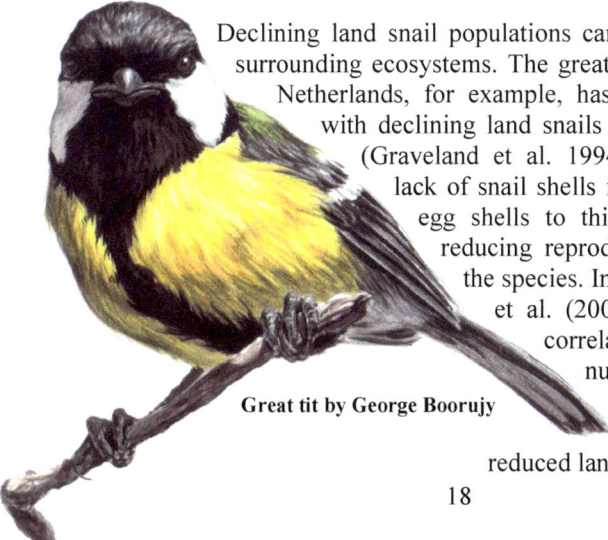

Declining land snail populations can have ripple effects to surrounding ecosystems. The great tit, *Parus major* in the Netherlands, for example, has declined precipitously with declining land snails as a result of acid rain (Graveland et al. 1994; Graveland 1996). A lack of snail shells in the bird's diet causes egg shells to thin and break, therefore reducing reproductive success rates of the species. In North America, Hames et al. (2002) have documented a correlation between a reduced number of wood thrushes and acid rain, hypothesizing a connection to reduced land snail populations.

Great tit by George Boorujy

18

Sensitive to changes in the environment, native land snails could provide an early warning to impending habitat deterioration, similar to the way that fresh-water mussels found in streams and rivers are used to determine the quality of waterways. For example, research has shown that when snails feed on various foods such as mushrooms, green vegetation or forest litter (detritus), environmental contaminants present are ingested and sequestered in their tissues (Dallinger and Wieser 1984a), the midgut gland being the main accumulation site of these trace elements (Dallinger 1993). Further laboratory experiments by Dallinger and Wieser (1984) have shown that land snails that ate lettuce laced with zinc, cadmium, lead, and copper readily sequestered these elements. More concerning though, snails quickly become poisoned when simply raised on soils contaminated with cadmium, raising fear that toxins in polluted soils may be more bio-active than previously believed (Scheifler et al. 2003). The environmental concerns to ecosystems and the consequence to snails and higher organisms that feed on contaminated gastropods are valid. Rimmer et al. (2005) found elevated levels of mercury in the blood of Bicknell's thrush on mountains in New England and thrushes are reported to eat snails (Martin et al. 1951), presumably to obtain calcium for egg laying. Native land snails could, therefore, be used to forecast impending problems created by anthropogenic pollutants reported to be accumulating in forests of the GSMNP. This hypothesis however remains untested in the park.

Parasites in Land Snails

Although nearly every kind of mollusk is inhabited by some form of parasite, only a few gastropods are of medical or veterinary importance (Burch 1962). Of these, almost all live in fresh water environments. Snails are required hosts in the life cycles of parasitic trematode worms. A few land snails such as *Cochlicopa lubrica* are vectors of lancet liver flukes in sheep, cattle, deer, and groundhogs (Burch 1962). *Zonitoides arboreus* and *Anguispira alternata*, both native land snails in the GSMNP, are implicated in the spread of lungworms in domestic sheep (Burch 1962). Multiple species of land snails and white-tailed deer play a significant role in the transmission of a parasitic nematode known as brainworm, (*Parelaphostrongylus tenuis*). The initial host of the nematode is the white-tailed deer and the intermediate hosts are several species of snails and slugs. Interestingly, brainworm nematodes do not appear to significantly affect white-tailed deer populations yet may cause debilitation or even fatality in elk, moose, goats, and sheep. The life cycle of the nematode begins with adult worms normally located between the membranes (meninges) that cover both the brain and in the spinal cord of the white-tailed deer. Eggs are deposited either on these membranes or directly into blood vessels. Those deposited on the membranes hatch and the larvae enter small blood vessels to be carried to the lungs where they enter the alveoli. Eggs deposited into blood vessels are carried to the lungs and eventually hatch with larval penetration of the alveoli. Activity in the lung tissue produces an interstitial pneumonia. The larvae pass up the respiratory tract from the alveoli, are swallowed and then eliminated in the feces of the deer. Larvae appear in the feces about three months after the host becomes infected. The larvae then enters into a snail or slug (the intermediate host) through the gastropods foot while crawling over infected deer feces or direct ingestion of deer feces by the gastropod. Development of the larvae in the gastropod to a stage when they are infective to the vertebrate host takes about three weeks. Cervids become parasitized by ingesting infected gastropods that are clinging to grass or other browse foods.

In the final host, other cervids, development of the larva to the adult worm takes place in tissues of the central nervous system, particularly the spinal cord. Parasites leave the tissues of the spinal cord after about 20-40 days and locate between the spinal membranes where they mature. Subsequently, they tend to accumulate in the cranial region. The adult worms are about 50 mm in length and may be seen fairly readily when free in the cranial cavity. From one to 20 worms have been found in the crania of infected deer, yet as previously stated, *P. tenuis* seldom causes damage in white-tailed deer. In other cervids, there is often extensive damage to tissues of the brain and spinal cord. The resulting neurologic disease is characterized by weakness, fearlessness, lack of coordination of movement, circling, deafness, impaired vision, paralysis and subsequent death. When in moose this disease is often called "moose sickness" or, "moose disease". A correlation to prevalence of the disease in elk, moose, and other cervids has been linked to an increase in population of the white-tailed deer. In Canada, studies by Anderson and Prestwood (1981) have substantiated the importance of this problem in management of big game and have given some indication of the dynamics of the host/parasite relationships among wild populations. Land snail species that have been implicated in the spread of brainworm infestation include: *Anguispira alternata, Deroceras leave* (a slug), *Deroceras reticulatum* (a slug), *Discus catskillensis, Euchemotrema fraternum, Philomycus carolinianus* (a slug), *Neohelix albolabris, Pallifera dorsalis* (a slug), *Striatura exiguum, Triodopsis tridentata, Ventridens intertextus,* and *Zonitoides arboreus.* Any species of land snail that occurs where there is a high prevalence of white-tailed deer could potentially be a host and contribute to the spread of this disease (Anderson and Prestwood 1981).

Land Snails as Pests

Land snails, including slugs, can be agricultural pests. By and large however, snails and slugs that are problematic in gardens are those species that are non-native introductions. Most have been accidently released into North America by way of plants, potting soils, or shipping crates but a few species like *Cepaea nemoralis* were introduced for their colorful shells. Exotic gastropods naturalize quickly and multiply. Degraded native habitats only make things worse by providing the conduit for dispersal and movement into new areas. These exotics can carry molluscan diseases and problematic parasites that can effect native land snails and wildlife populations like elk. Exotic slugs can be especially damaging pests in greenhouses and agricultural lands, costing millions of dollars worth of crop damage every year. Gardeners are most familiar with those primarily exotic species, the slugs, due to their insatiable appetites for tender vegetables. But who could blame them? While foreign gastropods quickly become overpopulated without natural controls in place, native snails and slugs are rarely problematic and most species actually become scarce or disappear entirely in areas where the natural vegetation has been eliminated. The best defense against exotic gastropod infestations is to keep natural forests intact.

A Human Connection

Mollusks are providing a number of life-supporting contributions to humans. For example, marine mollusks may help fight cancer. The drug Kahalalide F, a protein extracted from a species of mollusk that eats sea slugs in the Pacific Ocean, has shown great promise as chemotherapy for the treatment of liver cancer (Satheeshkumar et al. 2010). Lethal toxins produced by cone snails are used to develop a drug called Ziconitide for patients with cancer and AIDS who suffer from chronic pain that cannot be relieved by opiates and are not addictive (Wallace et al. 2008). Slime from the land snail, *Cornu aspersum* (one of the commonly eaten snails referred to as escargot) is now used to treat many different types of skin disorders. The snail slime repairs skin damage from overexposure to the sun and reduces scarring caused by severe acne. Land snail mucus is also known to contain natural antibacterial properties and, some scientists speculate, the next generation of human antibiotics. In Central America, the Maya use gastropods to treat a number of ailments including skin disorders, glaucoma, and whooping cough (Dourson 2009).

The Diet of Land Snails

Most snails are dietary generalists (Burch and Pearce 1990) consuming a wide variety of herbaceous plant leaves or stems, decaying vegetation and leaf litter (detritus), wood or bark, and fungal fruiting bodies such as mushrooms, animal scat, carrion and wood-inhabiting shelf or bracket fungi (Burch and Pearce 1990; Dourson 2008). Land snails in the genus *Haplotrema*, *Ventridens*, *Vitrinizonites* and *Mesomphix* are documented carnivores, feeding on a variety of other gastropods (Atkinson 1998). Gastropods sample and judge potential food by using the chemoreceptors located on the lower two tentacles (Shearer and Atkinson 2001). Once a food source is discovered, the snail begins the feeding process by first touching the food with its foot and mouth followed by rasping (rasping signs can be seen on the fungi illustrated to the right) and dislodging bits of food with the radula structure located in the mouth (Machensted and Markel 2001). The meal is then swallowed and muscular contractions move the food along the esophageal tract mixing it with saliva. Feeding episodes can last anywhere from a few minutes to nearly an hour, depending on the durability of the food being consumed (Dourson 2008).

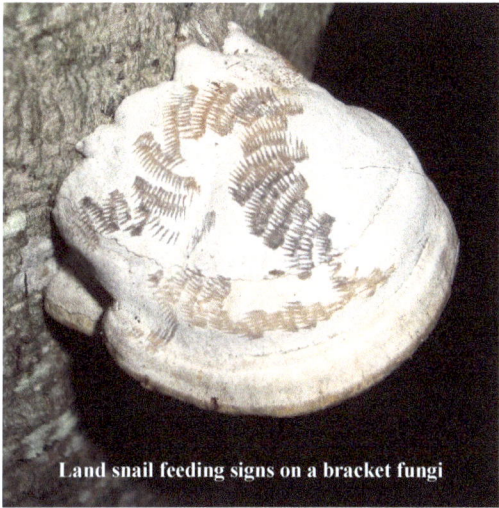

Land snail feeding signs on a bracket fungi

Fruiting bodies of fungi are a favorite food of native slugs. Below a brown-spotted mantleslug, *Philomycus venustus* feeds on the tissues of young golden trumpets, *Xeromphalina campanella*, Roan Mountain, North Carolina. As the slug consumes the tissues of mushrooms, they sequester toxins from the fungi suspected to bolster defensive properties of their mucus. Slugs in the genus *Philomycus* have been observed feeding on the destroying angel, *Amanita virosa* (pers. obs.). This is one of the most poisonous mushrooms known in North America and one that has caused the majority of mushroom fatalities in humans (Roody 2003), yet slugs seem impervious to the mushroom's lethal effects. Little is known about symbiotic relationships between mushroom and slug. What is known is that as slugs feed on the mushroom tissues, they no doubt ingest numerous spores in the process, which are carried for a time and excreted elsewhere, spreading the fungi spores to new locations.

Empty Snail Shells

When snails die, the shells do not immediately decompose. Some research suggests that shells can remain intact for years (Pearce 2008) and empty shells in some limestone locations can reach exceptional numbers. But these discarded shells are anything but vacant and actually provide secure and protected refuge for a whole host of living micro-invertebrates including pseudo-scorpions, ants, millipedes, tardigrades, and smaller land snails. Some invertebrates even deposit their eggs to be incubated in the security of shells. The translucent 3 mm wide shell of *Glyphyalinia indentata* (figure a) was the depository site for eggs of an unknown organism (pers. obs.).

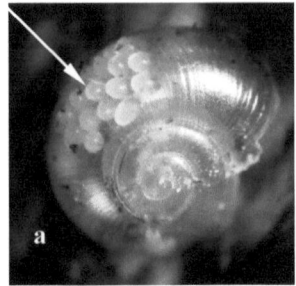

Fluorescence in Land Snails

Not to be confused with bioluminescence (the natural light observed in fireflies), fluorescence is the term used to describe the absorption of light at one wavelength and its emission in another. Only one species of land snail is known to produce bioluminescence; *Quantula striata* from Malaysia. Research has shown however that the mucus of several North American family groups of land snails including Discidae and Helicodiscidae have fluorescent slime under ultraviolet light. *Anguispira alternata* slime has a particularly brilliant, bluish fluorescence, while *A. kochi* has a similar fluorescence (Rawls and Yates 1971). Under laboratory conditions, the mucus on the foot, body and the mucus trails of *Anguispira* and *Discus* species glowed brightly when exposed to UV light. *Discus patulus* and *Helicodiscus parallelus* mucus appears to have fluorescence that is unique to each species. Other species of polygyrids were tested and all others failed to exhibit any sign of fluorescence. The fluorescence which was observed in the specimens of the three genera noted is extremely long-lived, being as bright and as distinctive in specimens preserved for twenty years or more as it is in living snails (Rawls and Yates 1971). I found that the crawling slime of *Anguispira jessica* (typically clear) did not fluoresce under UV exposure, but the defense slime produced by the species under attack (which is orange-yellow) did in fact fluoresce.

The function of fluorescence in land snails remains a mystery and there is some speculation that fluorescent slime has no real function at all, being simply a random act of evolution. I propose a functional hypothesis for the fluorescence in land snails. It turns out that moonlight has a component of UV light and *Anguispira* species are most active at night. What if certain nocturnal land snail predators (i.e. snail-hunting beetles) are avoidance-conditioned to the glowing defense slime the same way that predators are trained to steer clear of the bright colors of coral snakes? Snail slime, especially that of *Anguispira* species is pungent and distasteful to animals, including humans (*Anguispira* slime has a disagreeable-numbing affect in ones mouth, authors personal experience). Like the coral snake who uses color to save its venom, snails may be using florescence to save precious mucus reserves. The more snails are harassed, the more mucus is produced and losing copious slime (from an attack) would put the snail at risk of dehydration.

Fluorescent Land Snails

Above image of *Anguispira jessica,* taken with a flash and UV light and bottom picture of same animal with UV light only. Through the camera's lens, the defense-slime appears as a bluish glow, but to the human eye a fluorescent green. Images taken in the Doe River Gorge, Tennessee.

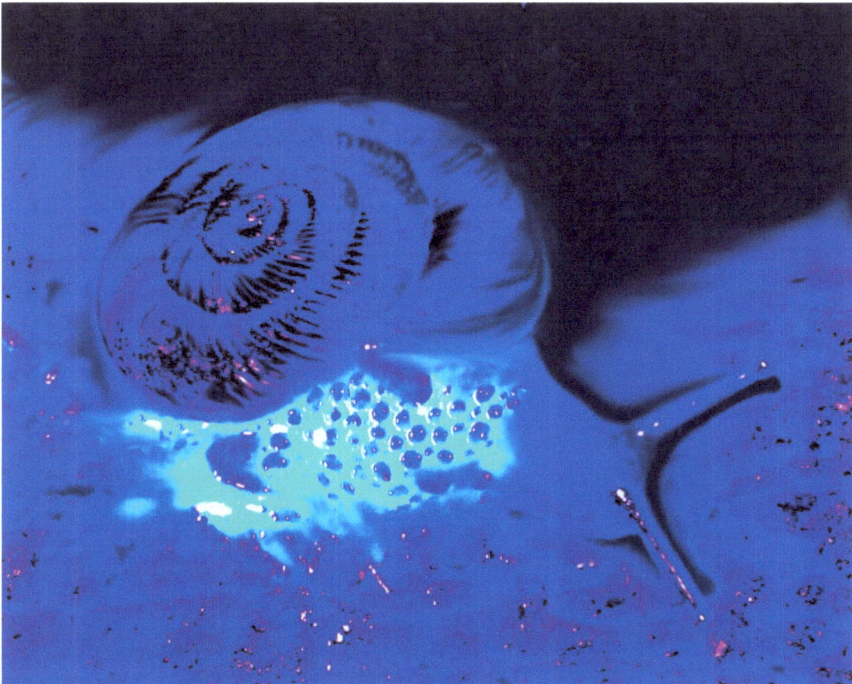

Shells, Teeth, Hairs and Viscous Slime

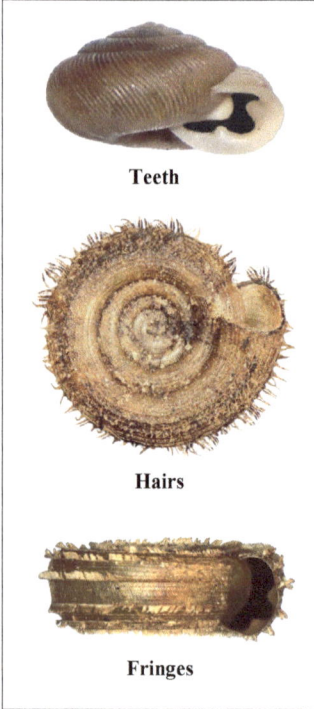

Teeth

Hairs

Fringes

Land snails use a variety of strategies to protect themselves from harm. The shell is the first line of defense and works well to ward off a number of predators. Opercula (snail doors) can prevent invertebrate attacks, but few snails in the park possess these protective structures. Shells and opercula also aid in the conservation of critical body moisture. Teeth and lamellae are believed to act as barricades, preventing the entry of predators such as predacious beetle larva. Some research suggests that aperture teeth may also provide a calcium storehouse used to repair damaged shells, act as pivotal points for balancing the shell during movement or trap air if the animal becomes immersed in water (Emberton 1994). The periostracal hairs and fringes found on *Inflectarius, Stenotrema* and *Helicodiscus* shells collect forest debris as snails inch through the leaf litter. Several theories have emerged for shell ornamentation, including increasing shell-crypsis or protecting shells from fracture during falls. Hairs, fringes and ribs function as water retention features, which hold and uniformly transfer (wick) water across the entire shell surface (pers. obs.). Although speculation, shells that contain these features may help facilitate snail movement across drier surfaces and or help keep the snail cooler. Several species of land snails such as *Anguispira* contain colorful defense mucus that has a anesthetizing affect in the mouth (personal experience), no doubt an effective predator-repellant. Although slugs are without protective shells, they are anything but defenseless. Slugs have copious, viscous, water insoluble mucus (a), which can gum-up the mandibles of beetles or cause antagonistic garter snakes to vomit (pers. obs.).

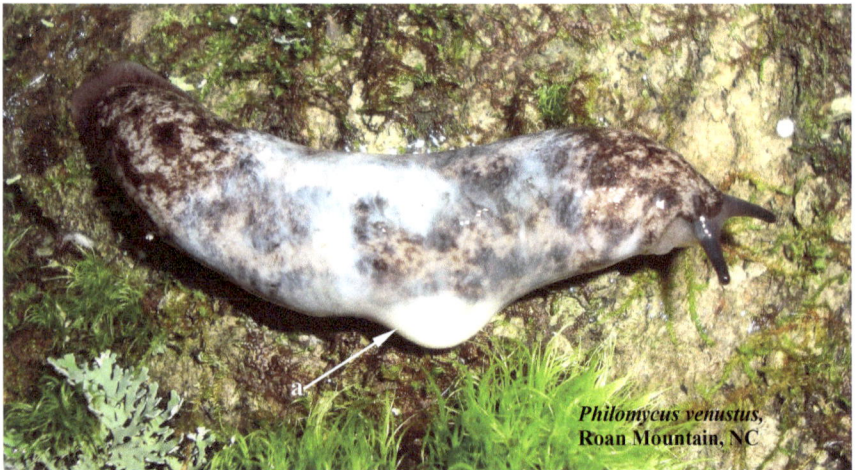

Philomycus venustus,
Roan Mountain, NC

Land Snails and Spider Webs

While crawling across a spider web, a flat bladetooth, *Patera appressa* appears to glean precious (condensed) water during a dry period in late July, Red River Gorge, Powell County, Kentucky. Land snails are often seen on spider webs, crawling effortlessly across the web's sticky surface without becoming entangled; an act that would imprison most other invertebrates of similar size. There is some evidence that snails are attracted to the silk nets (that trap condensed moisture) for a source of drinking water.

Land Snail Predators

With regard to land snail predators, small mammals such as shrews and moles are at the top of the list; above image of a hairy-tail mole preying on a Magnolia threetooth, *Triodopsis tridentata* Roan Mountain, NC. Smaller than moles, shrews are tiny venomous mammals that chew through the apex and sides of the shell quickly devouring the snail flesh within. Piles of empty shells under rocks and logs are a sure sign of small mammal feeding. Salamanders are also reported to eat a variety of land snails (Van Devender and Van Devender 2003). But precious little is known on the significance of land snails in the diets of salamanders, particularly Plethodontidae species. A Blue Ridge two-lined salamander (below image) eyes a highland slitmouth, *Stenotrema altispira*, Roam Mountain, NC.

Land Snail Predators

Thorax

Head

The formidable jaws
of the *Cychrine* beetle
used to grip and hold
slippery snail flesh

The head and thorax regions of *Schaphinotus* beetles are significantly smaller than other beetles of similar size, allowing the insect to enter through the aperture of the live snail to extract the slimy flesh with its serrated jaws. Below a snail-hunting *Cychrine* beetle dines on snail flesh of a gray-footed lancefoot, *Haplotrema concavum,* while a carrion-fly waits to deposit her eggs onto uneaten portions of the snail meat. In nature nothing goes to waste! Red River Gorge, Kentucky.

Carrion Fly

Reproduction

The majority of land snails found in the park are hermaphrodites, each individual having ovotestis in which both sperm and eggs are produced. When two individuals of the same species in search of propagation (through scent trails) find one another, they typically exchange sperm. Sperm can be store, for months to years, by each individual snail, until conditions are favorable for fertilization and egg laying, at which time eggs are deposited under logs or in moist leaf litter. Interestingly, many land snails, including several native slugs in the genus *Philomycus,* are characterized by the presence of "love darts".

Dr. Ron Chase of McGill University in Canada, has studied the complicated sex life of the European land snail most commonly found on people's plates, *Cornu aspersum,* that use calcareous love darts as part of their courtship. Actually, love darts are not shot but are forcefully expelled through the body wall of the partner. The dart itself appears in a variety of sizes and shapes, most species containing a single dart which is used only once. In his research on this fascinating sexual behavior, Dr. Joris Koene, learned that some land snails, employ multiple darts while others use the same dart to stab their partners repeatedly, as many as 3000 times. But what exactly is the function of the love dart? Further studies indicate that the dart functions after copulation to increase the reproductive fitness of the shooter. Snails are promiscuous and store sperm from multiple donors for several years before they use it to fertilize eggs. Thus, sperm donors ('males') must compete to fertilize eggs. Dart receipt promotes the safe storage of the shooter's sperm, so there will be more sperm from successful shooters available for fertilization than from unsuccessful shooters. Since the female function chooses the sperm by a lottery-like mechanism, successful dart shooters sire more babies than unsuccessful dart shooters. Chase and Blanchard (2006) tested whether the dart works by simply rupturing the skin or by injecting a bioactive agent. Just before the dart is thrust into a partner, it is covered with mucus (figure a) from a special gland located near the dart's launching site. Koene et al. (2013) conducted an interesting test in which needle stabbings were substituted for dart shooting. In one mating, saline was injected through the needle, in the other mating mucus was injected. They found that the matings that were associated with mucus injections were responsible for more than twice the number of offspring as were the matings associated with saline injections. Thus, mucus is the agent of the dart's effect on reproduction.

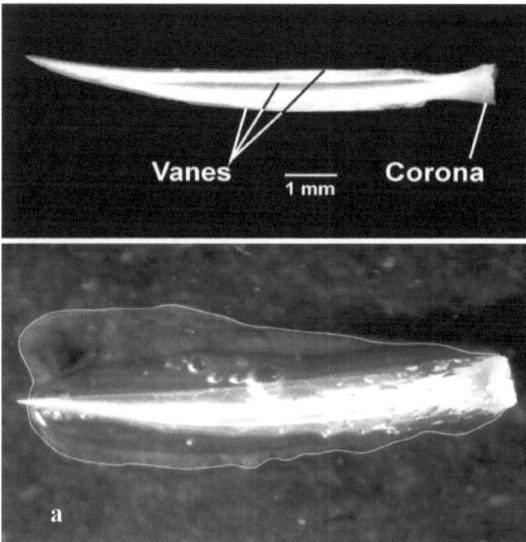

Vanes — 1 mm — Corona

Chase and Blanchard (2006)

29

Mating brown gardensnails,
Cornu aspersum

Love dart →

Chase and Blanchard (2006)

Snail eggs

What?
Wrong Brand?

Eros
Love
Darts

Other dart examples

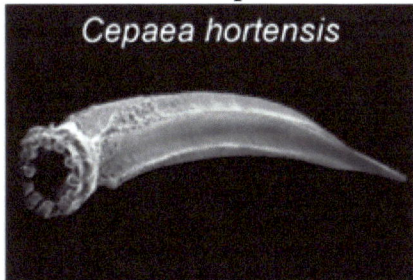

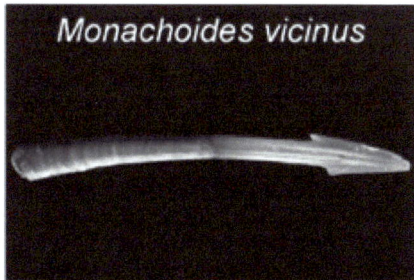

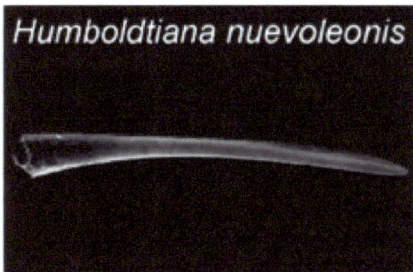

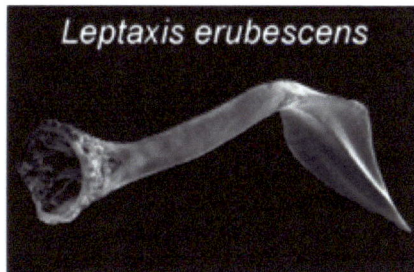

Cepaea hortensis

Monachoides vicinus

Humboldtiana nuevoleonis

Leptaxis erubescens

Koene and Schulenburg (2005)

How to Collect Land Snails

Collecting snails (shells or live animals) within the boundaries of the Great Smoky Mountains National Park is prohibited without a valid permit from the park. The park offers opportunities for participation in Citizen Science activities including snail searches. For further information, contact the park offices.

Selecting Sites to Survey

When surveying for land snails, there are specific habitats that should be targeted in order to comprehensively cover an area. These habitats include: under leaf litter, rocky outcrops, rock crevices, and logs; under exfoliating bark of standing and/or down dead trees; hollow trees like American beech and sycamore; damaged trees oozing sap, which attracts snails (Dourson pers. obs.); under and on top of caps of fungi; under moss mats and the flaps of rock tripe;lichen; bases of black walnut trees; crotches of trees; human-made features such as roadsides, steep banks, retaining walls, cement structures, spring houses, discarded bottles or other discarded refuse; cliffline features, caves, and rock talus.

Field and Lab Equipment Needed

Field equipment include:
- Ziploc bags
- Permanent marker
- GPS unit
- Field notebook
- Hand lens
- Quart-size drying bags

Collecting Methods

Samples of larger (macro) specimens from 5mm and greater should be collected and placed in Ziploc bags with date, site number, GPS coordinates, and collector name (preferably on the outside with a Sharpie). Do not put paper labels in bags with live snails. They will most likely eat the paper.

Samples of smaller (micro) specimens are best collected from leaf/soil collections; field sieves work well here. Sites that yield increased numbers of snails include the bases of black walnut and butternut trees, the bases of large mature hardwood trees, tree crotches and leaf litter along the edges of seeps. Optimal sites can be determined by collecting a handful soil/leaf litter then scanning the litter with a hand lens for evidence of micro specimens. If any snails are observed, a quart-sized cotton drying bag is filled with the material from the site, labeled with the date, site number, collector's name and GPS coordinates. These leaf samples are taken back to the lab and dried for approximately two weeks. Dried samples should be sifted through a series of sieves ranging from 4.76 mm down to 500 micrometers. The subsequent debris that remains after this sifting process is then searched with the aid of an Optivisor or other magnification device. It will be necessary to use a zoom microscope to determine the species of these small snails. Many of them have microscopic ornamentation that can be seen only under high magnification.

Field Notes and Labeling Samples

1). Date each daily entry and record the current weather conditions and last rain event.

2). Record for each area surveyed, the Forest Community Classification, main overstory tree species and dominant ground cover species, solar aspect, and elevation. Presence and abundance of rock or coarse woody debris (CWD) should be noted.

3). Record GPS position noting which datum used or mark precisely on a 7.5 minute quadrangle map.

4). Record the number of litter bags collected.

Preservation and Mailing Specimens

Generally, it is not necessary to collect live specimens due to the abundance of dead shells. If live snails are collected for the purpose of anatomy work, it is necessary to euthanize them in water for 24 hours so that they relax, then place them in a solution of 70-80% ethyl alcohol. It is best that the snail be preserved with the head and foot not retracted into the shell. Menthol or chlorotone crystals are sometimes used to relax land snails. Menthol cigarette tobacco can be soaked in water to provide a relaxant solution. Live snails can be sent through the mail by simply packing them in moist unbleached paper toweling. The paper will keep the snails hydrated and provide a food source. For experts involved in ongoing taxonomic studies, live snails provide valuable anatomical material.

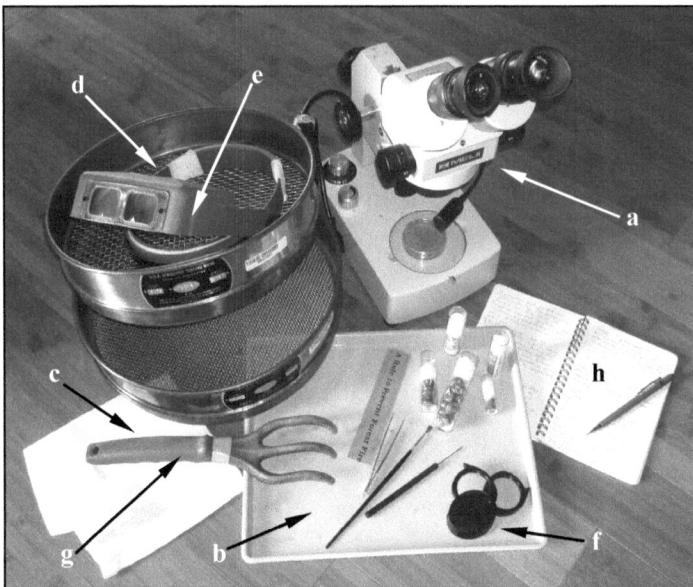

Necessary equipment for land snail work includes: a dissecting microscope (a), sorting tray (b), 1-liter litter bags (for the soil and leaf litter collections) (c), a series of soil sieves sizes #4 (4.75mm), #10 (2.00mm), #18 (1.00mm), #14 (1.40mm) (d), magnifying head loop (e), forceps, vials, labels, paintbrush to pick up and transfer small snails (all pictured on tray), hand lens (f), hand rake (g), and notebook (h).

Shell Morphology

Pictured (right) are three standard shell views (a, b, & c) used by malacologists to identify land snails. Always observe snails in these views. The frontal view (a) shows the shell's general form and aperture shape. The bottom view (b) shows one of the most important diagnostic features of any land shell, the umbilicus region. The top view (c) shows the apex or embryonic whorl, the number of whorls, and the width of the whorls. In general, the frontal and bottom views are the most important diagnostic views of land snail shells. While the top view has limited value for separating snails within the same genus, it is a reliable view for separating snails of different families. Short black arrows or lines indicate key features of shells. All measurements used in this book are for greater diameter or height of adult shells.

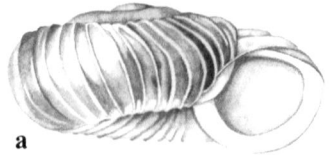

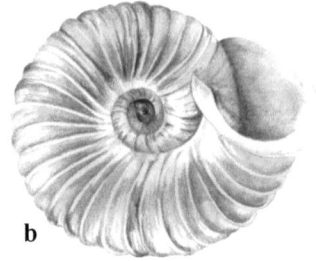

a

b

c

Terminology of the Shell

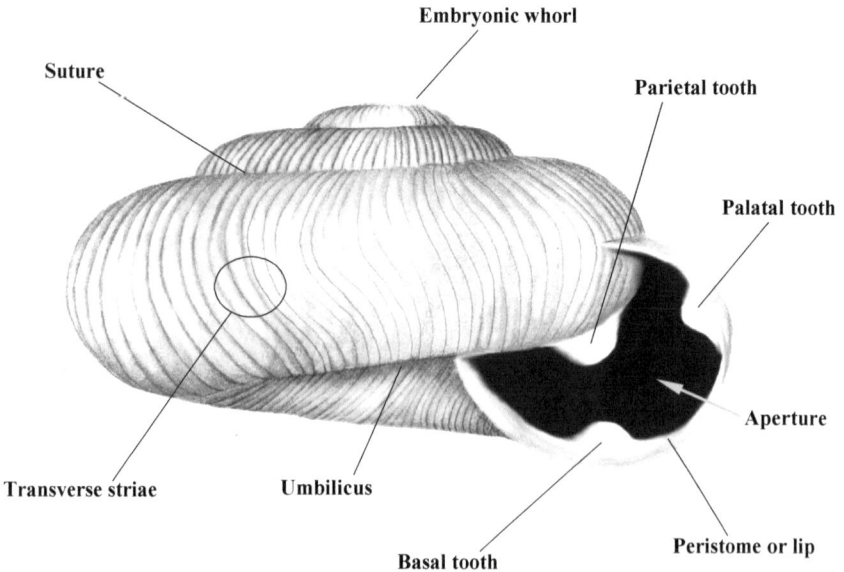

Embryonic whorl

Suture

Parietal tooth

Palatal tooth

Transverse striae

Umbilicus

Aperture

Basal tooth

Peristome or lip

0 cm 1 2 3 4 5 6 7 8 9 10 11

33

Umbilicus of the Shell (Burch, 1962)

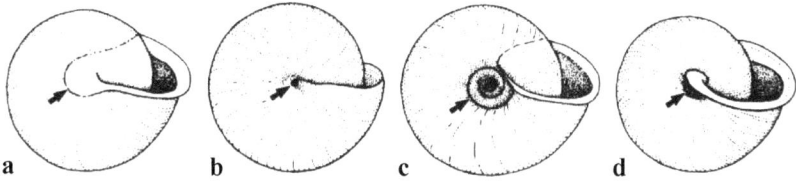

Figure A-Imperforate shell (closed umbilicus); Figure B-Perforate shell (small umbilical opening); Figure C-Umbilicate shell (wide umbilical opening); Figure D-Rimate shell (umbilical opening partially covered by aperture lip)

Shell Measurements

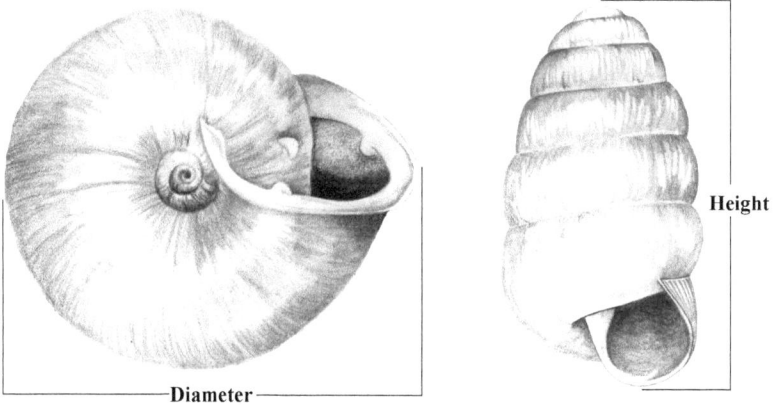

Diameter

Height

Shell Periphery	Peristome or Lip

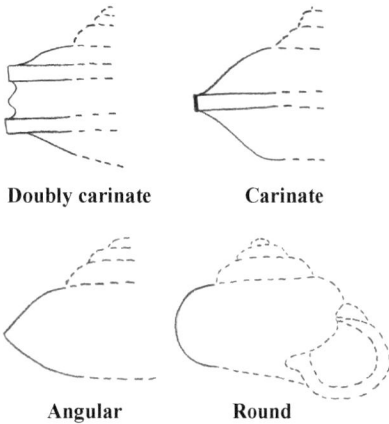

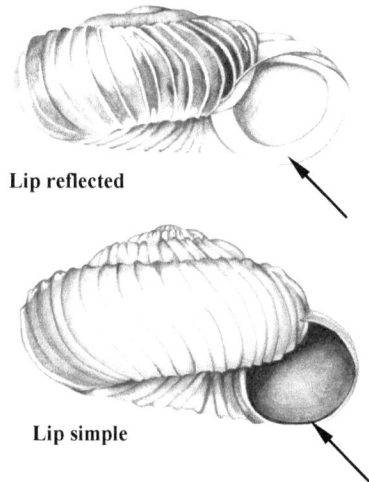

Doubly carinate   Carinate

Angular   Round

Lip reflected

Lip simple

Counting Whorls

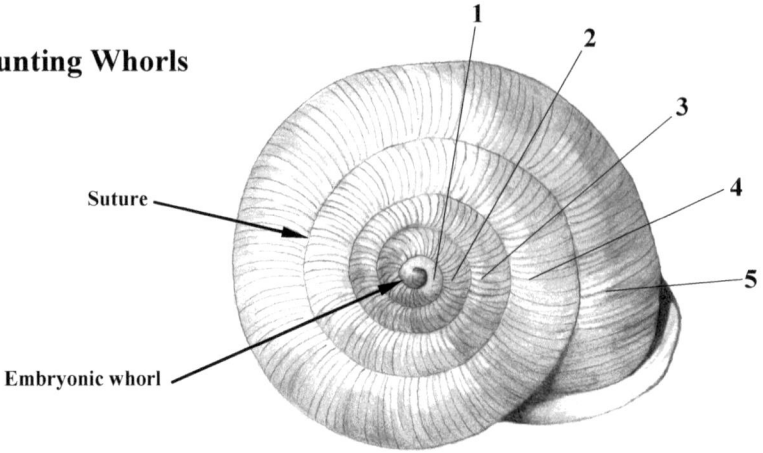

Suture

1
2
3
4
5

Embryonic whorl

Micro-features of the Shell Surface

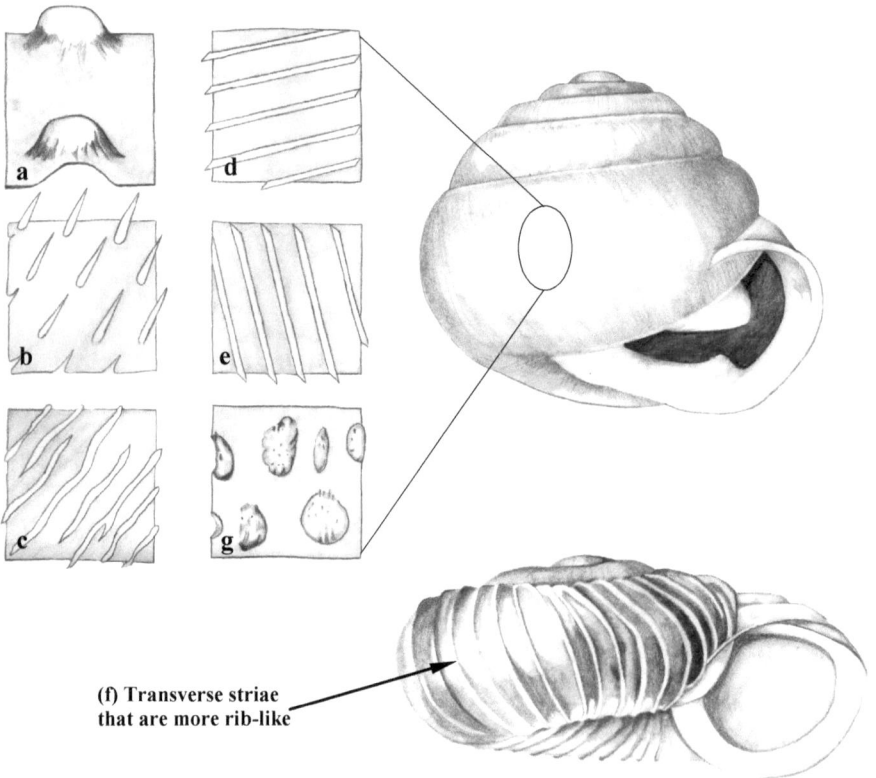

a
d
b
e
c
g

(f) Transverse striae
that are more rib-like

Papillae are raised bumps (a); hairs (b); wrinkles (c); spiral striae (can be in-dented or raised) run with the shell spire (d); transverse striae (can be indented or raised) are generally a micro-feature (e), but in some shells the striae are more rib-like (f); pits in the shell (g).

Basic Land Snail Anatomy

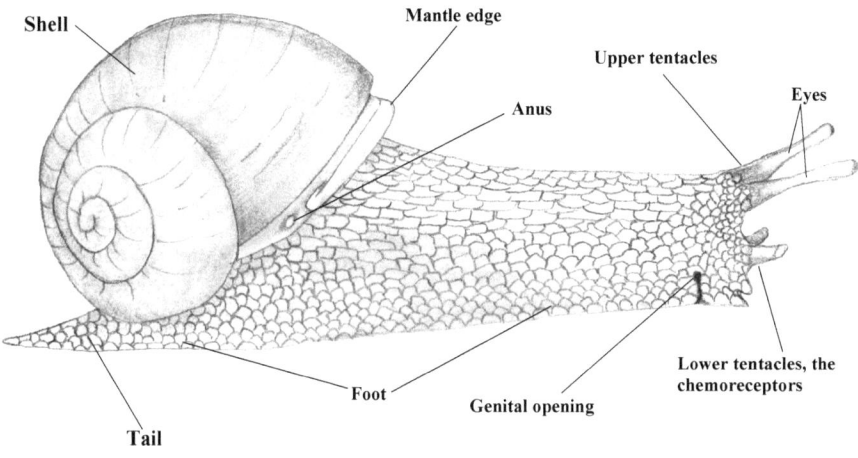

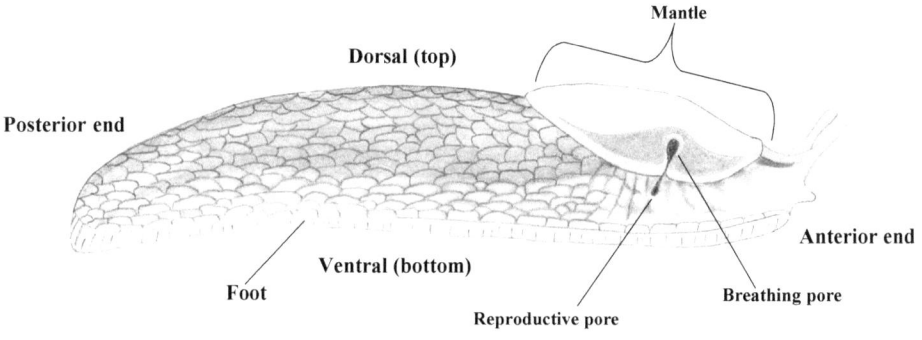

Shell

Mantle edge

Upper tentacles

Eyes

Anus

Lower tentacles, the chemoreceptors

Foot

Genital opening

Tail

Mantle

Dorsal (top)

Posterior end

Anterior end

Ventral (bottom)

Foot

Reproductive pore

Breathing pore

Internal Anatomy

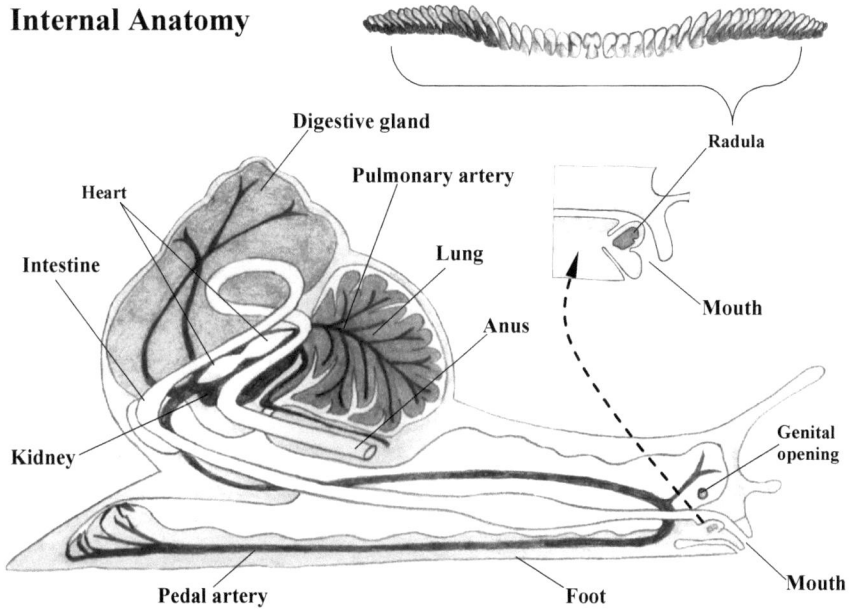

Digestive gland

Radula

Heart

Pulmonary artery

Intestine

Lung

Mouth

Anus

Genital opening

Kidney

Pedal artery

Foot

Mouth

Shell Growth (author)

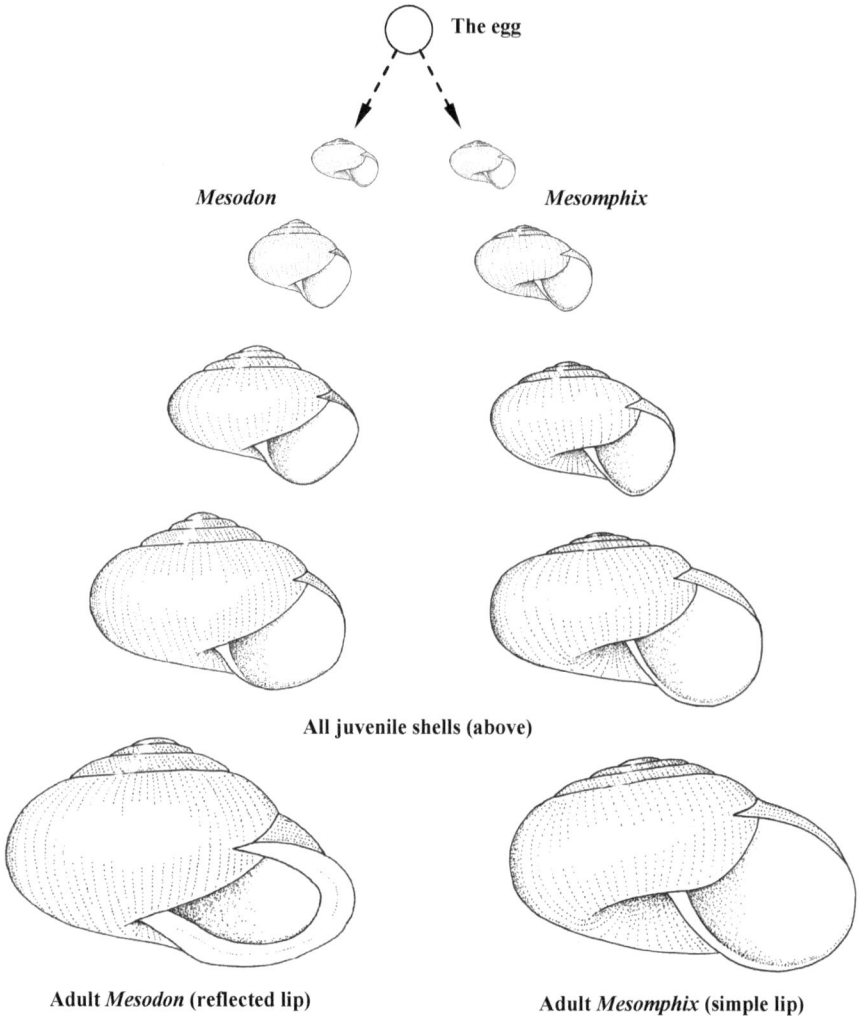

The egg

Mesodon

Mesomphix

All juvenile shells (above)

Adult *Mesodon* (reflected lip)

Adult *Mesomphix* (simple lip)

Immature shells of any species are difficult to identify. Determining the maturity of a shell can often be accomplished by examining the aperture and number of whorls. As shells mature, the shape of the aperture changes. Note the apertures of the juvenile shells of *Mesodon* and *Mesomphix*. The bottom of the aperture of each species appears to droop as if an invisible weight is attached. As shells of both species mature, the shape of the aperture changes to a more horizontally oval shape. For adult *Mesodon* species, a reflected lip forms and for adult *Mesomphix* species, the lip remains simple. Immature *Mesodon* species are easily confused with *Mesomphix* species. *Mesodon* and *Triodopsis* species do not form reflected lips until they reach maturity. Other species such as *Mesomphix, Anguispira* and *Ventridens* do not possess reflected lips at maturity. Finally, most mature shells should contain no less than 4 to 5 whorls at maturity.

Part I
Land Snails of the
Great Smoky Mountains National Park

With nearly 150 species, native land snails of the Great Smoky Mountains National Park are as varied and interesting as any group of animals found in the park.

Land Snails of the Great Smoky Mountains National Park

For its comparatively small size, the GSMNP has an exceptionally high number of land snail species, nine of which are endemic. One reason the park is so rich is largely tied to the great span of elevation, soils, geology and the age of the mountains. These conditions which are less dramatic elsewhere have fueled the engines of species diversity within park boundaries. Interestingly, many snails that were described as forms in early works by Pilsbry, Hubricht and others have been elevated to species by Emberton and Fairbanks. Currently, there are several interesting forms found in the GSMNP (page 256-257) which although synonymized with described species, may in fact represent speciation in its infancy. Further DNA and genital dissection research awaits these gastropods. Furthermore, these diverging snail forms may be key to bridging our better understanding of evolutionary biology. I have no doubt that additional species new to science await discovery in the GSMNP!

Species Accounts

The following Species Accounts lists every land snail reported within the boundaries of GSMNP. This includes all native and exotic snails and slugs. In addition, the text includes land snails not yet reported but known to occur close to the GSMNP and for that reason, have a reasonable probability of occurring in the park. Finally, land snails associated with the Southern Appalachian Mountains from around Mount Rogers, Virginia south to Chilhowee Mountain, Tennessee are found on pg 258-315. This information should make the text more useful to the Pisgah, Nantahala and Cherokee National Forests and the Nature Conservancy who manage lands near park boundaries. For each species illustrated the Common Name, Scientific Name, Description, Similar Species, Habitat, Status in the GSMNP (including rankings) and Specimen (providing the location of the particular specimen photographed) is included.

In addition, two plates are included (pages 317-318) illustrating the aquatic snails of the Great Smoky Mountain National Park with specimens provided by North Carolina Museum of Natural Sciences.

Maps

A hand-rendered map showing approximate, known locations (using yellow dots) for each species is positioned at the bottom of same page. If locations are not known, as is the case for several historical records, the maps will display a circle where the species is predicted to occur. For maps showing overall species distribution in the eastern USA, refer to Hubricht (1985), "The Distributions of the Native Land Mollusks of the Eastern United States", Fieldiana Zoology New Series, No. 24, Publication 1359: Field Museum of Natural History, Chicago, Illinois.

Abundance Categories of Land Snails in the GSMNP

The abundance of land snails within the GSMNP was determined by investigating records for each species and assigning a category based on that examination. The categories used in the book are as follows on next page.

1) Common: Species is found in some quantities in its habitat throughout the park.

2) Relatively Common: Species is usually encountered in its habitat in smaller numbers but with regularity.

3) Uncommon: Species not encountered frequently within its habitat but widespread in small numbers.

4) Rare: Rarely encountered even in suitable habitat (these snails have been furthered delineated based on factors discussed below).

Rare Status in the Park
Any native species in the GSMNP considered rare are of conservation concern to the National Park Service. This category of determining the status of land snails in the GSMNP is further categorized into two distinct groups. These groupings are critical to understanding the true status of the land snail species within the park and have implications for management and conservation of these rare species.

Globally Rare: Land snails that are globally rare and/or endemic to the GSMNP or the Southern Appalachians. These snails may have exceptionally limited habitat requirements making dispersal nearly impossible outside their range.

Locally Rare: May in fact be common and wide-ranging land snails outside the GSMNP but due to their limited distribution in the park are considered rare within park boundaries. The policy of the National Park Service is that native species should not be extirpated from a national park due to human impacts— even species that are relatively common in most places. These species have pushed their range limits at park borders, sometimes penetrating into the park's interior. Most of the land snails in the park that are considered **rare** will fall into this category.

Global Rankings: Global Rankings, provided by NatureServe (www.natureserve.org) and its network of natural heritage member programs, a leading source of information about rare and endangered species, and threatened ecosystems, is given for each land snail species are included under Status.

G1–Critically Imperiled—At very high risk of extinction or elimination due to very restricted range, very few populations or occurrences, very steep declines, very severe threats, or other factors.
G2-Imperiled—At high risk of extinction or elimination due to restricted range, few populations or occurrences, steep declines, severe threats, or other factors.
G3-Vulnerable—At moderate risk of extinction or elimination due to a fairly restricted range, relatively few populations or occurrences, recent and widespread declines, threats, or other factors.
G4-Apparently Secure—At fairly low risk of extinction or elimination due to an extensive range and/or many populations or occurrences, but with possible cause for some concern as a result of local recent declines, threats, or other factors.
G5-Secure—At very low risk or extinction or elimination due to a very extensive range, abundant populations or occurrences, and little to no concern from declines or threats.

Identifying Land Shells

Land snails are a study in subtlety and successful identification will depend on your aptitude for recognition of shell detail. The principal obstacle for beginners are the dichotomous keys. In an attempt to ascertain every species in one place, many keys overwhelm the reader with mind-numbing particulars. The reader can become frustrated quickly and lose interest. In this snail manual, the pictorial key takes you to genera only, keeping the finer details of each species in the text where I feel it belongs. Consequently, the keys are less cluttered with scientific muddle and hopefully easier to follow.

The keys are fundamentally based on two strategic characteristics: **shape** and **size** of the shell, with five additional shell supporting characters. These include: number of whorls, umbilicus, teeth or the absence of teeth, lip simple or reflected, and spiral striae or lack of this micro-feature. If you move through the keys but are uncomfortable with your final deliberation, go back and reconsider the shape (form) of the shell; this feature alone will most often lead to a misidentification. While most shells fit easily into certain shapes or categories, some will prove more difficult to discern, so allow for these discrepancies. Furthermore, it is not uncommon for the shell shape of some species to vary slightly from population to population and on occasion in the same population.

Although the colors of shells are used in this text, it is not generally a reliable consideration for their identification. Pictures of shells are from field specimens (not farm-raised), so many have flaws such as cracks, blemishes or are bleached from weathering. Minute species were photographed through a microscope and lighting was sometimes problematic. As a result, the natural color of some shells was compromised. Remember, shell shape or form and micro-features are the most important diagnostic features. In old shells that are bleached, slightly wetting the surface will often bring out the micro-features.

Record observations 1-7, then proceed with the shell identification.

(1) **Shape**-Determine the general shape of the shell (see next page); same-species shells will sometimes vary slightly from site to site, and occasionally, within the same population.

(2) **Greater Diameter or Height**-Measure the greater diameter for heliciform shells and the height for Succiniform, Conical and Pupa-shape shells. Sizes may sometimes vary slightly from measurements given in species accounts.

(3) **Whorls** - Count the number of whorls (refer to pg. 35)

(4) **Umbilicus** -Is the shell imperforate, perforate, umbilicate, or rimate?

(5) **Teeth**-Make note of any presence or absence of teeth in the aperture, their size and spatial arrangement.

(6) Lip – Is the lip simple or reflected?

(7) Spiral Striae-To determine if spiral striae exist, examine shell surface under a dissecting microscope.

This symbol indicates the use of a dissecting-scope is necessary or useful to view micro-features of the shell.

◆———◆ Scale bar is included throughout to present the actual size of the shell.

Basic Shell Shapes of Terrestrial Land Snails

Heliciform

Depressed heliciform

Shape of this shell lingers between depressed helici-form and heliciform; when this happens try both options.

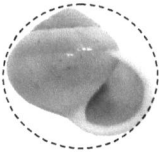

Roundish-shape

Dome-shape

Pill-shape

Vitriniform

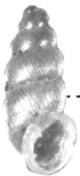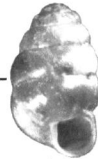

Pupa-shape

Conical

Succiniform

General groupings of land snails in this book are based primarily on shape, lip and sometimes teeth

*Shells taller than wide, simple lips (without teeth)	50
*Shells taller than wide, reflected lips (teeth present)	56
*Shells wider than tall, simple lip	76
*Shells pill-shape with narrow aperture	178
*Shells wider than tall, reflected lips	190
*Native slugs	232

Pictorial Key to GSMNP Land Snails

Keys work for adult shells only
(Shells are generally arranged from smallest to the largest)

---

1) Shell taller than wide, simple lip (without teeth) 2-20 mm

Pomatiopsis: shell conical, live snails with an operculum, 5-7 mm tall, page 52

Novisuccinea: shell succiniform and paper-thin, without an operculum, 10-20 mm tall, page 51

Cochlicopa: shell pupa-shape and translucent 4-7 mm tall, page 53

Columella: shell pupa-shape, 2-3 mm tall, page 55

2) Shell taller than wide, reflected lip (with teeth) under 4 mm

Carychium: shell pupa-shape, reflected lip, small parietal tooth 1.5-2 mm, page 57

Gastrocopta: shell pupa-shape, multiple teeth, lip notable reflected, 2-4.8 mm, page 63

Vertigo: shell pupa-shape, aperture containing multiple and sometimes crowded teeth, lip narrowly reflected with a slight dent on the right side of aperture, 1.5-2.3 mm, page 63

3) Depressed heliciform, simple lip (usually without teeth) under 4 mm

Punctum: shell minute with a strong spiral sculpture (scope required), umbilicate, 1.5 mm, page 77

Striatura: shell minute with weak spiral sculpture (scope required), umbilicate, 1.5-3.5 mm, Page 77

Hawaiia: shell minute with weak spiral sculpture (scope required), umbilicate, 2-3 mm; page 77

Lucilla: shell minute with a nearly smooth surface (scope required), umbilicate, 2-2.5 mm, page 77

Guppya: shell minute with a nearly smooth surface, but showing traces of spiral ornamentation, strongest around the apex of shell (scope required), perforate, 2 mm, page77

Pilsbryna: shell small with a nearly smooth surface, umbilicate, teeth in shells under 2.5 mm, adult shells around 4 mm without teeth or with reduced teeth, very rare, Part II, page 259

44

4) Depressed heliciform, simple lip (without teeth) 4 to 12 mm

Glyphyalinia: shell thin with distinctive indented transverse striae, imperforate to umbilicate, 4-12 mm, page 93

Zonitoides: shell surface smooth with transverse striae poorly developed, (except for *Z. patuloides* which has a well developed transverse striae) umbilicate, 4-8 mm, page 88

Discus: shell surface minutely ribbed (can be seen with hand lens of 10X), *Discus patulus* with a small tooth, other *Discus* species without teeth, widely umbilicate, 5-10 mm. page 120

5) Domed-shaped, simple lip (teeth usually absent) 2-8 mm

Euconulus: shell dome or bee-hive shaped and thin, *Euconulus dentatus* has one or two inner teeth, other *Euconulus* species without teeth, perforate, 2-3.5 mm, page 172

6) Heliciform to depressed heliciform, simple lip (teeth usually present, except for *Hendersonia*) 2-15 mm

Paravitrea: shell thin with either regular or irregular transverse striae on entire surface, strongest on the top, perforate to umbilicate, most species with internal teeth at some stage of growth, 2-8 mm, page 107

Showing teeth of *Paravitrea multidentata*

Gastrodonta: shell with two internal teeth, top surface ribbed and visible with a hand lens of 10X, perforate, 6-7.5 mm, page 88

45

Helicodiscus: shell flat with internal teeth, spiral striae are a constant and strong feature on *Helicodiscus* shells, widely umbilicate, 2-5 mm, page 120

Ventridens: shell can be flat or globose, glossy and with or without internal teeth, perforate to umbilicate, 5-15 mm, page 131

Hendersonia: shell heliciform, thick for its small size with a callus on outer lip, imperforate, operculate, 6-8 mm, page 172

7) Depressed heliciform, reflected lip (without teeth) under 3 mm

Vallonia: shell small with a widely reflected lip, widely umbilicate, 1.6-3 mm, page 191

8) Domed-shaped, reflected lip (teeth present) under 3 mm

Strobilops: shell small and beautifully sculptured with fine ribs, with two or more elongated teeth or lamellae, perforate, 2-2.8 mm, page 191

9) Pill-shaped, (aperture opening slit-like or nearly so) 5-15 mm

Stenotrema: shell small and pill-shape, lip opening restricted and narrow (b), imperforate, 5-15 mm, page 179

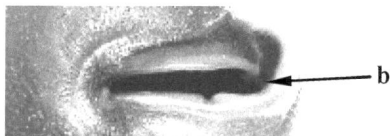

b

Euchemotrema: shell like *Stenotrema* but with a reflected lip, more open aperture and without a basal notch, rimate, 7-12 mm, page 195

10) Depressed heliciform, reflected lip (parietal tooth typically present) 5-14 mm (except *I. ferrissi*, 19-26 mm)

Inflectarius: shell like *Stenotrema* but with a more open aperture, with or without a basal and palatal tooth, imperforate, 7-14 mm, page 202

11) Heliciform to depressed heliciform, reflected lip, rimate shell (with or without teeth) 5-30 mm

Shells are rimate, <u>umbilicus never completely closed,</u> included here are *Praticolella, Euchemotrema, Inflectarius* (in part) and *Mesodon* (in part), page 195

12) Heliciform to depressed heliciform, simple lip (without teeth) 15-36mm

Mesomphix: shell generally glossy, without any teeth in all stages of growth, perforate to umbilicate, 15-36 mm, page 147

Anguispira: shell surface generally dull, without teeth in all stages of growth, color features always present, widely umbilicate, 15-31 mm, page 160

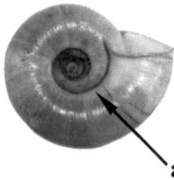

Haplotrema: shell glossy, without teeth, lip slightly reflected in old adults, umbilicate, umbilicus the widest of any large snail found in the GSMNP (a), 16-22 mm, page 169

a

13) Vitriniform, simple lip (without teeth) 16-20 mm

Vitrinizonites: shell glossy, paper-thin and flexible, without teeth in all stages of growth, perforate, 16-20 mm, page 171

14) Mostly depressed heliciform, reflected lip (long, low basal tooth always present) 13-22 mm

Patera: shell usually depressed (except for *P. clarki*) with a large parietal tooth, imperforate, 13-23 mm, page 207

15) Mostly heliciform, reflected lip, imperforate, (without a long, low basal tooth) 8-20 mm

Fumonelix: shell small, globose, usually with medium size parietal tooth, imperforate, 12-18 mm, page 211

16) Depressed heliciform, reflected lip (three distinct teeth) 9-25 mm

Triodopsis: shell more or less depressed, containing three teeth in the aperture, the parietal tooth the largest of the three, <u>umbilicate</u>, 9-25 mm, page 225

Xolotrema: shell more or less depressed, containing three teeth in the aperture, the parietal tooth the largest of the three, <u>imperforate</u>, 19-25 mm, page 225

17) Shell large, mostly heliciform, reflected lip, 15-45 mm, but most shells are greater than 20 mm

Mesodon: shell mostly large, with or without a small parietal tooth, 15-38 mm, imperforate to perforate, page 218

Appalachina: shell very large, thin and fragile, typically without teeth, 28-42 mm, umbilicate, page 218

Neohelix shell very large, thick and solid, without a parietal tooth, 25-45 mm, imperforate, page 218

18) Native slugs, page 232

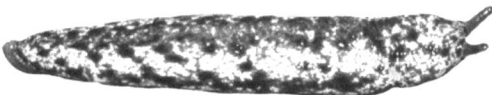

Philomycus species, over 80 mm

Megapallifera species, 40-60 mm

Pallifera species, under 30 mm

48

~SPECIES ACCOUNTS~

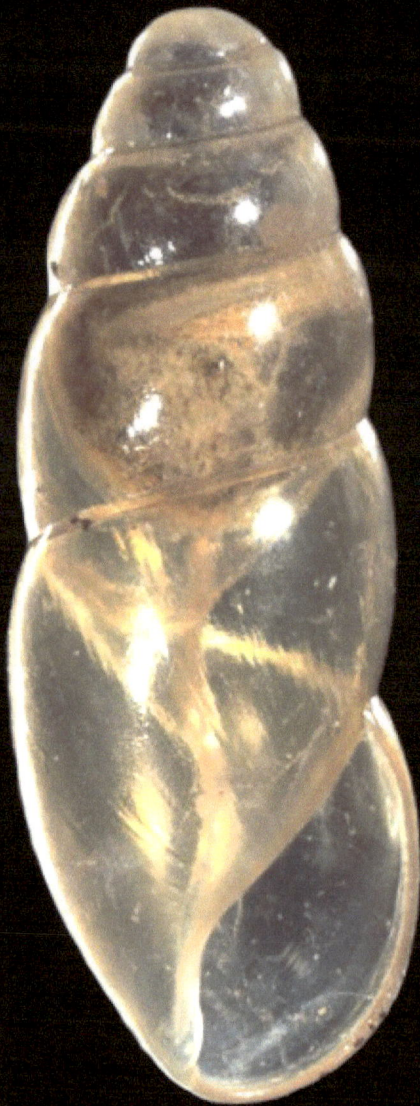

The study of snails is an inquiry into subtlety and masterful sculpture.

Cochlicopa morseana (Doherty, 1878)
"Appalachian pillar"

Land snails in this section are species that have shells taller than wide (Conical, Succiniform, or Pupa-shape) and that have simple lips; although some may have a portion of the lip (usually on the left side) that is slightly reflected. They include small (under 5 mm) but mostly mid-sized (10-20 mm) snails and are found throughout the GSMNP. At first glance, many of these snails look aquatic and some of these species are even referred to as amphibious, living close to, but not in water. Many have a glass-like luster and a near paper-thin but rigid shell. Consequently, these shells after death are quick to break down and disappear. All species in this section are without internal lamellae or teeth in the aperture.

Genera Included:
(in order of appearance in text)

Novisuccinea
Pomatiopsis
Cochlicopa
Columella

I hope there's no salt in here!

Oval ambersnail   Succineidae

Novisuccinea ovalis (Say, 1817)

Height: 10-20 mm

Description: Succiniform; lip simple; shell with 2.5-4.5 whorls; imperforate; no teeth or lamellae present in the aperture at any stage of growth; shell greenish-yellow to amber, weakly striate and near paper-thin; sole of foot bluish-gray, shading to orange at the edges.

Similar Species: *Catinella oklahomarum* (page 261) is smaller and more compact in form.

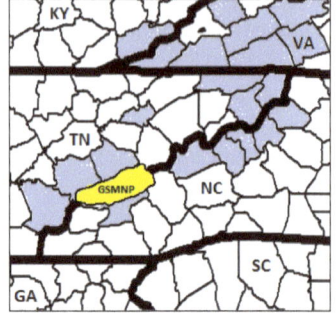

Habitat: This is the ambersnail of the southern mountains, low wet areas, hillsides and high mountain tops up to 2000 m; on jewelweed and stinging nettle.

Status: G5; Uncommon, first documented in the park by Pilsbry at Thunderhead Mountain and Mount LeConte; also documented along the ridge from Clingman's Dome to Starkey Gap (Dourson 2006).

Specimen: North Carolina, Mitchell County, Roan Mountain (author's collection).

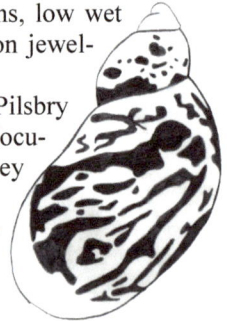

Above illustration showing the live animal as seen through the translucent shell

Roan Mountain, NC

51

Slender walker

Pomatiopsidae

Pomatiopsis lapidaria (Say, 1817)

Height: 5-7 mm

Description: Conical; lip simple; shell with 6-7 whorls; no teeth present at any stage of growth; fresh shells are a rich brown; live animal with an operculum (a), a calcareous door.

Similar Species: *P. cincinnatiensis* is notably smaller, has a more inflated profile and has one less whorl; not known from the GSMNP, but found in scattered eastern Tennessee counties.

Habitat: A calciphile often referred to as amphibious or even aquatic; common on wet limestone rock faces, in dripping seeps, and in mats of algae along small streams but also found in drier and more acidic situations.

Status: G5; Locally Rare, known only from the White Oak Sinks and along Little River near the Foothills Parkway.

Specimen: *P. lapidaria* from Kentucky, Pike County, Breaks Interstate Park (author's collection); *P. cincinnatiensis* from Kentucky, Edmonson County, Mammoth Cave NP (FMNH 229105).

P. lapidaria

a

P. cincinnatiensis

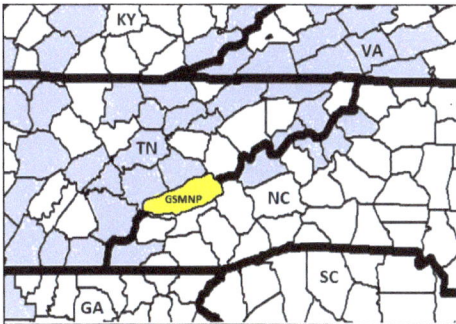

Appalachian pillar

Cochlicopa morseana (Doherty, 1878)

Height: 6.6-7.2 mm

Description: Pupa-shape; shell with 5.5 whorls; sutures slightly impressed; no teeth present at any stage of growth; shell smooth, glossy, translucent with a glass-like surface; foot of live animal white with a grayish head.

Similar Species: *Cochlicopa lubricella* is similar in shape, differing mainly in its smaller size, broader stature and by its slightly wider and thicker callus rim of the outer lip; *Gastrocopta* and *Vertigo* species are much smaller and contain teeth in their aperture; *Columella simplex* is around 4mm smaller and does not have translucent shells.

Habitat: A species that hides well in moist upland woods among deep leaf litter; it is rarely found on top of the litter even in wet weather, staying below or between the layers of leaves.

Status: G5; Relatively Common at elevations below 1200 m.

Specimen: Kentucky, Powell County, Furnace Mountain (author's collection).

Cochlicopidae

C. morseana   *Carychium exile*

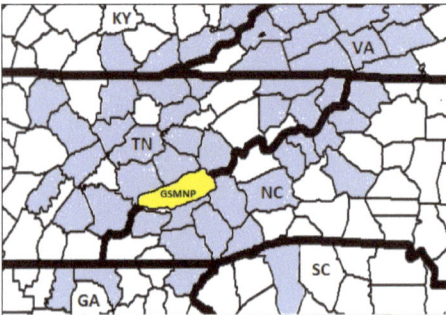

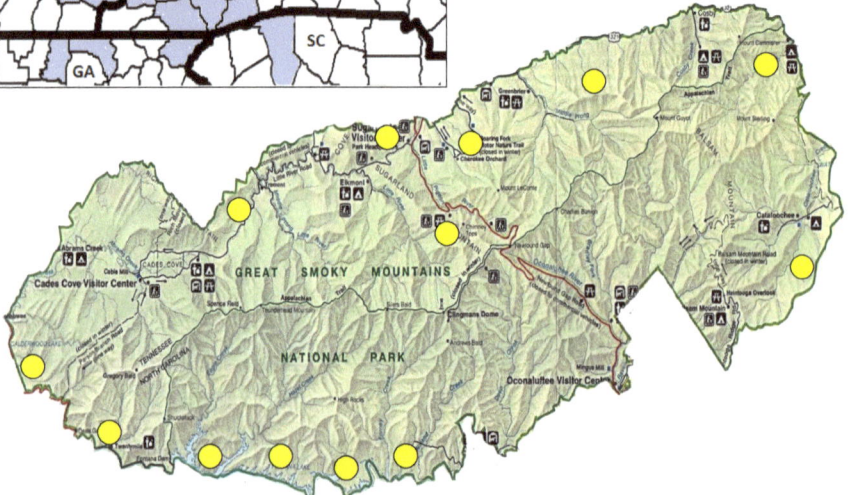

53

Thin pillar

Cochlicopidae

Cochlicopa lubricella (Porro, 1838)

Height: 5-6 mm

Description: Pupa-shape; shell with 5.5-6 whorls; sutures slightly impressed; no teeth present at any stage of growth; shell smooth, glossy, translucent with a glass-like surface.

Similar Species: *Cochlicopa morseana* is larger, narrower in form, its aperture more elongate with a thinner callous rim of the outer lip and is a lighter, more translucent honey color; *Gastrocopta* and *Vertigo* species are much smaller and contain teeth in their aperture.

Habitat: Almost certainly an **EXOTIC** in North America, prefers disturbed sites (yards and road verges) and hard clay soils with thin litter (Nekola 2003).

Status: G5; Locally Rare; currently known only from the Purchase Knob area of the park where it was likely introduced, the snail possibly hiding in transported soils around apple trees that were planted in the past.

Specimen: Kentucky, Wolfe County, yard in Zachariah (author's collection).

Edge of lip thickened

C. morseana   *C. lubricella*

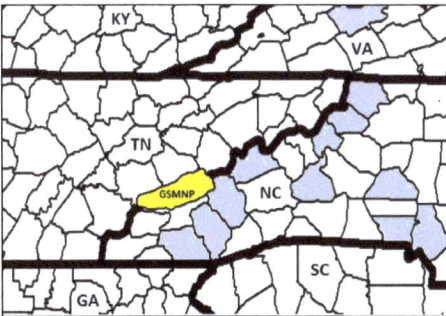

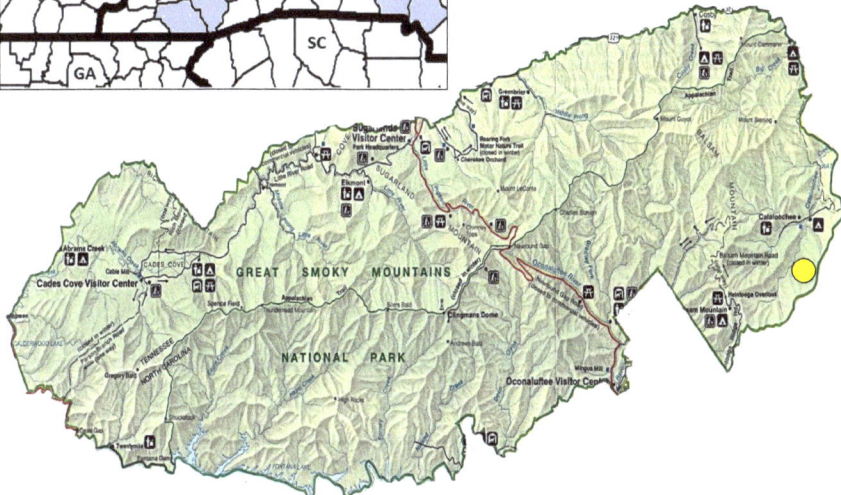

Toothless pupa

Columella simplex (Gould, 1841)

Height: 1.75-2.5 mm

Description: Pupa-shape; shell with a slight taper; left side of lip slightly reflected (a), while the right side is simple; shell with 5-7 whorls that taper over the first 4-5 whorls; perforate; no teeth present in aperture; transverse striae weakly developed; two forms are illustrated, the bottom shell more common in the GSMNP.

Similar Species: *Vertigo* and *Gastrocopta* species have teeth in their aperture, have notably reflected lips; *Carychium* are smaller, more narrow in form and with small teeth.

Habitat: Found across a wide range of open and forested habitats, ranging from subtropical to taiga, xeric to wet, and acidic to calcareous and on ferns and low shrubs, ubiquitous in most upland woods habitats (Nekola and Coles 2010).

Status: G5; Uncommon but found throughout the GSMNP; the classification of this species remains uncertain (Nekola and Coles 2010).

Specimen: Nekola and Coles image (2010).

Truncatellinidae

Two forms illustrated

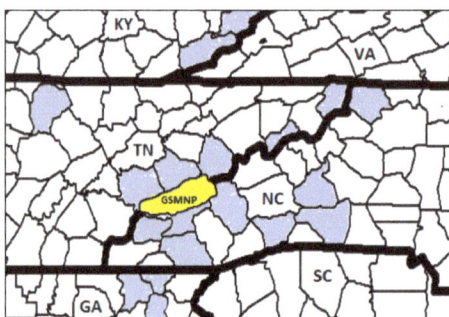

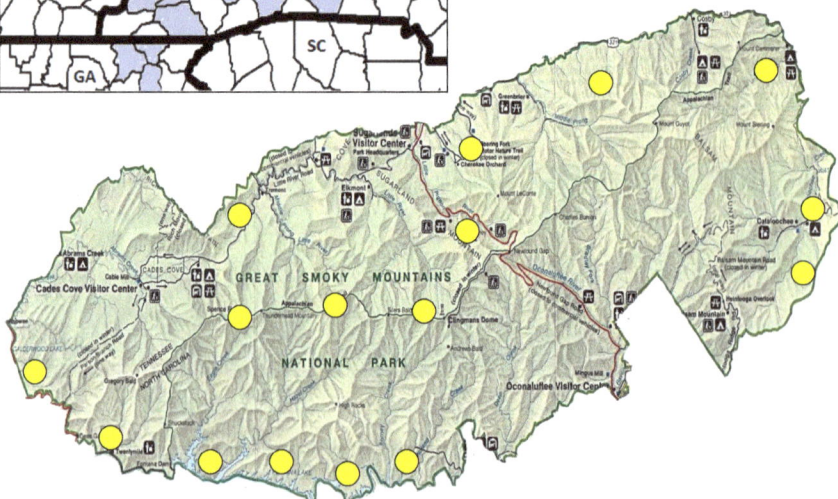

a

This section includes land snails that have shells that are taller than wide (Pupa -shape) and have reflected lips. These are the miniatures of terrestrial gastropods, most under 3 mm as adults. In some habitats such as glades, dry outcrops of limestone or wet seeps, they may represent as much as 75% of the species found. In other nonspecific locations, these small gems (including *Punctum*, *Striatura* and *Guppya*) typically add as much as 45% to the total land snail fauna (Dourson & Beverly, 2009). This is why gathering leaf-litters at every site is so important. In general, fresh dead and live shells of this group are translucent and dead shells becoming bleached white with age. *Vertigo* and *Gastrocopta* species are cryptic, remaining well-hidden in the detritus and leaf litter. These two genera also have teeth in their apertures which are thought to prevent attacks from predators like predacious beetle larva. Although tiny, shells of *Carychium* species tend to be more easily spotted a result of their lighter color.

Genera Included:
(in order of appearance in text)

Carychium
Gastrocopta
Vertigo

Don't say SNAIL MAIL like it's a bad thing!

Tree thorn   Ellobiidae

Carychium arboreum (Dourson, 2012)

Height: 1.9-2.1 mm

Description: Pupa-shape; lip reflected; shell tiny with 4.5 whorls; as in all *Carychium* species, there are two horizontal entering lamellae that can be seen on the left side of the aperture (strong lens required); transverse striae are well-developed and widely spaced (see below); no spiral striae; shells are translucent in live individuals and fresh shells.

Similar Species: *Carychium exile* is shorter in height but is wider in stature; *C. clappi* has finer transverse striae (see below comparison); *C. nannodes* is smaller with a smoother surface.

Habitat: Found in elevated tree crotches that contain ample leaf litter fill and on occasions, well-developed rock talus in lower elevation mixed hardwood.

Status: G1; Globally Rare; Endemic to the GSMNP.

Specimen: North Carolina, Swain County, Forney Creek, GSMNP (GSMNP collection).

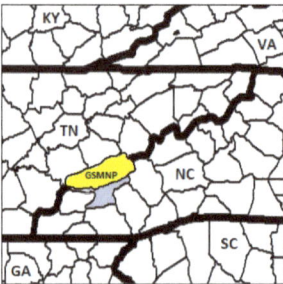

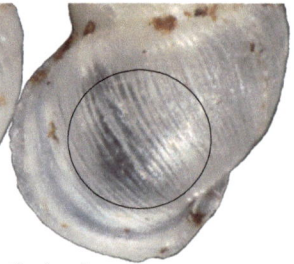

C. arboreum   *C. clappi*

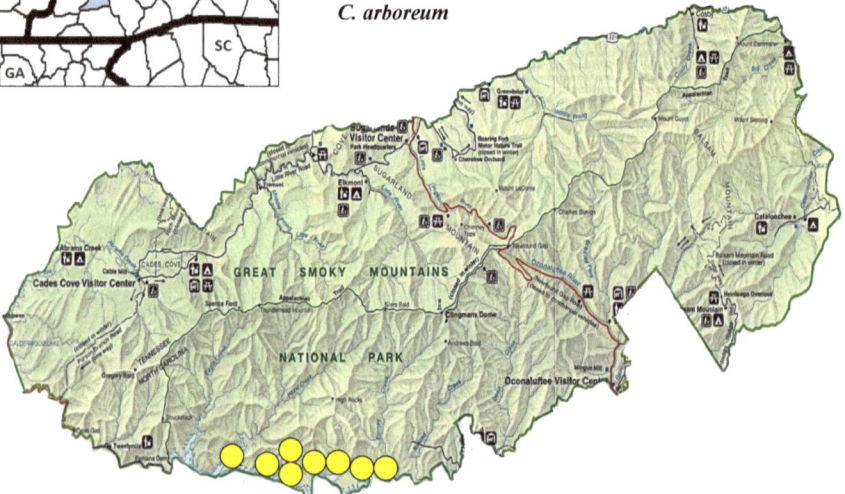

Appalachian thorn

Ellobiidae

Carychium clappi (Hubricht, 1959)

Height: 1.7-1.8 mm

Description: Pupa-shape; lip reflected; shell tiny with 4.5 whorls; two horizontal entering lamellae that can be seen on the left side of the aperture; transverse striae well-developed and closely-spaced; shells translucent in live individuals, becoming bleached white with age.

Similar Species: *Carychium exile* is shorter in height but is wider in stature and has less distinct transverse striae on the first three whorls; *C. nannodes* is smaller and has a smoother surface.

Habitat: Found living between layers of moist leaves located in moist depressions on hillsides; it is often found with *C. exile* and *C. nannodes* and on occasion with *C. arboreum*.

Status: G5; Common throughout the park at all elevations.

Specimen: Tennessee, Blount County, White Oaks Sinks, GSMNP (GSMNP collection).

Three most common *Carychium* in the park (proportionate)

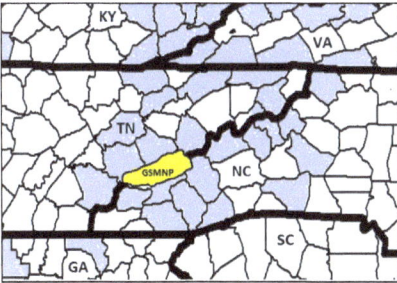

C. nannodes   *C. exile*   *C. clappi*

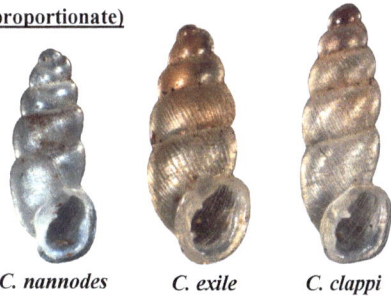

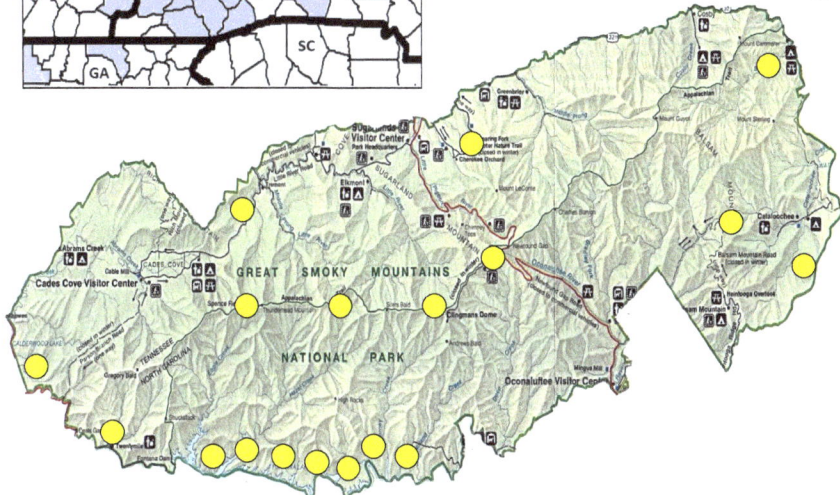

Ice thorn   Ellobiidae

Carychium exile H. C. Lea, 1842

Height: 1.5-1.8 mm, average size 1.75 mm

Description: Pupa-shape; lip reflected and thickened; shell with 4.5-5 whorls; there are two horizontal entering lamellae on the left side of the aperture (see below enlarged aperture view); transverse striae are relatively well-developed; no spiral striae; shells are translucent in live individuals and fresh shells.

Similar Species: *C. clappi* is taller, more narrow in form and the transverse striae are more pronounced on the first three whorls; *C. nannodes* is smaller and lacks distinct transverse striae.

Habitat: In pockets of moist leaves, often found living in low depressions and around small seeps.

Status: G5; Common; the most frequent *Carychium* in the park and found at nearly all elevations below 1200 m.

Specimen: Tennessee, Blount County, White Oak Sinks, GSMNP (GSMNP collection).

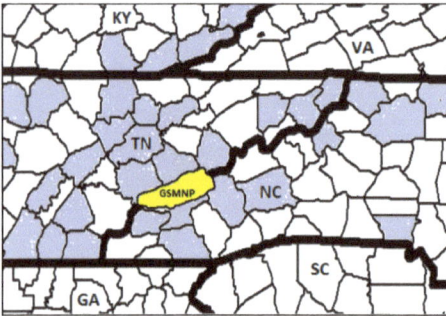

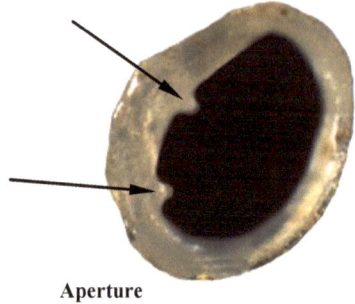

Aperture

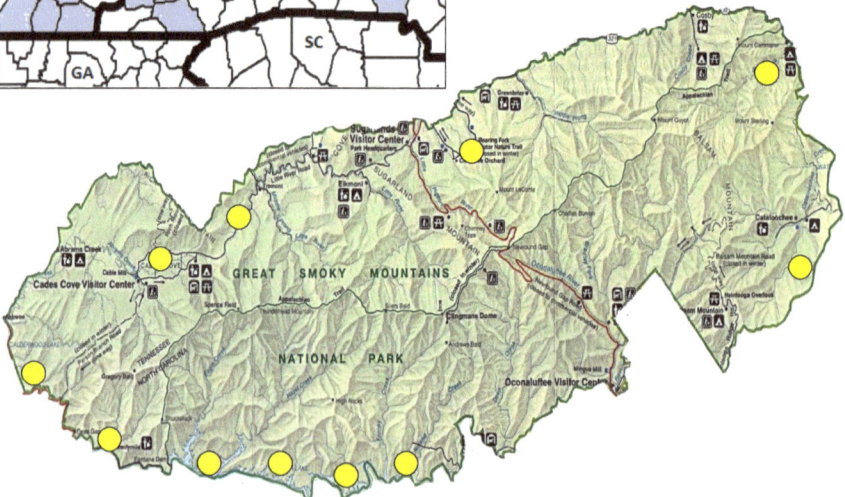

Obese thorn

Ellobiidae

Carychium exiguum (Say, 1822)

Height: 1.5-1.6 mm

Description: Pupa-shape; lip widely reflected; shell with 4.5-5 whorls; there are two horizontal entering lamellae seen on the left side of the aperture; transverse striae are poorly developed and at times, hardly perceptible; no spiral striae; shells are translucent in live individuals.

Similar Species: *Carychium clappi* is slightly larger, more narrow in form, with transverse striae that are more distinct; *C. exile* is around the same height but is narrower (the most important feature separating the two species) and has more distinct transverse striae.

Habitat: Found in pockets of moist decaying leaves found in sinkholes or around the entrances of caves.

Status: G5; Locally Rare; most populations are known from Cades Cove, White Oaks Sinks and several locations above Fontana Lake, expected to occur elsewhere in the park.

Specimen: Tennessee, Blount County, Cades Cove, GSMNP (GSMNP collection).

Aperture

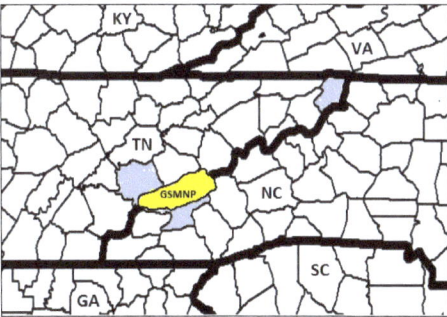

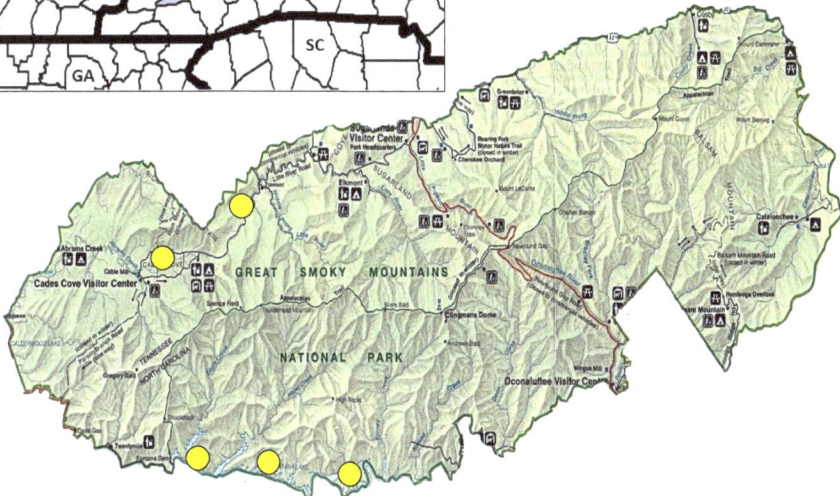

File thorn   Ellobiidae

Carychium nannodes (Clapp, 1905)

Height: 1.3-1.5 mm

Description: Pupa-shape; lip reflected; shell with 4.5-5 whorls; sutures notably deep; there are two horizontal entering lamellae that can be seen on the left side of the aperture (strong lens required); transverse striae are poorly developed and at times, hardly perceptible; shells are translucent when fresh but become bleached with age. This is the smallest *Carychium* species in the GSMNP and southern Appalachian Mountains

Similar Species: *Carychium exile* and *C. clappi* are both larger with distinct transverse striae.

Habitat: Found in pockets of moist decaying leaves in low depressions, also common around walnut, butternut, basswood and buckeye trees.

Status: G5; Common throughout the GSMNP in elevations below 900 m.

Specimen: Kentucky, Powell County, Furnace Mountain (author's collection).

Aperture

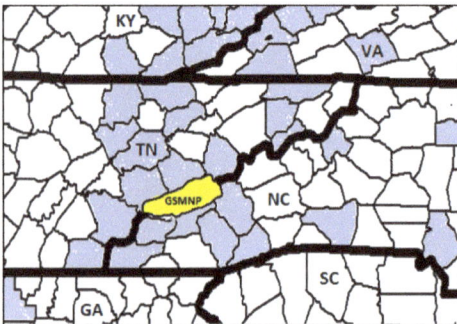

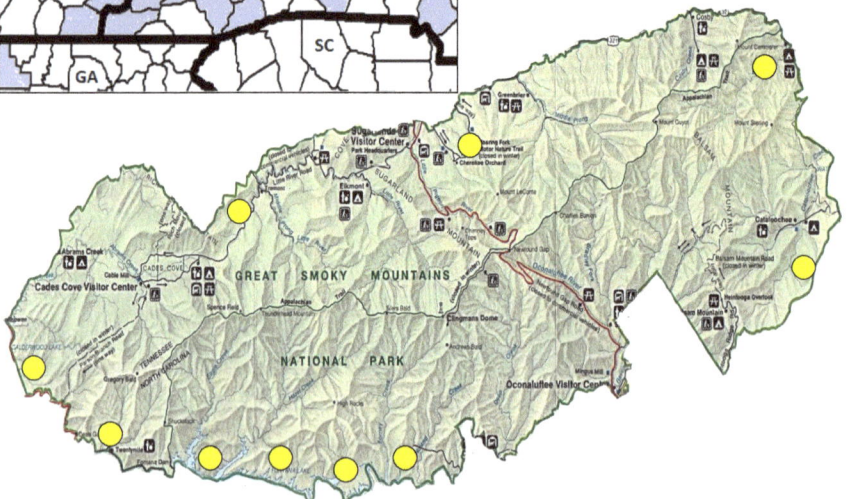

Carychium exile H. C. Lea, 1842
"Ice thorn"

Key to *Gastrocopta* and *Vertigo* Species

Gastrocopta and *Vertigo* are tiny snails with Pupa-shape shells typically with well-developed teeth in the aperture. In general, they differ from *Carychium* by their larger, more obese and toothier shell. *Gastrocopta* species are generally larger than *Vertigo*, have wider, more reflected lips (a) and are without a small dent (b) in the right side of the aperture. Shell surface for either group can be smooth or with fine transverse striae but are typically without spiral striae. To better illustrate the teeth and their spatial arrangement in the aperture of these small gastropods both illustrations and photographs are used.

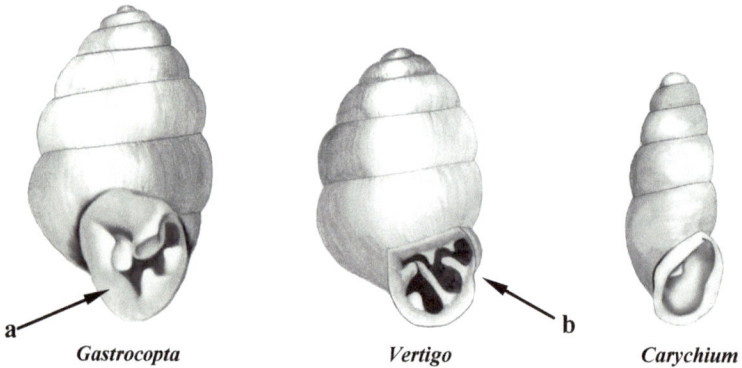

Gastrocopta   *Vertigo*   *Carychium*

Terminology Used in Pupillidae Identification (Nekola and Coles 2010).

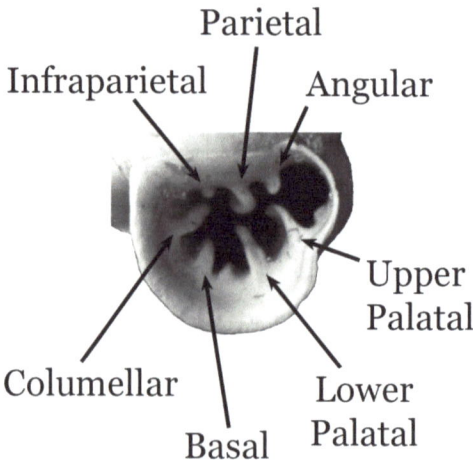

Parietal

Infraparietal   Angular

Upper Palatal

Columellar   Lower Palatal

Basal

Location of major apertural lamellae used in Pupillid identification

63

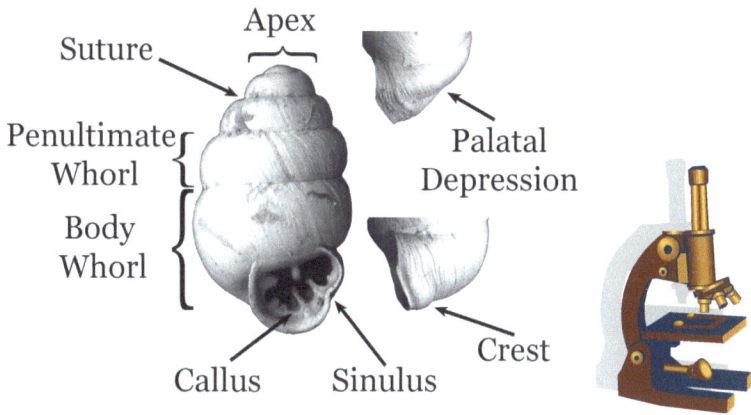

Major shell features used to identify Pupillid species

Keys to *Gastrocopta* and *Vertigo* Species in the GSMNP

(modified from Nekola and Coles (2010) "Annotated Keys To Eastern North America Pupillids")

Gastrocopta

1. Angulo-parietal lamella a single large, folded sheet ....*G. contracta*

2. Columellar lamella with both forward and basally pointing components, appearing more or less pyramidal in apertural view; shell usually >4 mm tall ....
....*G. armifera*

3. Shell narrowly conical, with height more than 1½ times width; lower palatal lamella deeply entering the aperture
....*G. pentodon*

4. Basal and palatal lamellae absent; shell >2.4 mm tall..........................*G. corticaria*

Vertigo

1. Strong apertural lamellae and sinulus; shell translucent, aperture as tall as wide; columellar lip of aperture rounded, not markedly broad; infra-parietal lamella often present ....*V. ovata*

2. Lower palatal lamella curved and deeply entering aperture; shell height <1.9 mm.
....*V. milium*

3. Body whorl narrower than penultimate whorl, making shell bluntly pointed at both top and bottom; four lamellae, with an elongate vertical columellar
....*V. oscariana*

4. Shell height >1¾ mm; a weak upper palatal lamella often present.
....*V. tridentata*

5. Shell height <1¾ mm; upper palatal lamella absent
....*V. parvula*

6. Striae indistinct, with shell often appearing smooth under low (×10) magnification; single deep depression over both palatal lamellae;~1¾ mm
....*V. bollesiana*

7. Striae distinct, with shell not appearing smooth under low (×10) magnification; palatal depression weak or absent; most forms >1¾ mm tall, with small southern Appalachian forms being ~1¾ mm tall... *V. gouldii*

Bottleneck snaggletooth

Gastrocoptidae

Gastrocopta contracta (Say, 1822)

Height: 2.2-2.5 mm

Description: Pupa-shape; lip widely reflected; shell with 5.5 whorls; with large teeth, the parietal tooth the largest (a); spiral of shell tapering; aperture more or less triangular in shape; shells are somewhat translucent in live individuals and fresh shells are frequently covered with soil; old shells quickly become bleached.

Similar Species: *Gastrocopta armifera* is larger, has 6-7 whorls, the aperture teeth are less crowded; the lamella configuration of *G. contracta* is also very different than *G. armifera* and has a tapered (not ovate) shell shape.

Habitat: This species is found in nearly all terrestrial habitats including but not limited to low wet places to dry mountainsides and acidic forests.

Status: G5; Common and widespread; along with *G. pentodon,* this is one of the most common *Gastrocopta* species in the GSMNP.

Specimen: Nekola and Coles image (2010).

a

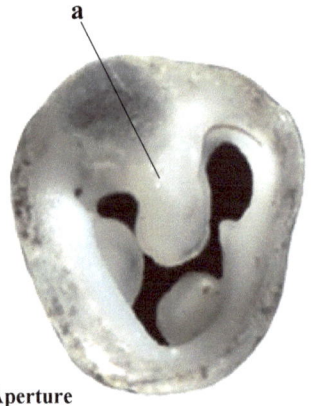

Aperture

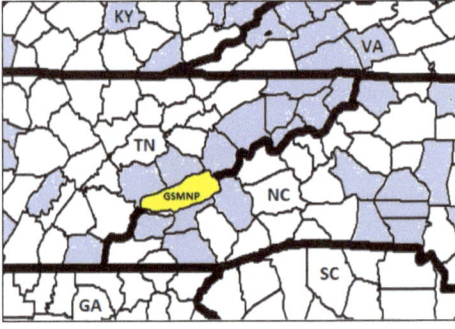

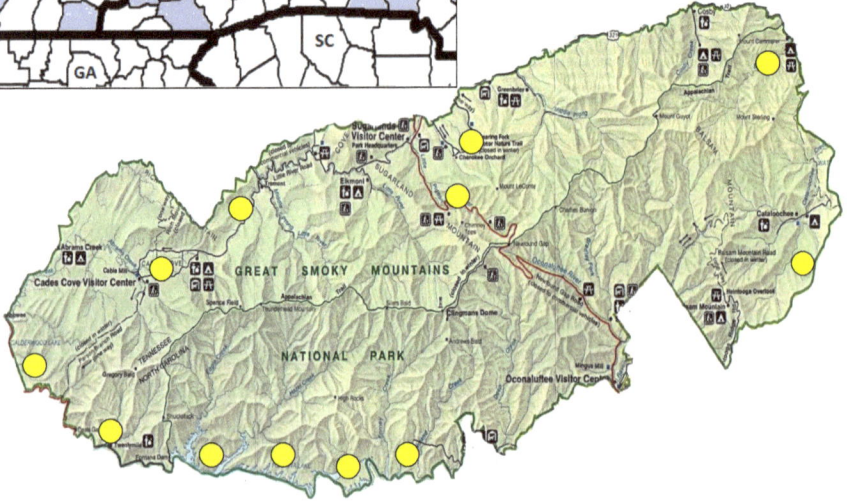

Armed snaggletooth

Gastrocoptidae

Gastrocopta armifera (Say, 1821)

Height: 3-4.8 mm

Description: Pupa-shape; lip reflected; shell with 6.5-7.5 whorls; aperture crowded with large teeth, the parietal tooth large and forked (b); columellar lamella with both forward and basally pointing components, appearing more or less pyramidal in apertural view (Nekola and Coles 2010).

Similar Species: *Gastrocopta contracta* is smaller, having only 5.5 whorls and the aperture teeth are notably larger, choking the aperture; other *Gastrocopta* species are notable smaller.

Habitat: A calciphile found in more open, sunny areas such as cedar glades and limestone outcrops.

Status: G5; Locally Rare; known only from the western side of the park above the Little Tennessee River, this species has apparently remained on the fringes of the park.

Specimen: Nekola and Coles image (2010).

b

Aperture

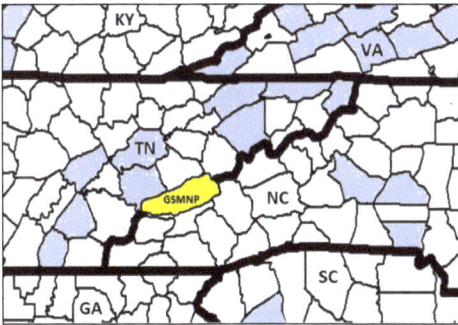

Comb snaggletooth

Gastrocopta pentodon (Say, 1821)

Height: 1.5-1.8 mm

Description: Pupa-shape; lip narrowly reflected; shell with 5 whorls; parietal tooth is the largest (a); outer lip with a distinct palatal callus (b) or ridge usually containing 5 or 6 teeth but sometimes up to 9 teeth; shells are somewhat translucent in live individuals and fresh shells are typically covered in soil.

Similar Species: *Gastrocopta contracta* and *G. armifera* are both larger, containing less but larger teeth in their aperture.

Habitat: A species that is found in a variety of habitats including dry upland hardwood forests, acidic forests such as pine savannas, around limestone outcrops and occasionally found in low wet places.

Status: G5; Common and widespread; the most common *Gastrocopta* species in the GSMNP, found at practically all but the highest elevations.

Specimen: Nekola and Coles image (2010).

Gastrocoptidae

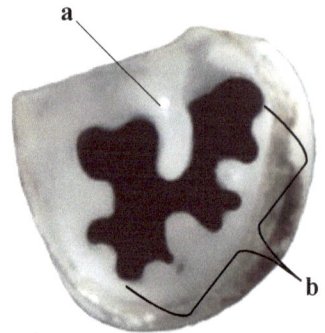

Aperture

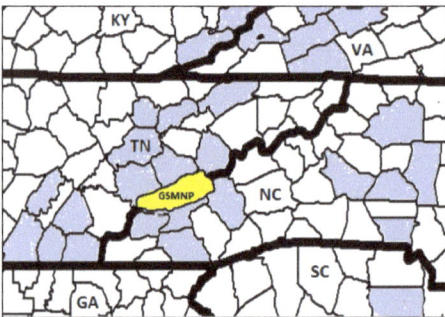

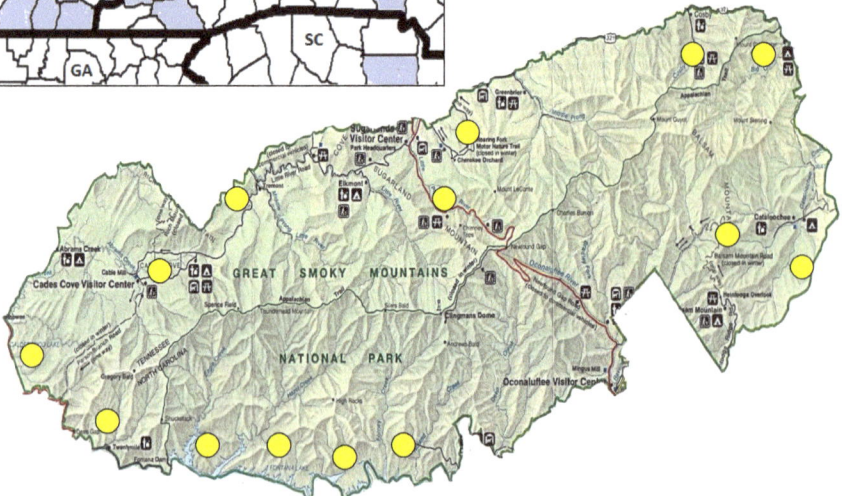

Bark snaggletooth

Gastrocoptidae

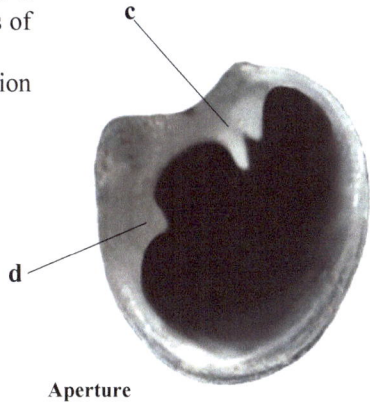

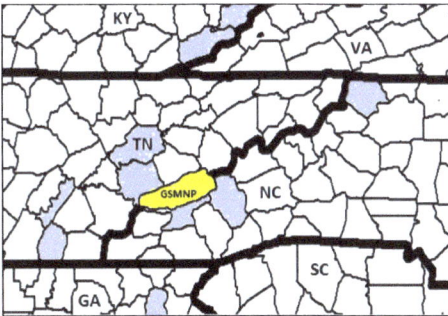

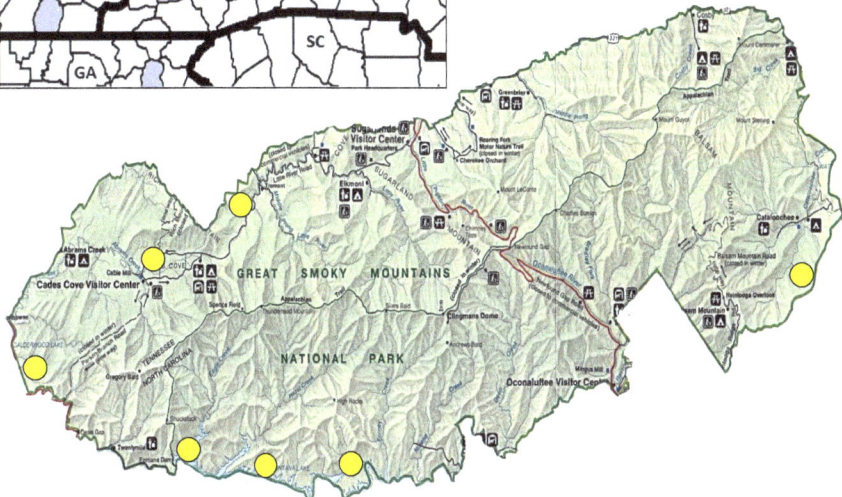

Gastrocopta corticaria (Say, 1816)

Height: 2.5 mm

Description: Pupa-shape; lip reflected; shell with 5.5 whorls; one small forked tooth present on the parietal wall (c) and one small columellar tooth (d); outer palatal wall without teeth; shells are somewhat translucent in live individuals and fresh shells.

Similar Species: This is the only *Gastrocopta* species in the GSMNP to possess only two teeth, lacking any lamellae or teeth on the outer (palatal) wall of the aperture; *Columella simplex* is without any teeth in the aperture and has a simple non-reflected lip.

Habitat: A calciphile found in mixed hardwood forests with limestone outcrops but also found in acidic forests where it may be restricted to mossy tree trunks; in wet weather it can be found crawling on logs and the trunks of trees (Hubricht 1985).

Status: G5; Uncommon with scattered location across the park.

Specimen: Nekola and Coles image (2010).

c

d

Aperture

68

Ovate vertigo

Vertigo ovata (Say, 1822)

Vertiginidae

Height: 2.2-2.3 mm

Description: Pupa-shape; left side of lip reflected while the right side of the lip is simple containing a small dent in the outer aperture wall, a common character of *Vertigo*; shell with 4.5-5 whorls; the 6-9 teeth are well-developed crowding the aperture.

Similar Species: *Vertigo ovata* does not have a long, curved lower palatal like *V. milium*; it is strongly ovate, shell volume easily 5x more than *V. milium* (Nekola and Coles 2010).

Habitat: Found on cattail leaves in swamps, sedge meadows, wet and mesic prairie, low calcareous meadows, river banks and upland forests (Nekola and Coles 2010).

Status: G5; Locally Rare; in the park known only from the Cades Cove area where it was obtained from ATBI (Discover Life in America) pitfall traps.

Specimen: Nekola and Coles image (2010).

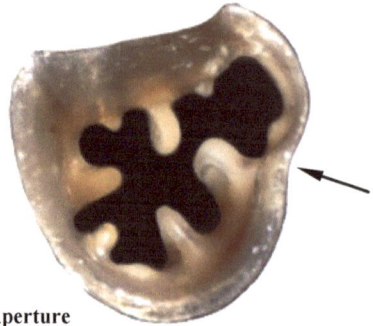

Aperture

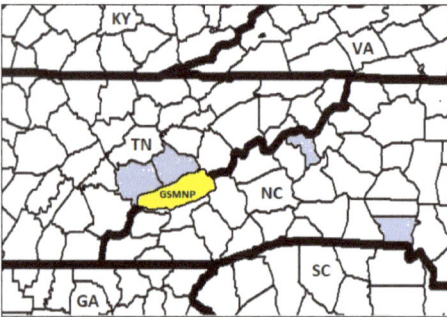

Blade vertigo

Vertiginidae

Vertigo milium (Gould, 1840)

Height: 1.4-1.8 mm

Description: Pupa-shape; small dent in the outer aperture wall; shell with 4.5-5 whorls; usually with 6 teeth that crowd the aperture, one of the palatal teeth especially long (a); shell surface glossy, weakly striate; shells are somewhat translucent in live individuals and fresh shells but become bleached white with age.

Similar Species: The inner teeth are relatively long and fill the aperture; the long, curved lower palatal and the broad inward-sloping columellar plate of *V. milium* are very different from the peg-shaped columellar of *V. ovata*.

Habitat: Found on cattail leaves in swamps, sedge meadows, wet and mesic prairie, low calcareous meadows, river banks, and wooded wetlands (Nekola and Coles 2010).

Status: G5; Locally Rare; not a common species in the southern mountains; known only from Cades Cove and Twin Creeks.

Specimen: Nekola and Coles image (2010).

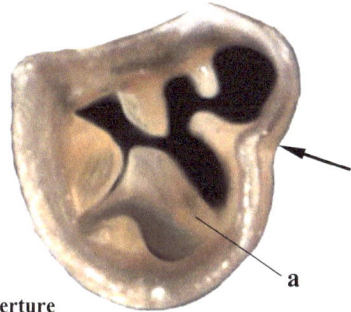

Aperture

a

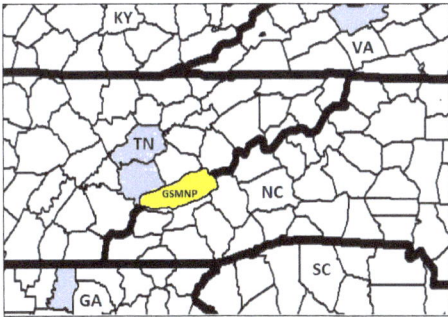

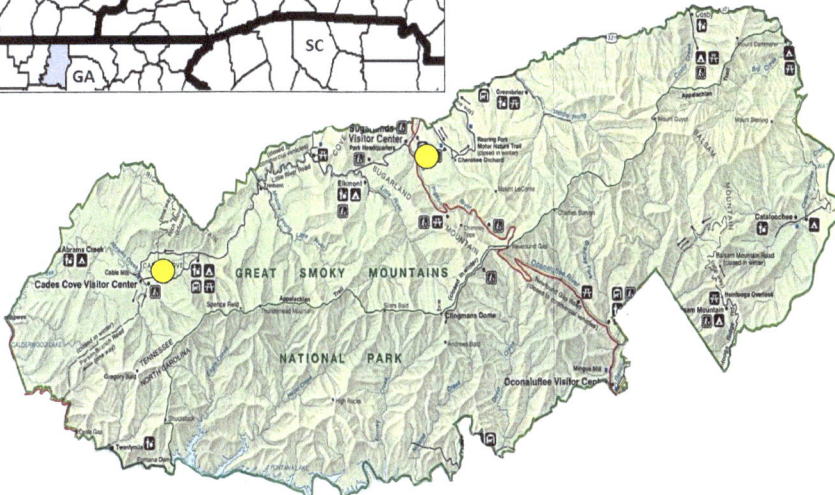

Capital vertigo

Vertiginidae

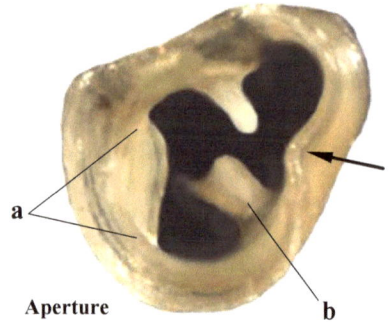

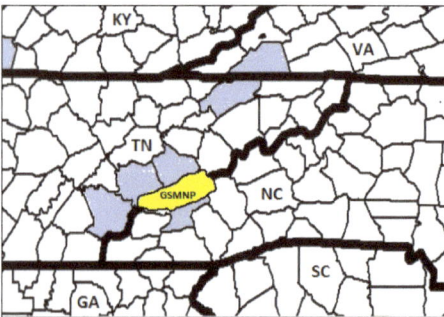

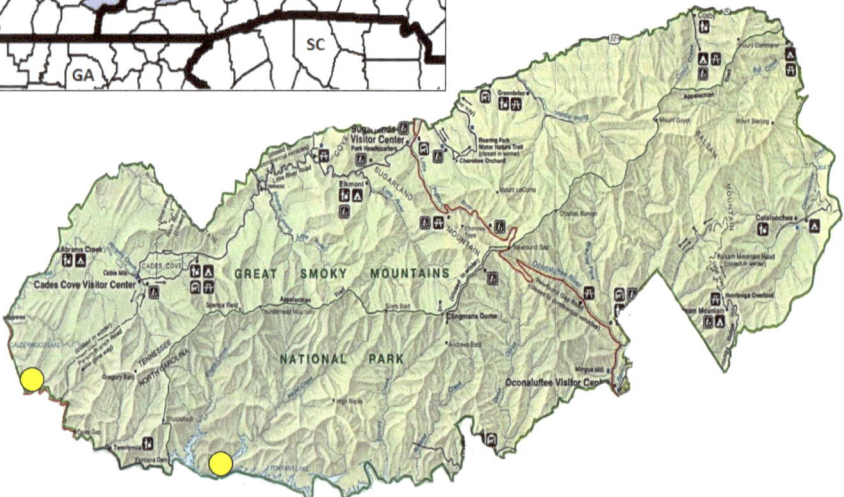

Vertigo oscariana (Sterki, 1890)

Height:1.4-1.6 mm

Description: Pupa-shape; small dent in the outer aperture wall; shell with 4.5-5 whorls; 3 medium-sized teeth, columellar lamella blunt and wide (a); palatal fold (tooth) short, thick and set deeply within the aperture (b); shells are translucent in live individuals; old shells become bleached white.

Similar Species: *Vertigo parvula* has a peg-shaped, not sheet-like columellar; *V. oscariana* is the only *Vertigo* which has a body whorl much less wide than the penultimate (Nekola and Coles 2010).

Habitat: Found in well-decomposed accumulations of broadleaf and pine litter in mesic-wet woodlands and shaded rock outcrops (Nekola and Coles 2010).

Status: G4; Locally Rare; known only from around Fontana Lake and the western edge of the park above the Little Tennessee River.

Specimen: Nekola and Coles image (2010).

a

b

Aperture

Honey vertigo

Vertiginidae

Vertigo tridentata (Wolf, 1870)

Height: 1.8-2.3 mm

Description: Pupa-shape; outer lip containing a small dent in the outer aperture wall (c); shell with 5 whorls; 3 or 4 short teeth, aperture semi-circular in shape; shells are somewhat translucent in live individuals, but like most *Vertigo* species becoming bleached with age.

Similar Species: The shell color of *V. tridentata* is honey-brown whereas *V. gouldii* is a dark-brown; the parietal lamella is pointing at the lower palatal in *V. tridentata*, not the upper palatal as seen in *V. gouldii* and *V. bollesiana* (Nekola and Coles 2010).

Habitat: A species found climbing on herbs in the Lamiaceae family (mint family) in low sunny places (Hubricht 1985), bedrock glades, in well decomposed leaf litter accumulations on shaded cliff ledges and talus (Nekola and Coles 2010).

Status: G5; Locally Rare; currently known from White Oak Sinks only.

Specimen: Nekola and Coles image (2010).

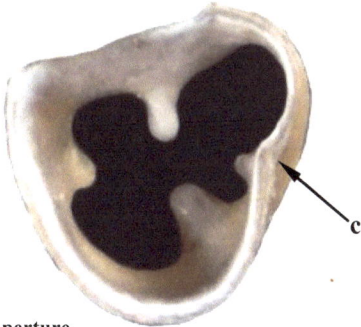

Aperture

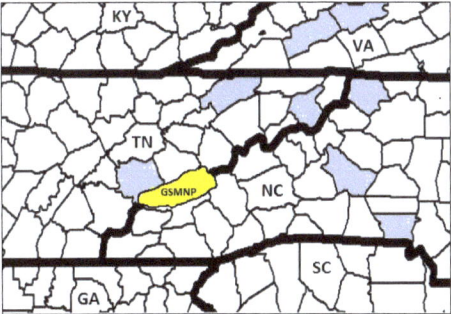

Smallmouth vertigo

Vertiginidae

Vertigo parvula Sterki, 1890

Height: 1.4-1.6 mm

Description: Pupa-shape; lip dent weakly developed; shell with 5 whorls; 3 rather large teeth; columellar lamella small; as with most Vertiginidae and other small species, the shells are somewhat translucent in live individuals.

Similar Species: *V. parvula* stands closest to *V. tridentata*, but is consistently lacking the upper palatal, has a short, elongated lower palatal (a) and is also much smaller than *V. tridentata* (Nekola and Coles 2010); *V. oscariana* has a columellar lamella that is notably wider and the palatal tooth is shorter and sits a little deeper in the aperture.

Habitat: Found in accumulations of well-decomposed leaf litter in base-rich cove forests, rock outcrops, and talus slopes at mid-low elevations (Nekola and Coles 2010).

Status: G3; Locally Rare; known from areas along the Foothills Parkway.

Specimen: Nekola and Coles image (2010).

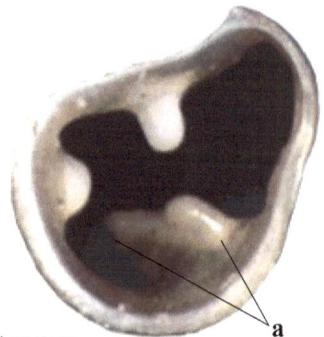

Aperture

a

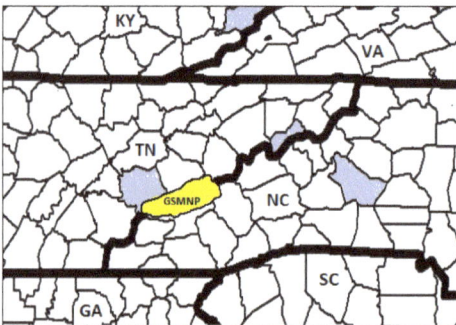

Delicate vertigo

Vertiginidae

Vertigo bollesiana Morse, 1865

Height: 1.5 mm

Description: Pupa-shape; the dent in the outer aperture wall is weakly developed or imperceptible; shell with 4.5-5 whorls; 5 teeth, parietal lamella and palatal folds well-developed; shell surface has poorly developed transverse striae; this species always shows a strong depression on the outside of the shell over both palatal lamellae (Nekola and Coles 2010).

Similar Species: *Vertigo gouldii* is larger and in general, better developed transverse striae; *V. bollesiana* has a large depression (outside of shell) over the palatals (see page 64), while *V. gouldii* is without this feature; also *V. bollesiana* tends to be a bit wider for its height than *V. gouldii.*

Habitat: Found in leaf litter under shrubs, on cliff face ledges and boulder tops in mesic upland forests and white cedar wetlands (Nekola and Coles 2010).

Status: G4; Locally Rare; Clingman's Dome.

Specimen: Nekola and Coles image (2010).

Aperture

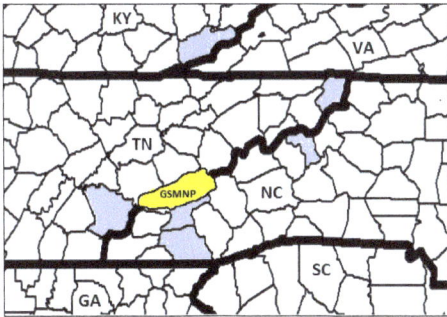

Variable vertigo

Vertigo gouldii (A. Binney, 1843)

Vertiginidae

Height: 1.5-2.1 mm

Description: Pupa-shape; the dent in the outer aperture wall is weakly developed or imperceptible; shell with 4.5-5.5 whorls; 4 to 5 teeth, all around the same size; parietal lamella and palatal folds well-developed; shell surface has well-developed transverse striae; shells are somewhat translucent in live individuals and fresh shells but become bleached with age.

Similar Species: *Vertigo bollesiana* is smaller, the transverse striae are poorly developed and most importantly *V. bollesiana* has a large depression on the outside of the shell over the palatal lamellae (see page 64) whereas *V. gouldii* has a small depression (Nekola and Coles 2010).

Habitat: The species is limited to forested sites and is most common on wooded bedrock outcrops.

Status: G5; Common; the most common *Vertigo* species in the GSMNP.

Specimen: Nekola and Coles image (2010).

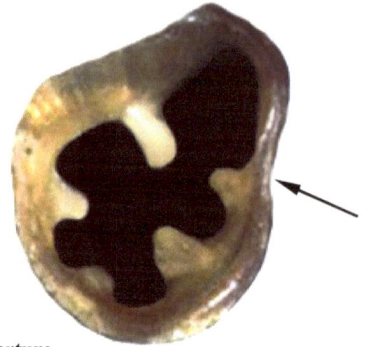

Aperture

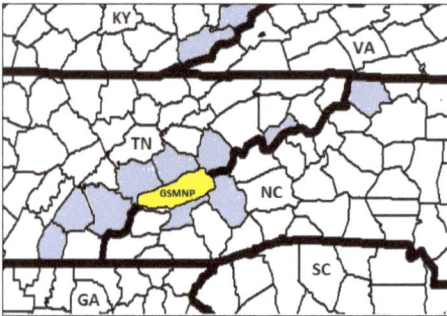

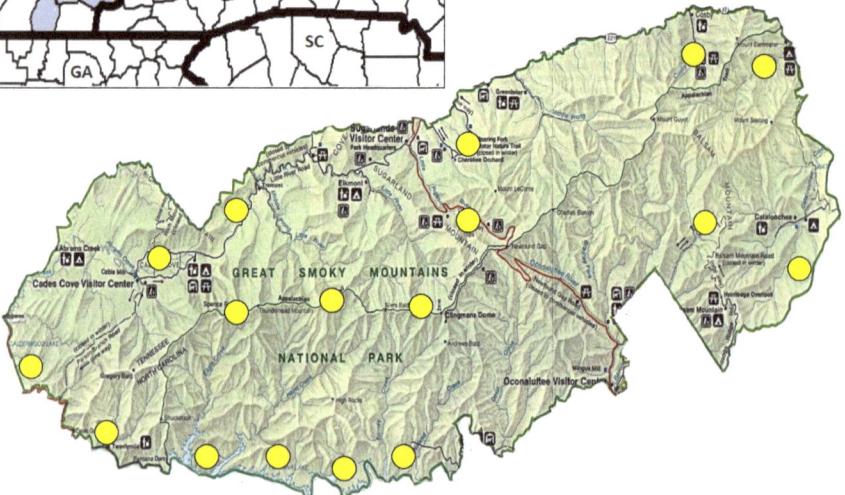

Land snails in this section have shells that are wider than tall, either heliciform or depressed heliciform and have simple lips (although a few may have slightly reflected lips). They include minute snails from 1.5 mm to snails well over 35 mm including many of the mid-size snails occurring in the GSMNP. Several of the smaller species in this section such as *Glyphyalinia, Helicodiscus,* and *Striatura* are in company with some of the more beautifully sculptured gastropods in North America. Many have glasslike, transparent shells while others display some of the most extraordinary micro-ornamentation found on any organism. This outstanding craftsmanship, however, can only be appreciated under the lens of a strong scope.

Genera Included:
(in order of appearance in text)

Punctum
Striatura
Hawaiia
Lucilla
Guppya
Zonitoides
Gastrodonta
Glyphyalinia
Paravitrea
Discus
Helicodiscus
Ventridens
Mesomphix
Anguispira
Haplotrema
Vitrinizonites
Euconulus
Hendersonia

Simple lip

Does my helix make me look fat?

Land snails under 3.5 mm, umbilicate (except *G. sterkii*), simple lip, typically without teeth or lamellae

Land snails in this section are less than 3.5 mm in diameter, without reflected lips and generally found in moist leaf litter and detritus. To separate these woodland gems, it is essential that a strong hand lens or dissecting microscope be employed, otherwise a correct ID will be nearly impossible. Each genus has its own unique ornamentation; some species like *Punctum* and *Striatura* are some of the most beautiful land snails known! These microscopic features are key for a successful identification. If you have small specimens collected from leaf litter that fit the above discussed categories, then proceed with the following separations below.

Grouping 1: shells under 1.5 mm, umbilicate, whorls tightly coiled (b), with crisscrossing striae (a), pg 78

Punctum minutissimum
Punctum vitreum
Punctum blandianum
Punctum smithi (small tooth)

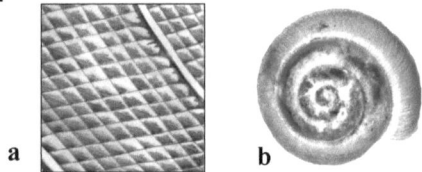

a   b

Grouping 2: shells between 1.5 mm to 3.5 mm, umbilicate, whorls loosely coiled (c), with crisscrossing striae (a), pg 83

Striatura meridionalis
Striatura ferrea

c

Grouping 3: shells between 1.5 mm to 2.8 mm, umbilicate, surface without crisscrossing striae, with a dull-sheen, pg 85

Hawaiia minuscula

Grouping 4: shells between 2.2 mm, umbilicate, surface without crisscrossing striae, smooth and polished, pg 86

Lucilla scintilla

Grouping 5: shells between 1.2 to 1.3 mm, perforate, surface without crisscrossing striae, smooth, translucent and very glossy, pg 87

Guppya sterkii

Small spot

Punctidae

Punctum minutissimum (I. Lea, 1841)

Diameter: 1.1-1.3 mm

Description: Depressed heliciform; lip simple; aperture roundish, not reflected; shell with 3.5-4.5 whorls; umbilicate; pale brown to corneous; no teeth present; transverse and spiral striae that crisscross each other, this amazing micro-feature is so small it can only be seen under the lens of a strong microscope; fresh and live shells are thin and translucent.

Similar Species: *Punctum vitreum* has a smaller umbilicus and has more widely-spaced ribs; *P. blandianum* has a larger umbilicus and flatter shell, its micro-surface features however are the same as in *P. minutissimum*.

Habitat: Found in deep moist pockets of leaf litter in depressions or around logs; between the layers of moist, matted leaves around small seeps in mature hardwood forests.

Status: G5; Common, the most common *Punctum* species in the GSMNP, occurring at all elevations.

Specimen: Kentucky, Letcher County, Bad Branch NP (author's collection).

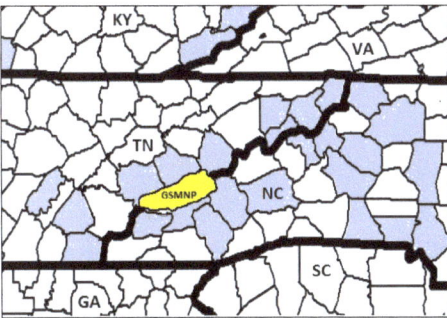

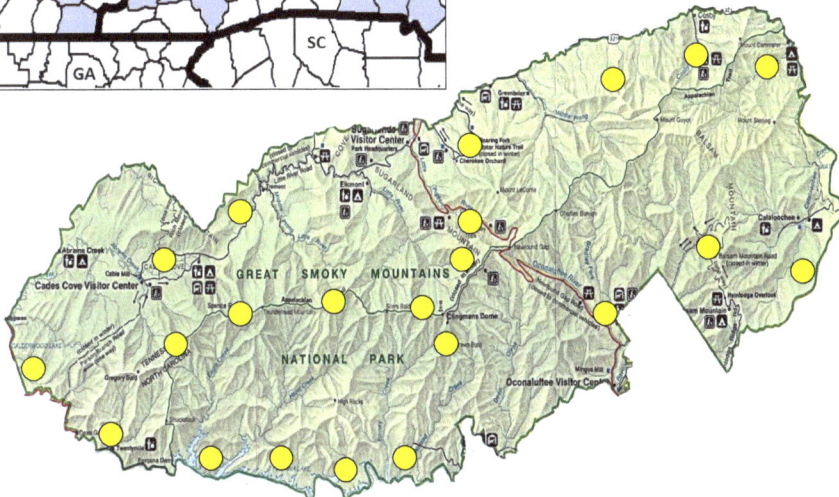

Glass spot

Punctidae

Punctum vitreum (H. B. Baker, 1930)

Diameter: 1.2-1.4 mm

Description: Depressed heliciform; lip simple; aperture roundish; shell with 4-4.5 whorls; umbilicate; corneous to colorless; no teeth present; transverse and spiral striae that are more rib-like and crisscrossing each other (strong microscope required to see these microscopic features); fresh shells are thin, fragile and more or less translucent.

Similar Species: *Punctum minutissimum* has a slightly wider umbilicus and has less widely-spaced ribs; *P. blandianum* has a notably larger umbilicus but its micro-surface features are more or less the same as seen in *P. minutissimum*.

Habitat: Found between the moist, matted leaves around small seeps in mixed hardwoods.

Status: G5; Uncommon; records include areas along Forney Creek, Big Creek, White Oak Sinks, Foothills Parkway and the west end of TN side around the Calderwood area.

Specimen: North Carolina, Haywood County, Purchase Knob (GSMNP collection).

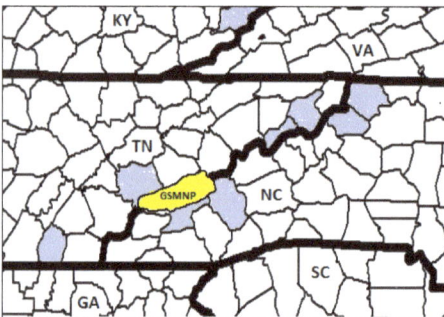

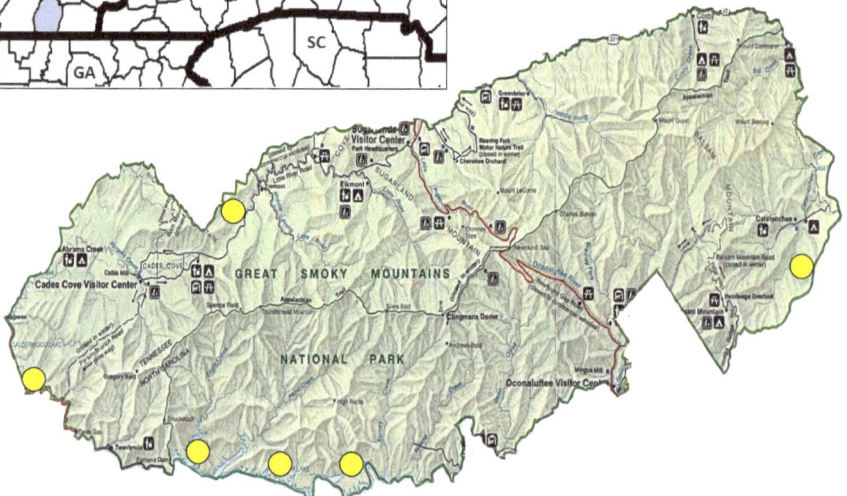

Brown spot

Punctidae

Punctum blandianum (Pilsbry, 1900)

Diameter: 1.1-1.3 mm

Description: Depressed heliciform; lip simple; aperture roundish to slightly oval-shaped; shell with 4 whorls; widely umbilicate; pale brown; no teeth present; transverse and spiral striae that are rib-like and well-developed (strong microscope required); fresh and live shells thin and translucent, but older shells will bleach with time.

Similar Species: *Punctum minutissimum* is the same size but has a notably smaller umbilicus; *P. vitreum* has more widely spaced ribs than is seen in *P. blandianum,* a smaller umbilicus and a more elevated, compact shell.

Habitat: Found in deep moist pockets of leaf litter resting in depressions or around logs but also between the layers of moist, matted leaves found around small seeps.

Status: G5; Uncommon; as leaf litter surveys continue more sites will no doubt be found.

Specimen: Tennessee, Rutherford County, Murfreesboro, (FMNH 235261)

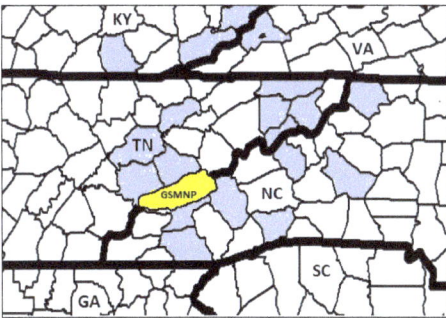

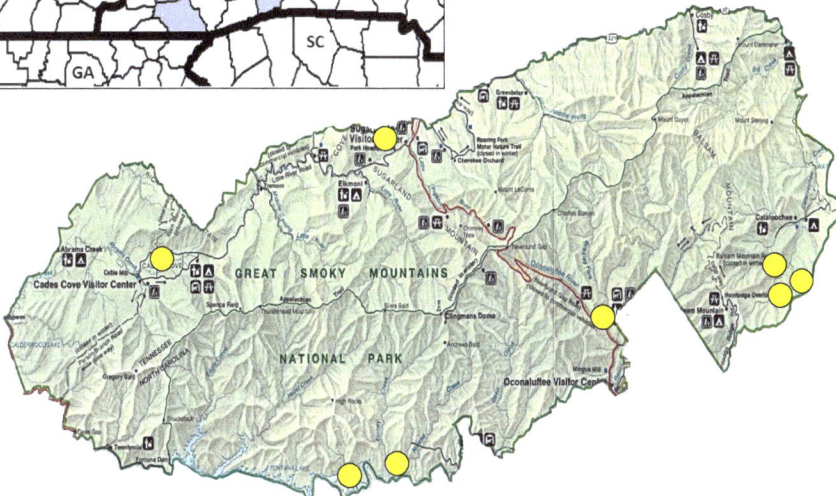

Lamellate spot

Punctidae

Punctum smithi Morrison, 1935

Diameter: 1.1-1.2 mm

Description: Depressed heliciform; lip simple; aperture roundish; shell with 4-4.5 whorls; umbilicate; pale tan or brown; one basal tooth or lamella present just inside the lower aperture (a); transverse striae are poorly developed (microscope required); periphery rounded.

Similar Species: No other small species having the shape and size of *P. smithi* has a single tooth in the aperture; *Paravitrea* species typically have multiple teeth or lamellae; other *Punctum* species have more notable microsculpture and are without a tooth or lamella in their apertures.

Habitat: The species resides in deep moist pockets of leaf litter found in depressions or around rotting logs in ravines of mixed hardwood forests.

Status: G4; Uncommon in western portions of the park around the Little Tennessee River.

Specimen: Kentucky, Jefferson County, near McNeely Lake Dam, Okolona, (FMNH 234970)

a

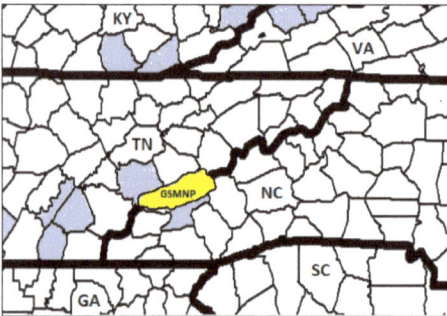

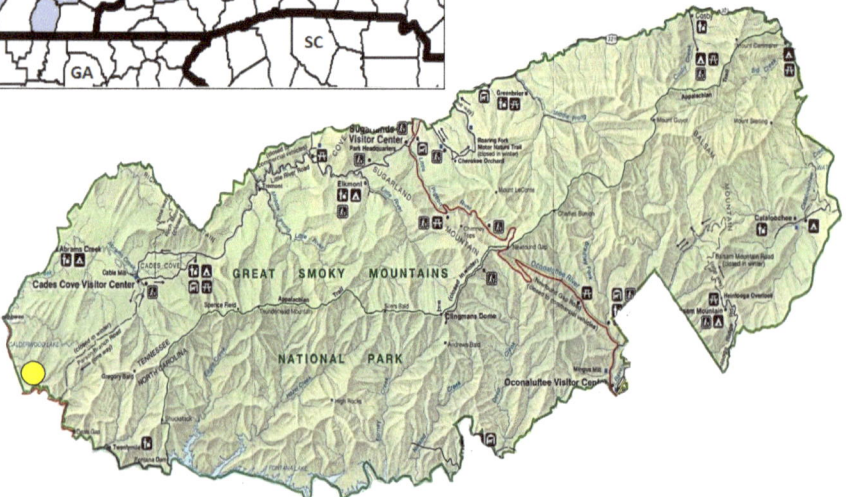

Punctum species Compared

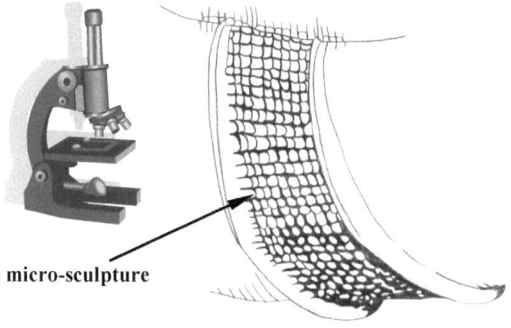

Above showing the unique micro-sculpture
found in *Punctum* species

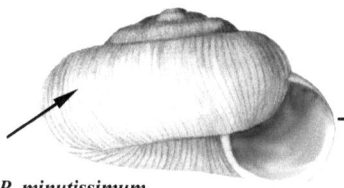

P. minutissimum

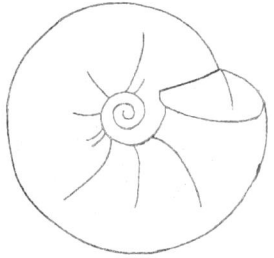

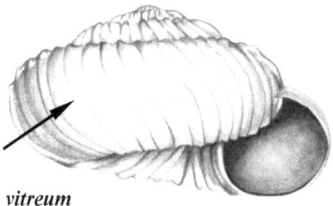

P. vitreum

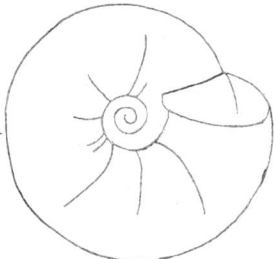

P. blandianum

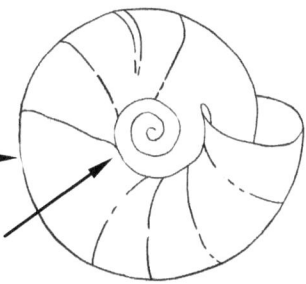

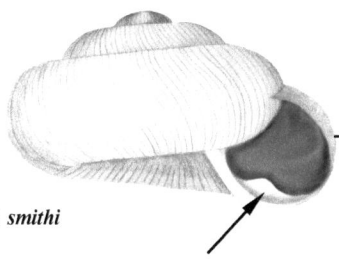

P. smithi

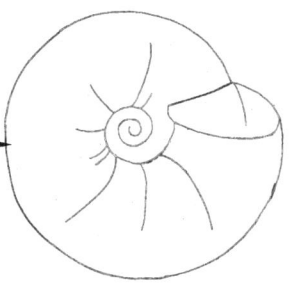

Median striate

Gastrodontidae

Striatura meridionalis Pilsbry and Ferriss, 1906

Diameter: 1.7-1.8 mm

Description: Depressed heliciform; aperture roundish; lip simple; shell with 3-3.5 whorls; widely umbilicate; corneous with a greenish cast; shell translucent in live snails and fresh dead; the apex with spiral but not transverse striae; transverse striae becoming well-developed on later whorls, forming minute riblets that are not parallel with the peristome (lip) and thus the growth lines but cross them at an angle, this character alone distinguishes them from *Punctum* species; without teeth.

Similar Species: *Striatura ferrea* is larger in diameter, has an oval-shaped aperture and a notably smaller umbilicus.

Habitat: Found in mixed hardwood forests on hillsides and ravines living between the layers of moist leafs; also found at the base of black walnut trees.

Status: G5; Common throughout the park.

Specimen: North Carolina, Macon County, Nantahala National Forest (author's collection).

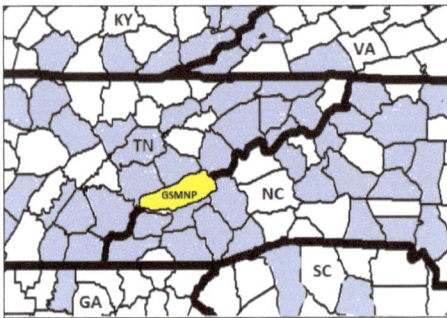

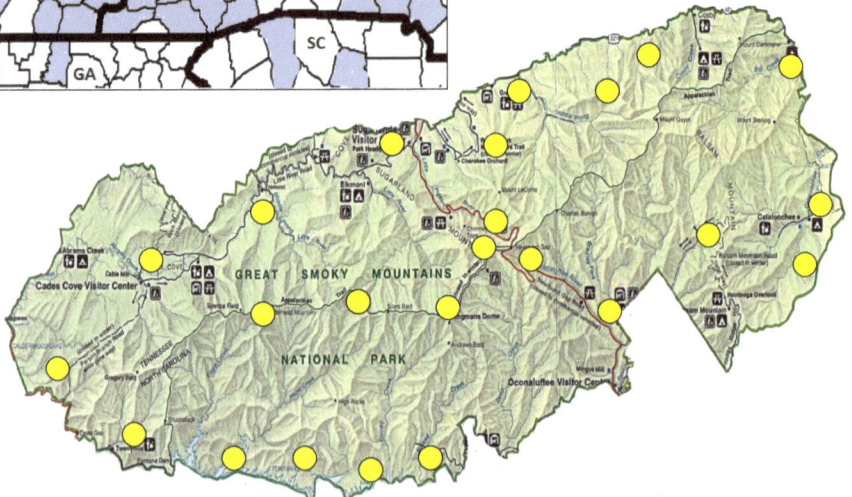

Black striate

Gastrodontidae

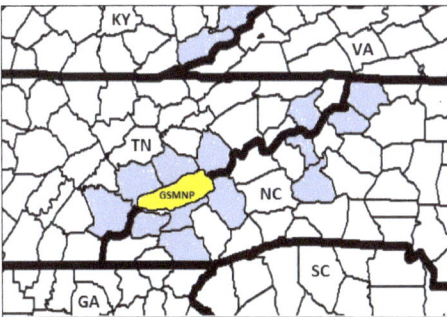

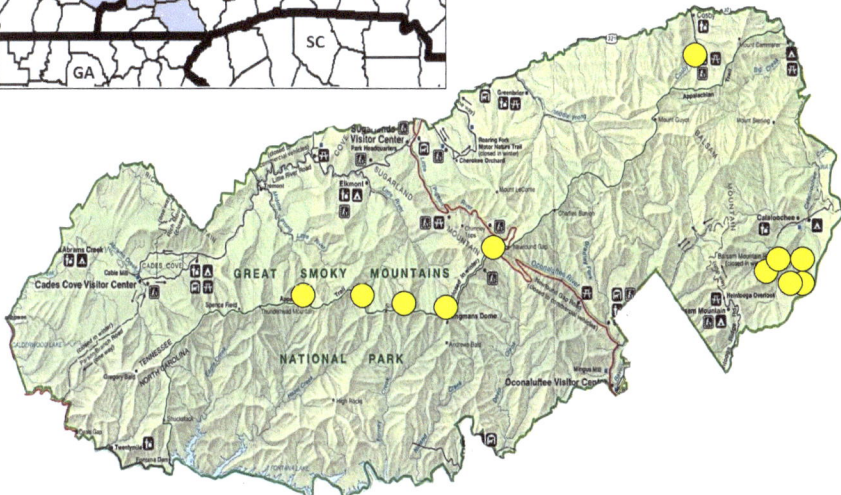

Striatura ferrea Morse, 1864

Diameter: 2.5-3.4 mm

Description: Depressed heliciform; aperture oval; lip simple; shell with 3.5-4 whorls; narrowly umbilicate; shell grayish and rather dull; transverse striae not as well-developed as in *S. meridionalis*; spiral striae not well defined but are a constant and reliable diagnostic feature; without teeth; deceased and dried animal can be seen through the translucent shell giving it a darker appearance in some spots.

Similar Species: *Striatura meridionalis* is smaller in diameter, has a rounded aperture and a notably larger umbilicus; *Punctum* species are smaller and have a wider umbilicus.

Habitat: Found in mixed hardwood forests in upper elevation mountainsides and ravines living under layers of moist leaf litter.

Status: G5; Uncommon to Common; leaf litter sampling will no doubt turn up additional sites.

Specimen: North Carolina, Cherokee County, Nantahala National Forest (author's collection).

84

Minute gem

Pristilomatidae

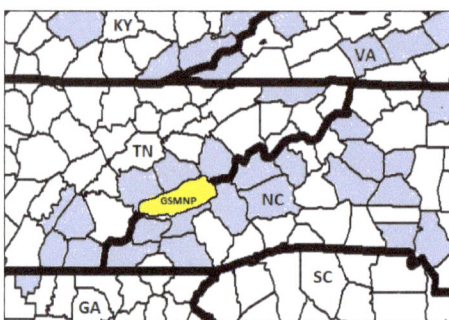

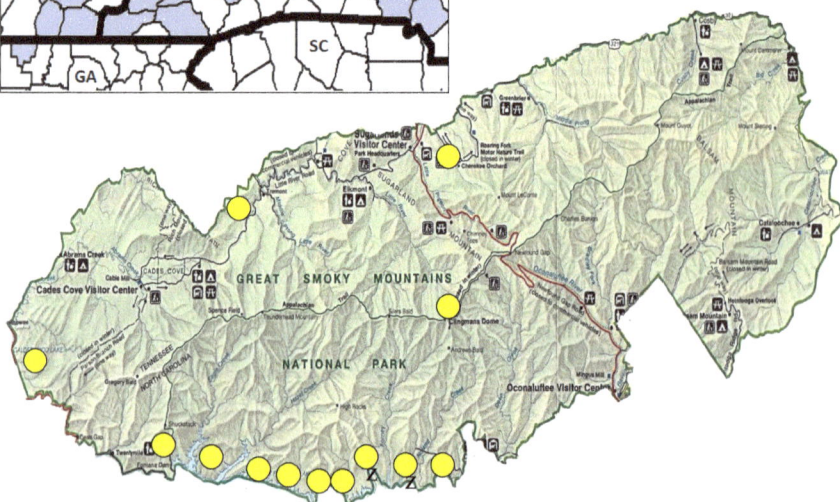

Hawaiia minuscula (A. Binney, 1840)

Diameter: 2-2.8 mm

Description: Depressed heliciform; lip simple; shell with 3.5-4.5 whorls; live shells thin, pale gray and translucent, dead shells quickly turning white with age; widely umbilicate; transverse striae are irregularly spaced and there are always faint spiral striae present.

Similar Species: Both *Lucilla scintilla* and *L. singleyana* have a much less distinctive sculpture and most importantly, a smaller umbilicus; *Punctum* species have a crisscrossing sculpture; the spiral striae of *Striatura* species is better developed.

Habitat: A snail of bare ground occurring in floodplains, meadows, roadsides and in urban areas; the snail also occurs in the mountains in mixed hardwood forests under leaf litter and detritus.

Status: G5; Locally Rare to Uncommon across the park, mostly at lower elevations.

Specimen: Kentucky, Letcher County, Bad Branch (author's collection).

85

Oldfield coil

Helicodiscidae

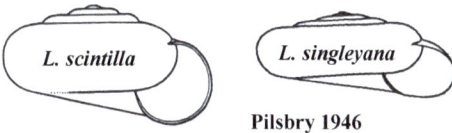

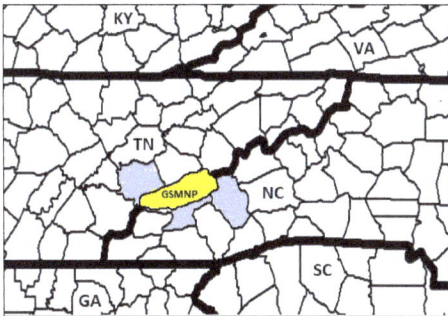

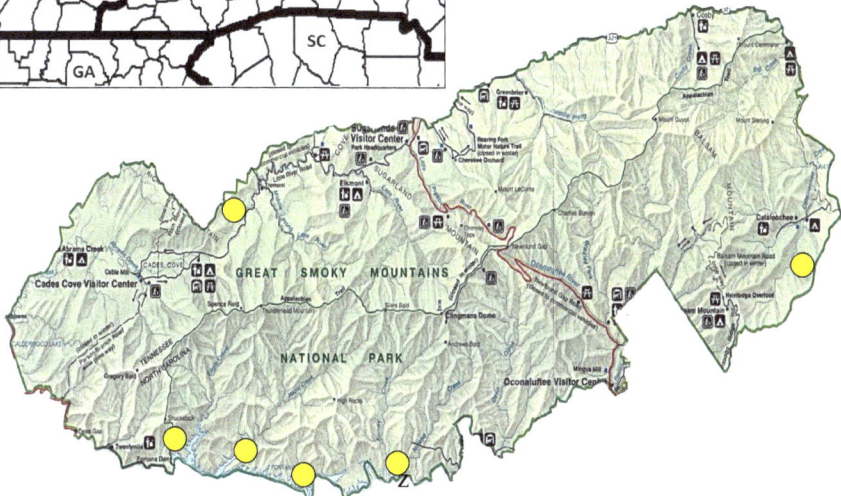

Lucilla scintilla (Lowe, 1852)

Diameter: 2.2 mm

Description: Depressed heliciform; lip simple, rounded; shell with 3-4 whorls; fresh shells are yellowish corneous with a dull sheen and translucent but turn white with age (a bleached specimen shown here); widely umbilicate; no traces of spiral ornamentation.

Similar Species: *L. singleyana* displays fine microscopic spiral lines, is more compressed and has an oval-shaped aperture (see below); separating *Lucilla* species is no picnic.

Habitat: Usually a species of open grassy areas, meadows, old fields and roadsides.

Status: G4; Locally Rare to Uncommon across the GSMNP at lower elevations.

Specimen: Texas, Val Verde County, Pecos River and US Route 90 (FMNH 240318)

L. scintilla   *L. singleyana*

Pilsbry 1946

86

Brilliant granule

Guppya sterkii (Dall, 1888)

Diameter: 1.2-1.3 mm

Description: Depressed heliciform; lip simple; shell with 3.5-4 whorls; minutely perforate, the umbilicus not completely closed; shell yellowish and highly translucent in live snails and fresh dead shells; transverse striae weak (strong light and scope required); spiral striae always present but sometimes a hard feature to detect, best seen around the first whorls but also on the base of shell; tilting the shell slightly will bring out this minute feature.

Similar Species: This species differs from other minute snails by having a nearly closed umbilicus.

Habitat: Found in mixed hardwood forests at all elevations; this is likely one of the most common minuscule snail residents in leaf litter.

Status: G5; Common; one of the most frequent tiny leaf litter species in the park.

Specimen: North Carolina, Swain County, Nantahala National Forest (author's collection).

Euconulidae

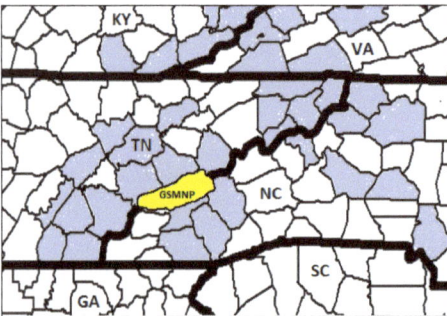

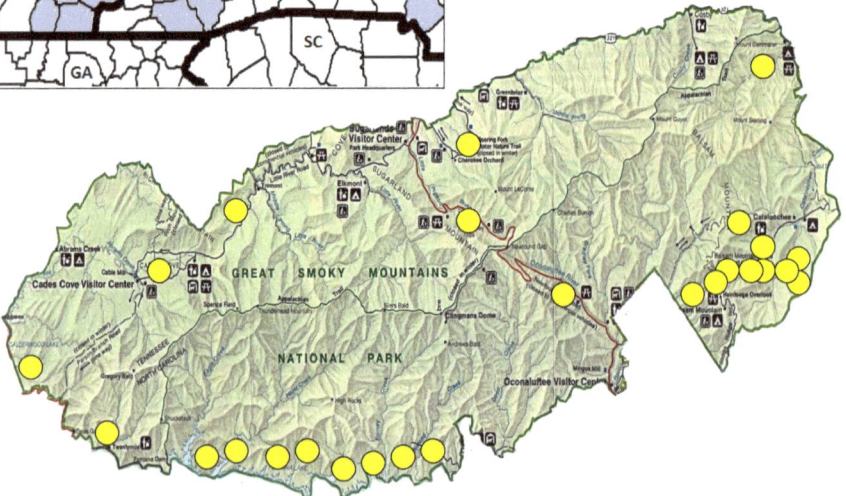

Shells 4-8.4 mm, umbilicate or perforate, simple lip, the last whorl relatively narrow, shell rather solid

Zonitoides species are small colonial snails often associated with rotting hardwood in advance stages of decay, standing or downed trees. They are sometimes found in large numbers. Key features to this group of gastropods are the open umbilicus and absence of any teeth in all stages of growth. They are occasionally confused with *Glyphyalinia* and *Paravitrea* species but, in general *Zonitoides* are larger, have much thicker shells and are less finely sculptured. Pay close attention to any spiral ornamentation or the absence of it under the scope. At times, this micro-feature is hard to see, depending on the lighting source and angle of the shell. It is best viewed on the top surface and will take some experience to key in on this finer feature. Slightly tilting the shell (while under the scope) will sometimes help reveal this micro-feature, wetting the surface of the shell will also bring out these features, especially if the shells are bleached or badly weathered. There are three species reported from the park.

Gastrodonta species differ from *Zonitoides* by being more tightly coiled (d), perforate (c) and are always with well developed teeth. One species is reported from the park.

Grouping 1: shells 5-8.4 mm, umbilicate (a), whorls moderately to loosely coiled (b), spiral striae usually a weak feature or absent, pg 89

Zonitoides elliotti
Zonitoides arboreus
Zonitoides patuloides

Grouping 2: shells 6-7 mm, perforate (c), with tightly coiled whorls (d) and strongly developed ribbing on top of the shells (e), pg 92

Gastrodonta interna

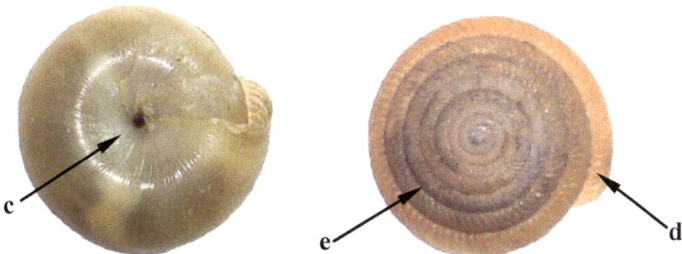

Green gloss   Zonitidae

Zonitoides elliotti (Redfield, 1856)

Diameter: 7.5-8.4 mm

Description: Depressed heliciform; lip simple with a slightly reflected (a) and square-shape; shell with 5-6 whorls; umbilicate; greenish-horn color and glossy; transverse striae are poorly developed; spiral striae present but are so faint that detection is often difficult; without teeth in any stages of growth; small callus within the aperture above the reflection.

Similar Species: The slightly reflected lower lip and thicker shell of *Z. elliotti* is a key characteristic feature in separating it from all other snail shells found in this size class.

Habitat: Found in mixed hardwood and pine woods on hillsides and ravines in well-rotting wood, partly embedded in the soil or under rotting hardwood logs; the species is often found in colonies, upwards of twenty-five on one log (pers. obs.).

Status: G4; Common and widespread.

Specimen: North Carolina, Swain County, Nantahala National Forest (author's collection).

a

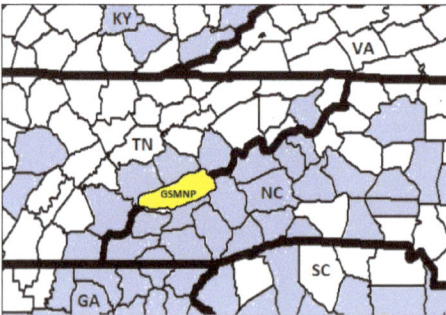

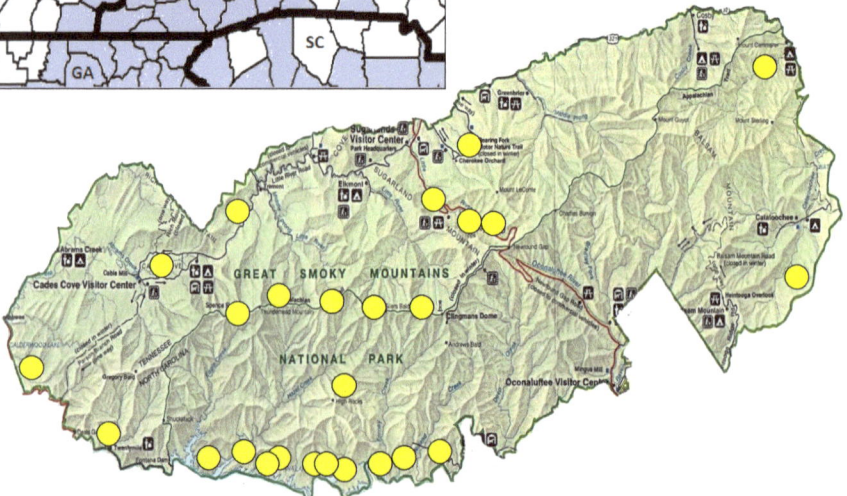

Quick gloss

Zonitidae

Zonitoides arboreus (Say, 1816)

Diameter: 5-6 mm

Description: Depressed heliciform; lip simple not reflected; shell with 4.5–5 whorls; umbilicate; olive buff and glossy; transverse striae are poorly developed; spiral striae present but a weak feature; without teeth; periphery round.

Similar Species: *Z. elliotti* is larger, has a thicker shell and a slight reflection of the lower lip; *Z. nitidus* (not yet recorded from the GSMNP) is slightly larger and taller with a darker shell, no spiral striae and has a more roundish aperture.

Habitat: A common species found on or under exfoliating bark of standing or down rotting trees in advanced stages of decay; very common on old timbers inside abandoned coal mines (MacGregor pers. comm. 2010); usually found in small colonies.

Status: G5; Common throughout the GSMNP at all elevations; this is the most common and wide ranging land snail in North America.

Specimen: Kentucky, Powell County, Furnace Mountain (author's collection).

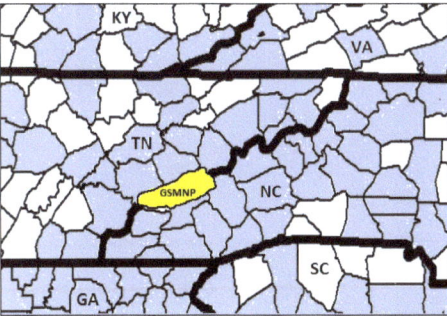

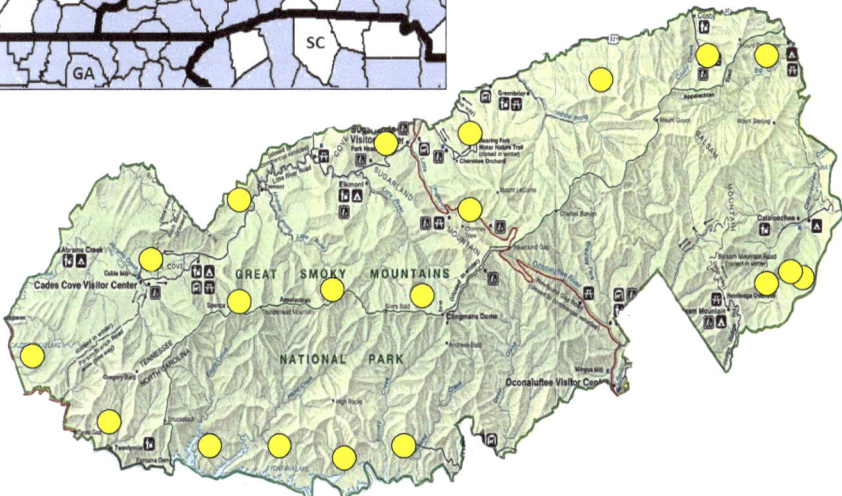

Appalachian gloss

Zonitidae

Zonitoides patuloides (Pilsbry, 1895)

Diameter: 5-5.8 mm

Description: Heliciform and compact; lip simple and rounded, not reflected; shell with 4.5–5 whorls; umbilicate; the transverse striae are strongly developed on the entire shell surface; without spiral striae; no teeth in any stages of growth; periphery well rounded.

Similar Species: *Z. elliotti* is larger, has a thicker shell and a slight reflection of the lower lip; *Z. arboreus* has a notably thinner shell, is depressed heliciform and has a much smoother surface; *Gastrodonta interna* is larger, containing internal teeth and is perforate.

Habitat: Found in pockets of deep, moist leaves on mountainsides and in ravines of deciduous forests (Hubricht 1985).

Status: G3; Locally Rare; reported by Pilsbry in Cades Cove and around Thunderhead Mountain; this is a rare species, the anatomy of which is unknown (Pilsbry 1946).

Specimen: North Carolina, Cherokee County (author's collection).

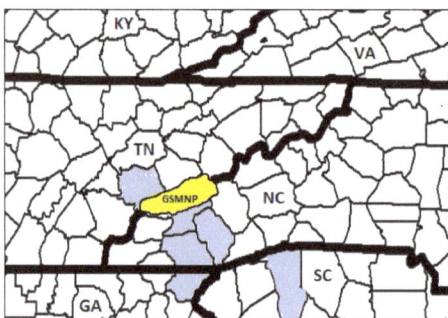

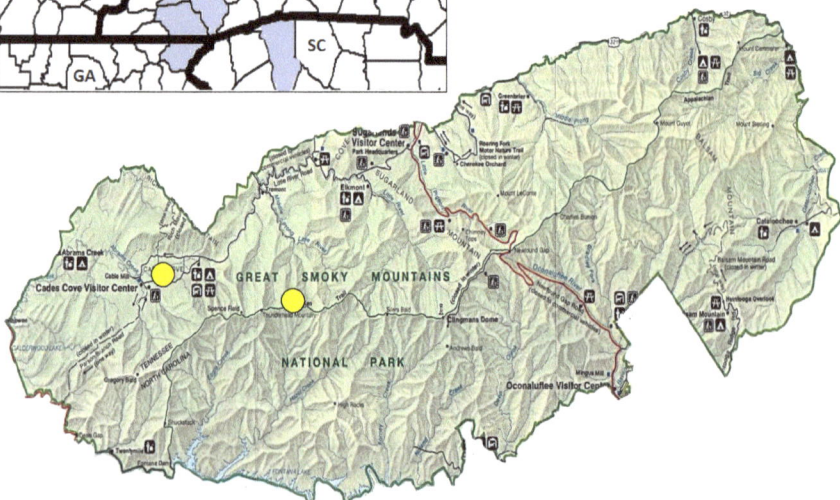

Brown bellytooth

Gastrodontidae

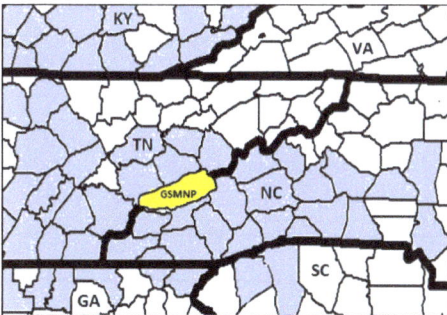

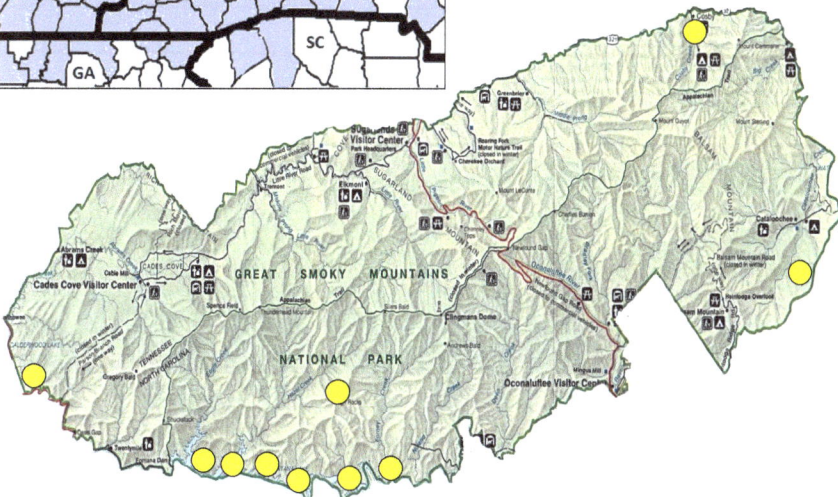

Gastrodonta interna (Say, 1882)

Diameter: 6.5-7.4 mm

Description: Heliciform; lip simple; shell with 8-9 tightly coiled whorls; perforate; cinnamon-brown; shell solid (thick) for its small size; transverse striae are distinct, rib-like and regularly spaced on top but much weakened on the base; two teeth are present at all stages of growth and can be seen through the bottom of live and fresh shells (a).

Similar Species: *G. fonticula* (not recorded in the GSMNP) has a larger umbilicus; *Z. patuloides* is umbilicate, less tightly coiled and without teeth.

Habitat: A colonial species found under leaf litter, often found under loose bark of decaying wood on hillsides and ridge-tops of mixed hardwood forests.

Status: G5; Uncommon; currently reported from the North Shore of Fontana Lake and Cosby, but it likely occurs in more interior locations of the park.

Specimen: Kentucky, Letcher County, Bad Branch (author's collection).

a

Shells 4-12 mm, without teeth, simple lip, last whorl wide, shells thin, translucent and fragile (*Glyphyalinia*)

Glyphyalinia species can be a troublesome group of land snails for the novice. Even for the experienced malacologist, they are challenging. Key features for this group of gastropods are the small size, the loosely-coiled whorls, the last whorl usually widely expanded (a), the glossy and translucent shells and most importantly, the indented transverse striae (c), which can vary from closely-spaced (d) to widely spaced (e). The somewhat unique indented or etched sculpture on the shell surface is the signature of this genus and is easily viewed under a hand lens of 10X. Although *Paravitrea* species have indented sculpture as well, they are tightly coiled, the last whorl not expanded (b) and are usually with internal armature at some stage of growth. *Glyphyalinia* species never contain internal teeth. Give special attention to any spiral sculpture under the scope. At times, this micro-feature is hard to see, depending on the light source and angle of the shell. It is best viewed on the top surface and will take some experience to key in on this fine feature. Slightly tilting the shell (while under the scope) will sometimes help reveal this micro-feature. Wetting the surface of the shell will also bring out these features, especially if the shells are bleached or badly weathered. *Glyphyalinia* snails here are arranged in three groups; **(1)** shells that are imperforate, **(2)** shells that are perforate or rimate and **(3)** shells that are umbilicate.

Glyphyalinia rhoadsi **illustrating the remarkable indented striae**

Glyphyalinia

Paravitrea

a

b

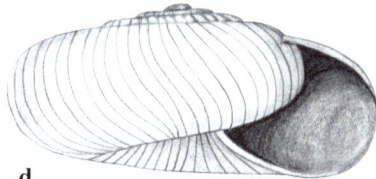

d

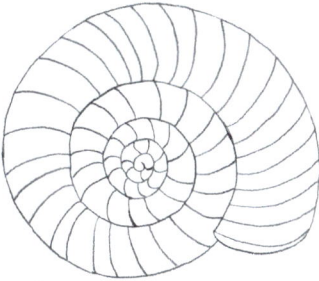

c

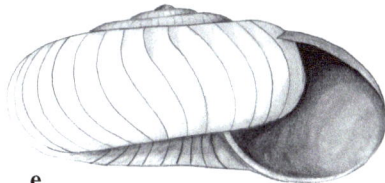

e

Grouping 1: shells that are imperforate, pg 95

G. cryptomphala

Grouping 2: shells that are perforate or rimate, pg 96

G. praecox
G. junaluskana
G. sculptilis
G. caroliniensis
G. indentata

Grouping 3: shells that are umbilicate, pg 102

G. rhoadsi
G. wheatleyi
G. cumberlandiana
G. pentadelphia

Thin glyph

Gastrodontidae

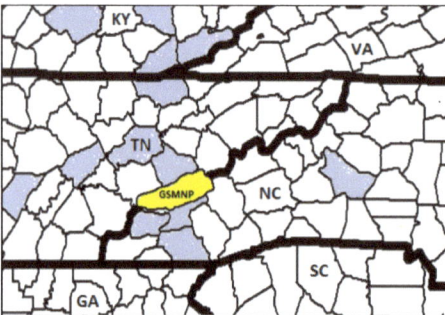

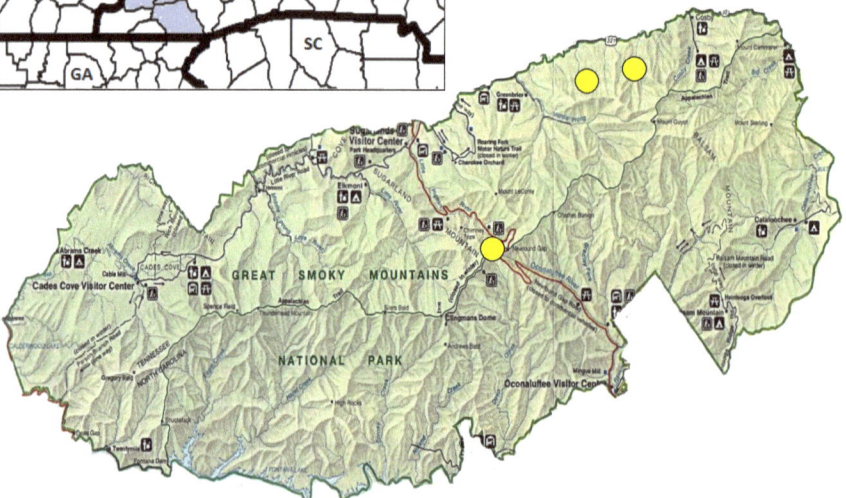

Glyphyalinia cryptomphala (G. H. Clapp, 1915)

Diameter: 5.1-6 mm

Description: Depressed heliciform; lip simple; shell with 5-5.5 loosely coiled whorls; imperforate, the umbilicus completely covered in all stages of growth; shell fragile, light horn to white, glossy and semi-transparent; indented transverse striae are well-developed, closely and nearly equally spaced and continue to the base; the spiral striae is weakly defined but is a constant feature; deceased and dried animal seen through bottom view of shell.

Similar Species: *Glyphyalinia praecox* is slightly larger and is perforate, its umbilicus not completely closed as in *G. cryptomphala*.

Habitat: Most often found in a variety of mixed hardwood forests under moist leaf litter along river bluffs and in ravines.

Status: G5; Uncommon;. known only from a few scattered locations in the park.

Specimen: Tennessee, Coffee County, Lusk Cave, 4 miles ENE of Hillsboro (author's collection).

Brilliant glyph

Gastrodontidae

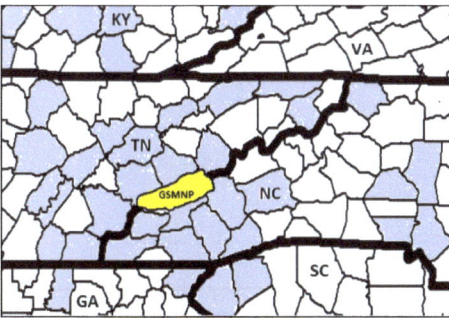

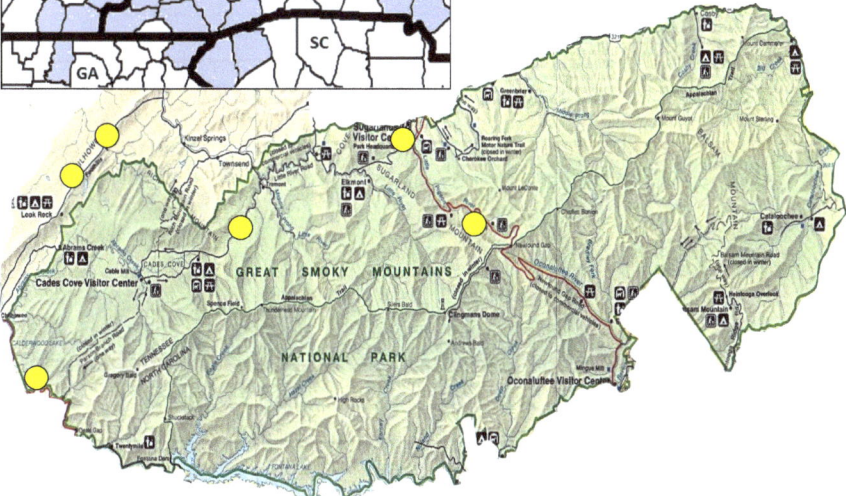

Glyphyalinia praecox (H.B. Baker, 1930)

Diameter: 6.2-6.3 mm

Description: Depressed heliciform; lip simple; shell with 4.5-5 loosely coiled whorls; perforate, the minute umbilical opening is partly covered by an expansion of the columellar lip; shell fragile, bronze-colored, glossy and transparent; indented transverse striae are well-developed, closely and nearly equally spaced and continue to the base; the spiral striae may be weakly defined but are a constant feature; deceased and dried animal seen through all three shell views

Similar Species: *Glyphyalinia indentata* has a slightly higher profile, the apical whorls are less tightly coiled and the last whorl is more capacious and is a corneous not a bronze color.

Habitat: Found under moist leaf litter in floodplains and on talus slopes in mixed hardwood forests.

Status: G4; Uncommon; in the park reported from scattered locations.

Specimen: North Carolina, Swain County, GSMNP (GSMNP collection).

Light glyph

Gastrodontidae

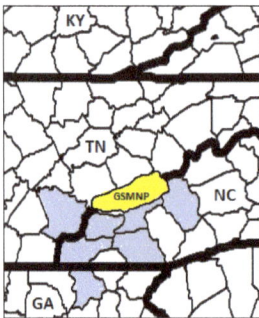

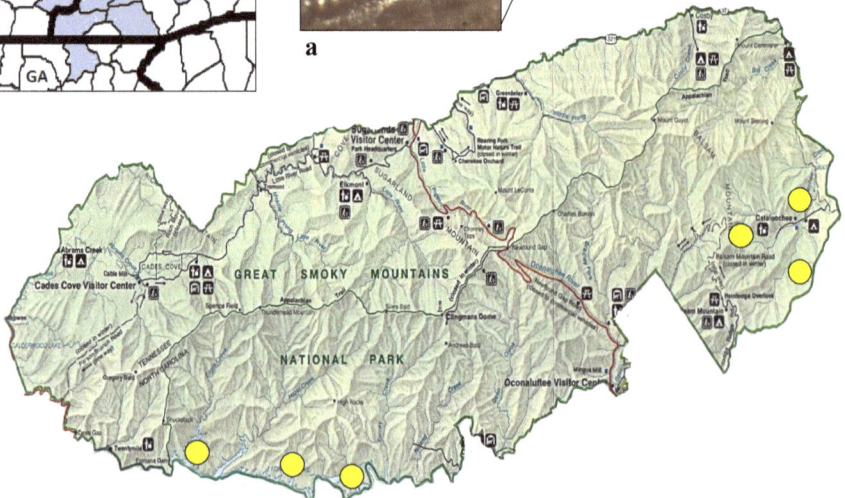

Glyphyalinia junaluskana (Clench and Banks, 1932)

Diameter: 8-12 mm

Description: Depressed heliciform; lip simple; shell with 6 loosely coiled whorls, perforate or rimate, the minute umbilicus opening is partly covered by an expansion of the columellar lip; glossy and translucent; indented transverse striae are well-developed, equally and closely-spaced, bordered by distinctive beading (a).

Similar Species: *Glyphyalinia indentata,* and *G. praecox* are smaller and the transverse striae are farther apart and more irregularly spaced; *G. sculptilis* does not have the distinct beading found along the transverse striae (figure a) of *G. junaluskana.*

Habitat: A snail of upland mixed hardwood forests under and among moist leaf litter.

Status: G2; Locally Rare; Endemic to the region of the GSMNP; reported from southern portions of the park and Purchase Knob.

Specimen: North Carolina, Swain County, Nantahala National Forest (authors collection).

a

The Sculpted glyph, *Glyphyalinia rhoadsi* showing the extraordinary internal and external architecture of the shell

Suborb glyph

Gastrodontidae

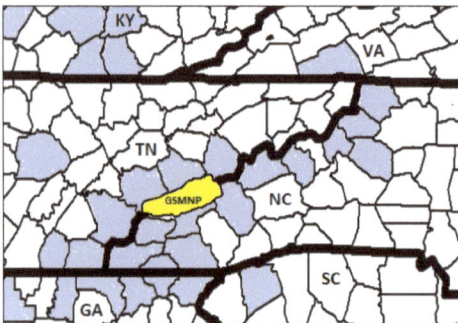

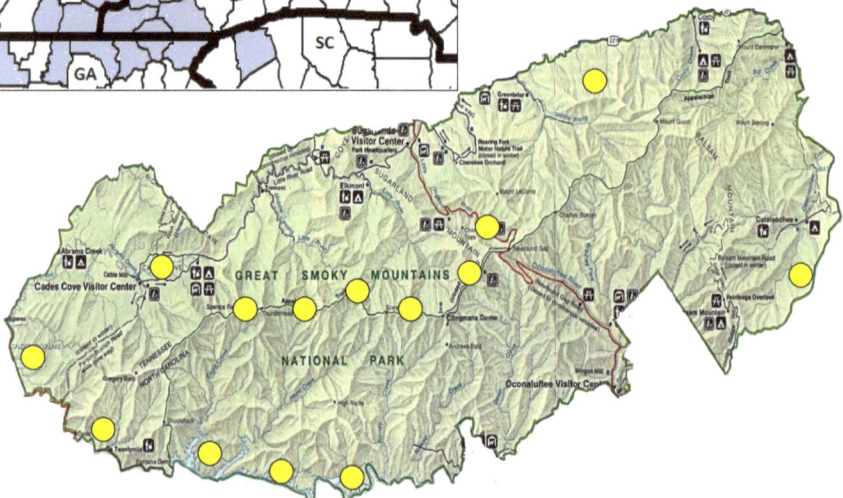

Glyphyalinia sculptilis (Bland, 1858)

Diameter: 6.5-12.7 mm

Description: Depressed heliciform; lip simple; shell with 7 loosely coiled whorls; perforate, the minute umbilical opening is partly covered; shell slightly thicker than other *Glyphyalinia* species, pale horn-colored, glossy and translucent; indented transverse striae are well-developed, equally and closely-spaced (about 0.5 mm apart at the periphery of the body whorl).

Similar Species: *Glyphyalinia indentata,* and *G. praecox* are smaller and the transverse striae are farther apart and more irregularly spaced; *G. junaluskana* has unique beading along its transverse striae.

Habitat: A snail of upland mixed hardwood forests under and among moist leaf litter and detritus.

Status: G4; Relatively Common but with scattered locations reported in the GSMNP.

Specimen: North Carolina, Swain County, Nantahala National Forest (author's collection).

Spiral mountain glyph

Gastrodontidae

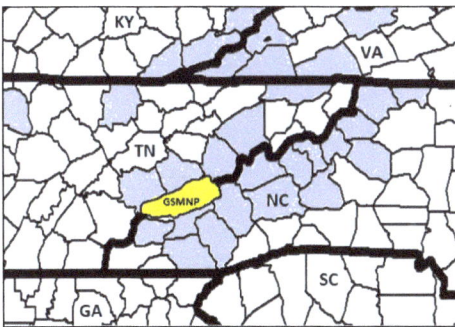

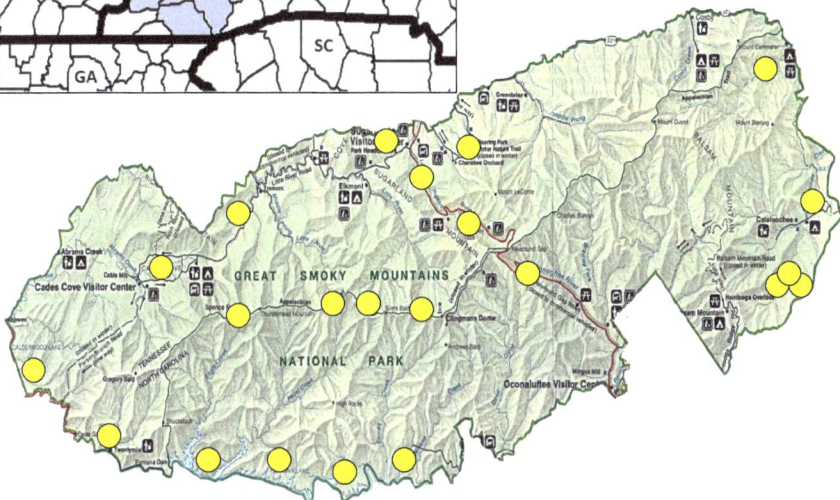

Glyphyalinia caroliniensis (Cockerell, 1890)

Diameter: 5-12 mm

Description: Depressed heliciform; lip simple; shell with 4.5-5.5 loosely coiled whorls; perforate or rimate; shell fragile, corneous, glossy and semi-transparent; indented transverse striae are well-developed, but are widely and nearly equally spaced; the spiral striae may be weakly defined but are a constant feature (a strong scope required to see this sometimes faint feature).

Similar Species: This species is most similar to *Glyphyalinia indentata* but is larger and has a more conspicuous spiral striae; *G. praecox* is bronze-colored; *G. sculptilis* has a thicker shell and closer set transverse striae.

Habitat: Mixed hardwood forests under moist leaf litter along river bluffs and also a species of mountainsides.

Status: G4; Common throughout the park at most elevations.

Specimen: North Carolina, Swain County, Nantahala National Forest (author's collection).

Carved glyph

Gastrodontidae

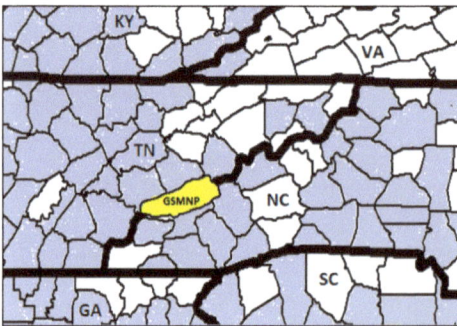

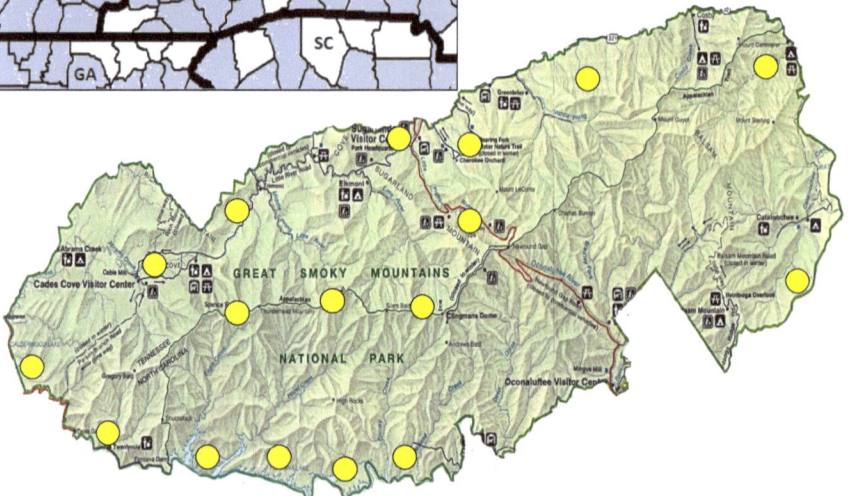

Glyphyalinia indentata (Say, 1823)

Diameter: 4.7-7.1 mm

Description: Depressed heliciform; lip simple; shell with 4.5-5 loosely coiled whorls; perforate or rimate; shell fragile, corneous, glossy and translucent; indented transverse striae are well-developed, widely but nearly equally spaced, about 28 on the last whorl; the spiral striae may be weakly defined but are a constant feature (scope required); periphery well rounded; deceased and dried animal seen through shells.

Similar Species: This species is most similar to *Glyphyalinia caroliniensis* but is fully 4 to 5 mm smaller when collected from the same region; *G. rhoadsi* has a wider umbilicus and stronger developed indented transverse striae.

Habitat: Found in a variety of mixed hardwood forests under leaf litter but also occasionally found along roadsides and in urban areas.

Status: G5; Common; found at all elevations throughout the park.

Specimen: North Carolina, Macon County, Nantahala National Forest (author's collection).

Sculpted glyph

Gastrodontidae

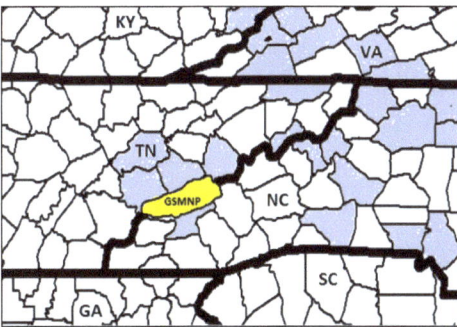

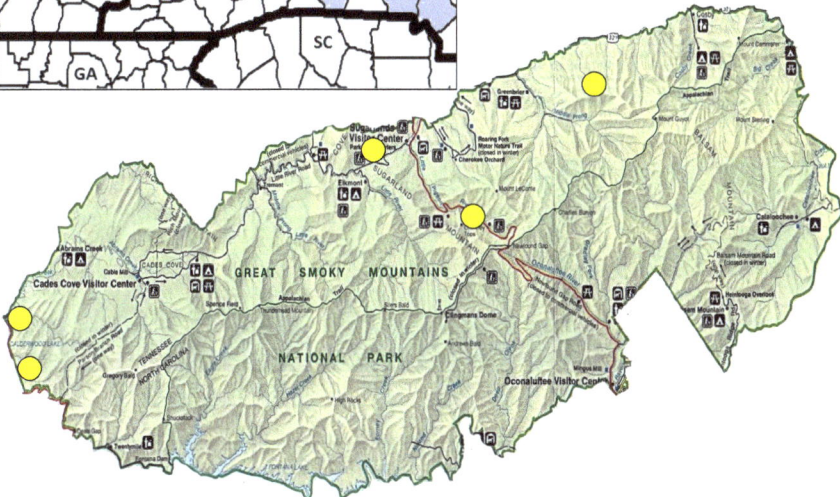

Glyphyalinia rhoadsi (Pilsbry, 1899)

Diameter: 4.5-5.3 mm

Description: Depressed heliciform; lip simple; shell with 4-5 loosely coiled whorls; umbilicate, umbilicus 1/12 the shell diameter; shell fragile, corneous, glossy and semi-transparent; indented transverse striae are well-developed and nearly equally spaced continuing to the base; the extremely weak spiral striae may be seen near the suture and umbilicus, but this feature is absent in some specimens (Pilsbry 1946).

Similar Species: This species is most similar to *Glyphyalinia indentata* but is slightly smaller, has better developed transverse striae and has a notably wider umbilicus.

Habitat: Found in a variety of mixed upland hardwood forests under moist leaf litter.

Status: G5; Uncommon; reported from the Foothills Parkway, Calderwood and several other scattered locations in the GSMNP.

Specimen: Virginia, Dickenson County, Garden Hole above river (author's collection).

Bright glyph   # Gastrodontidae

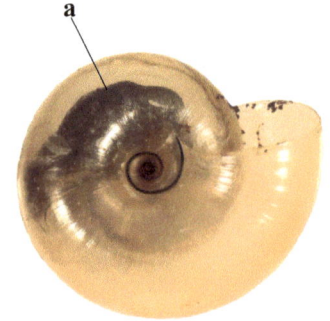

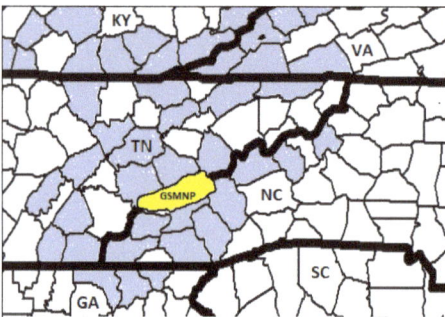

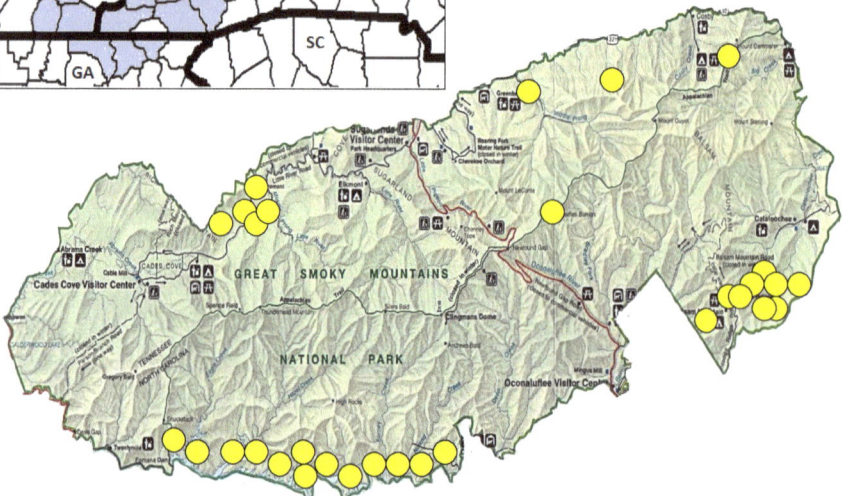

Glyphyalinia wheatleyi (Bland, 1883)

Diameter: 3.5-6 mm

Description: Depressed heliciform; lip simple; shell with 5-5.5 loosely coiled whorls; umbilicate (size varying); shell glossy; brownish horn; indented transverse striae (radiating lines) are moderately developed and irregularly spaced; spiral striae may be absent or only mere traces in some populations; deceased and dried animal (a) can be seen in all three views.

Similar Species: *Glyphyalinia cumberlandiana* is smaller and has a thinner, more fragile shell and is a light horn-colored; *G. rhoadsi* has a notably smaller umbilicus, a different color and stronger developed transverse striae.

Habitat: Found in a wide range of habitats including under moist leaf litter and deep detritus deposits located from valleys to mountaintops in mixed hardwood forests.

Status: G5; Common; this wide-ranging and somewhat variable species is likely more widespread in the park than current records indicate.

Specimen: Kentucky, Fayette County, Ravens Run Nature Reserve (author's collection).

Hill glyph

Gastrodontidae

Glyphyalinia cumberlandiana (G. H. Clapp, 1919)

Diameter: 2.5-3 mm

Description: Depressed heliciform; lip simple; shell with 4-4.5 loosely coiled whorls; umbilicate; shell exceedingly fragile, glossy, light horn-colored and semi-transparent; indented transverse striae are faint and irregularly spaced; the weak spiral striae are most notable on the first and second whorls; like a miniature *G. wheatleyi;* deceased and dried animal seen through the translucent top view of the shell, giving the first few whorls a darker color.

Similar Species: *Glyphyalinia wheatleyi* is similar in form but is larger.

Habitat: Reported as a calciphile of rocky limestone areas in shaded mixed hardwood forests but also a snail of acidic habitats especially in and around rock talus.

Status: G4; Uncommon; found along the Foothills Parkway, Purchase Knob, White Oak Sinks and other scattered locations in the park.

Specimen: Kentucky, Powell County, Furnace Mountain (author's collection).

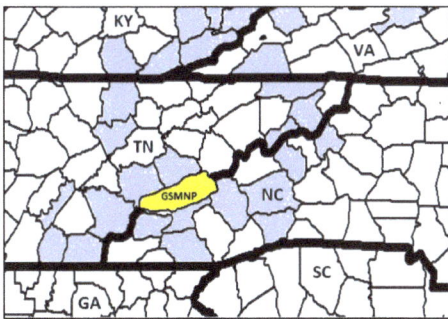

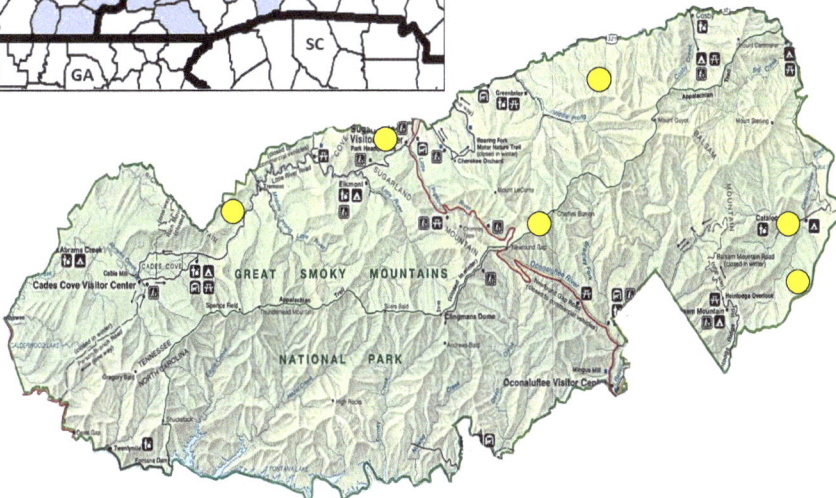

Pink glyph

Gastrodontidae

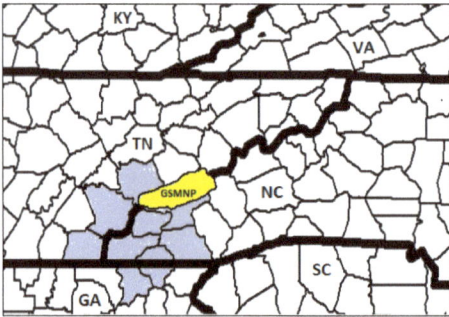

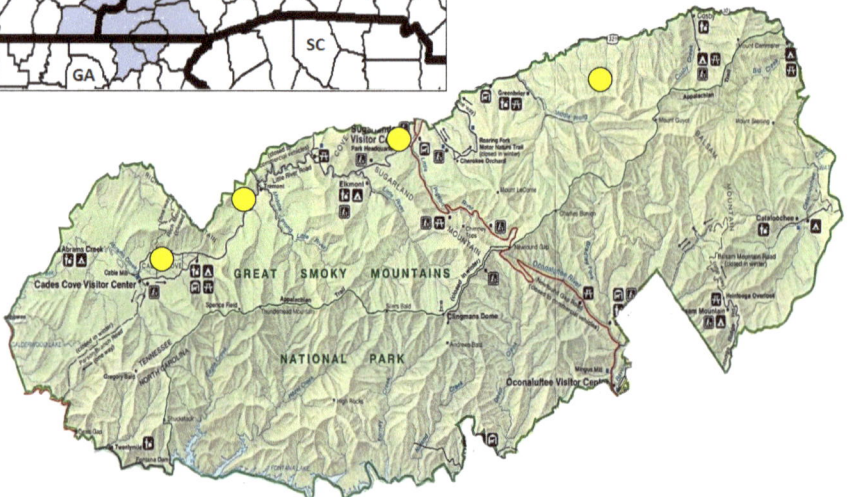

Glyphyalinia pentadelphia (Pilsbry, 1900)

Diameter: 5 mm

Description: Depressed heliciform; lip simple; shell with 4.5 moderately coiled whorls; widely umbilicate; glossy, with a pinkish cast; semi-transparent; indented transverse striae are irregularly spaced; the spiral striae are most notable on the first and second whorls, becoming somewhat obsolete on the last whorl (microscope required).

Similar Species: *Glyphyalinia wheatleyi* is similar in form but has a notably smaller umbilicus and the last whorl is more loosely coiled.

Habitat: Found under moist leaf litter in upland mixed hardwood forests.

Status: G2/G3; Uncommon; Endemic to the region; reported from Cades Cove, White Oak Sinks, Sugarlands and the Albright Grove.

Specimen: Tennessee, Polk County, Ocoee Gorge (author's collection).

Note the indented (engraved) transverse striae and translucent shell that are symbolic of *Glyphyalinia* species.

Glyphyalinia pentadelphia X30

Shells 2-8 mm, teeth typically present, simple lip (*Paravitrea*)

Paravitrea species are undoubtedly one of the most difficult groups of land snails to differentiate. Most are under 6 mm (a few species slightly larger), have more than 5 or 6 tightly coiled whorls, the last whorl not greatly expanded (a) and most species at some stage of development are with internal armature; *Glyphyalinia* species have loosely coiled whorls, the last whorl greatly expanded (b) and are without teeth. *Paravitrea* have either teeth (c and d) or lamellae (e). Without a good series of *Paravitrea* shells at various stages of growth, taxa recognition will be difficult. Immature specimens are essential for the identification of the genus. If adult *Paravitrea* species are encountered during collections, it is imperative that leaf litter samples (always a good idea anyway) also be secured at the same locations. This will likely insure that young *Paravitrea* specimens are included during the survey. Many *Paravitrea* snails contain teeth at the juvenile stage that are reabsorbed (for the calcium content) in the adult stage (see next page, figures f, g, h and i). *Paravitrea* species in this book are arranged in three groups: **(1)** juvenile and adult shells without teeth or lamella barriers, **(2)** juvenile shells with teeth or lamella barriers but as adults are usually without these protective structures and **(3)** juvenile and adult shells with teeth or lamella barriers.

Grouping 1: juvenile and adult shells without teeth or lamella barriers, pg 109

*P. capsella
*P. petrophila
*P. clappi

Grouping 2: juvenile shells with teeth or lamella barriers, adults without, pg 112

*P. placentula
*P. variabilis

Grouping 3: juvenile and adult shells with teeth or lamella barriers, pg 114

*P. andrewsae
*P. lamellidens
*P. multidentata
*P. umbilicaris
*P. umbilicaris form *dentata*

Paravitrea multidentata

Shell Morphology & Armature of *Paravitrea* Species

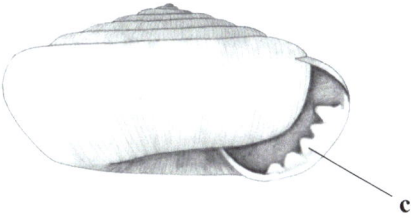

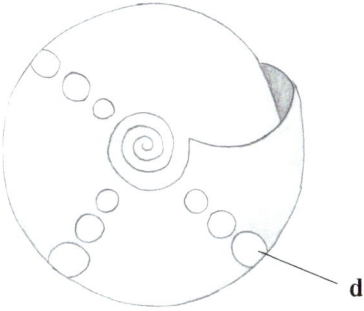

Paravitrea

a

Glyphyalinia

b

c

d

f

g

h

Figures (f-i) are examples of *Paravitrea placentula* showing the first three juvenile shells (f-h) containing teeth. As the shells mature, teeth become smaller, but in the last shell, an adult (i), the teeth are completely reabsorbed for the calcium content.

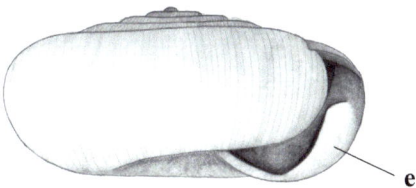

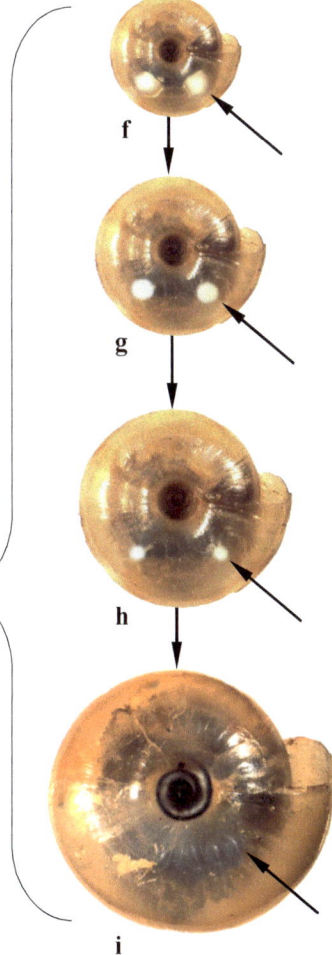

e

i

Dimple supercoil

Pristilomatidae

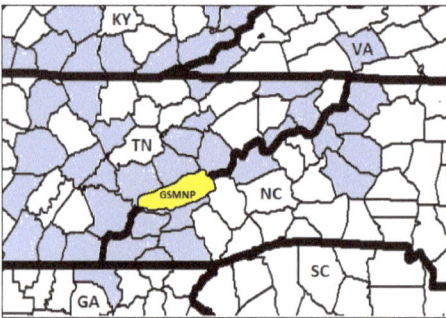

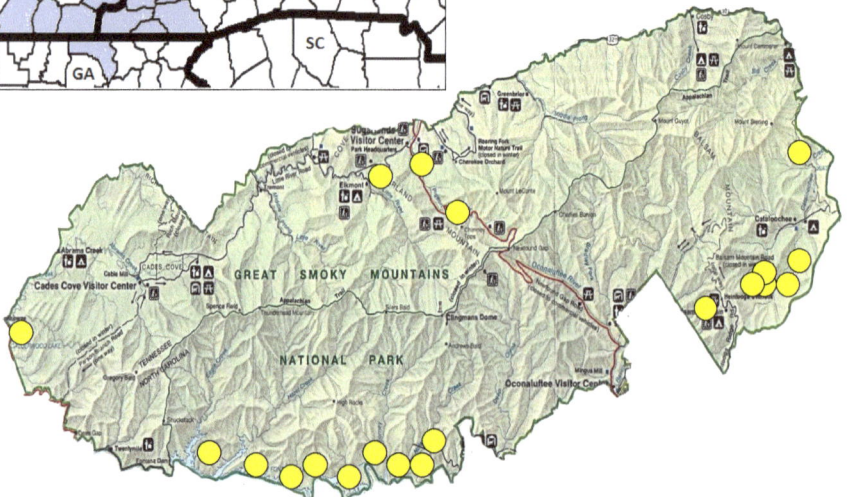

Paravitrea capsella (Gould, 1851)

Diameter: 4.8-6.2 mm

Description: Depressed heliciform; lip simple; shell with 6.5-7.5 tightly coiled whorls; umbilicate; aperture roundish; amber color; live animal pale-slate colored; glossy and translucent; transverse striae moderately developed on top of shell but nearly smooth on the base; without teeth in adults; paired teeth rarely seen in juvenile shells.

Similar Species: *Paravitrea placentula* is larger and has paired teeth in young shells; *P. petrophila* has a notably wider umbilicus.

Habitat: A habitat generalist found in a wide range of mixed hardwood forests.

Status: G5; Common in lower elevation forests throughout the GSMNP; this snail is part of a complex of anatomically distinct but confusing species with little or no shell differences; DNA will be necessary to sort out this composite group of snails (Hubricht 1985).

Specimen: North Carolina, Graham County, Nantahala NF (authors collection).

Cherokee supercoil

Pristilomatidae

Paravitrea petrophila (Bland, 1883)

Diameter: 5.7-6 mm

Description: Depressed heliciform; lip simple; shell with 5.5-6 loosely coiled whorls, the last whorl notably expanded (a), widely umbilicate; more than is seen in other *Paravitrea* species; fresh shells are shining, translucent and whitish, live animal whitish; transverse striae widely and irregularly spaced; without teeth; deceased and dried animal can be seen in all three views, making the shell darker in places.

Similar Species: This species is often confused with *Glyphyalinia* snails but primarily differs in its tighter coils; it stands closest to *Glyphyalinia pentadelphia* in terms of shape but has one more whorl in shells of the same size and is ivory white, not pinkish.

Habitat: A habitat generalist found in a wide range of mixed hardwood forests under leaf litter and detritus.

Status: G4; Uncommon; reported from the White Oaks Sinks area only.

Specimen: North Carolina, Macon County, Tate Gap (authors collection).

a

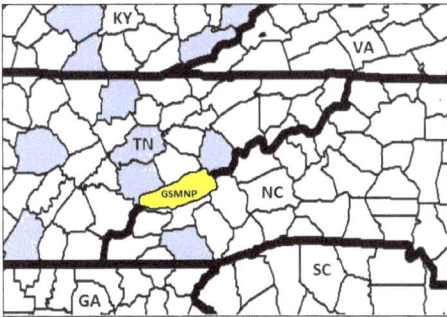

Mirey Ridge supercoil

Paravitrea clappi (Pilsbry, 1898)

Pristilomatidae

Diameter: 5.8 mm

Description: Depressed heliciform; lip simple; shell with 6.5-7 tightly coiled whorls, the last whorl not notably expanded; perforate; fresh shells; shell glossy, thin and translucent; transverse striae are well developed, closely and evenly spaced, continuing to the base; no teeth in any stage of growth.

Similar Species: This species differs from other *Paravitrea* by its well developed transverse striae, a perforate umbilicus, and a glossy chestnut-amber (Pilsbry, 1946) or pink shell (Van Devender, 1985).

Habitat: A habitat specialist usually found in well-developed, higher elevation (over 1500 meters) rock talus where it burrows deep into leaf litter that accumulates between the interstitial spaces of boulders.

Status: Provisionally ranked G1; Globally Rare, Endemic to the GSMNP; not ranked in Natureserve, this form is included and may represent a new species. Further study is needed to make this determination.

Specimen: Tennessee, Blount County, Thunderhead Mountain, GSMNP (FMNH 5847).

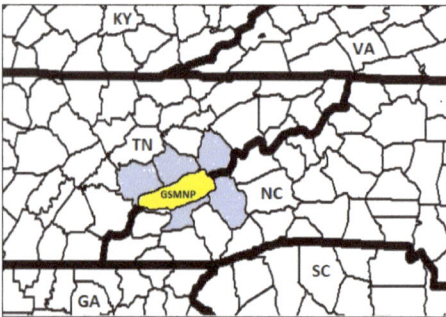

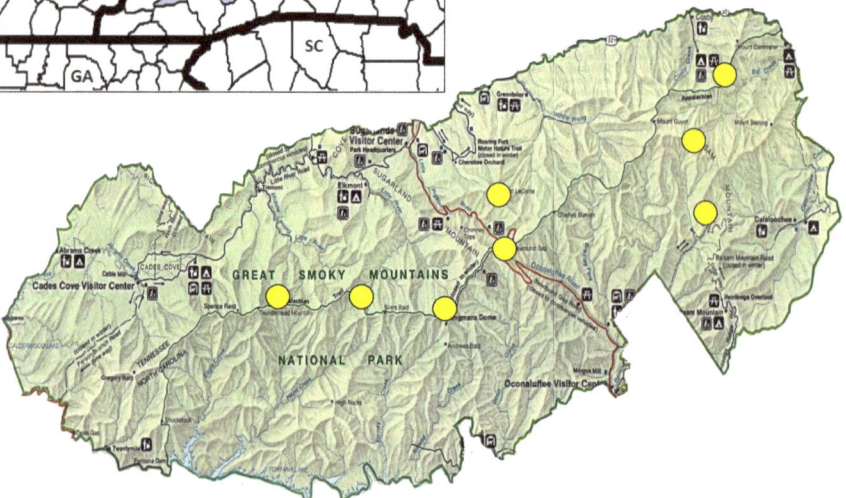

Glossy supercoil

Paravitrea placentula (Shuttleworth, 1852)

Pristilomatidae

Diameter: 7.2-7.8 mm

Description: Depressed heliciform; lip simple; shell with 5-6 tightly coiled whorls, umbilicate; the last whorl slightly expanding; corneous; shell thin, glossy and translucent in fresh and live shells; indented transverse striae are faint and irregular; two pairs of rather large teeth that are arranged close together can be seen through the bottom of fresh and live shells of juveniles up to 3.5 mm in diameter (a); adults are without teeth.

Similar Species: Most similar to *Paravitrea capsella* but larger, and contains teeth in juvenile shells.

Habitat: Acidic coves, rich woods, upper elevation northern red oak and montane oak hickory forests.

Status: G3; Relatively Common; reported from along the Foothills Parkway, White Oak Sinks, North Shore and other scattered locations.

Specimen: Tennessee, Blount County, White Oak Sinks, GSMNP (GSMNP collection).

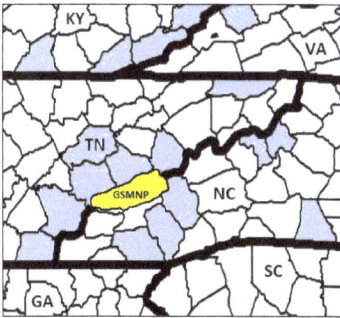

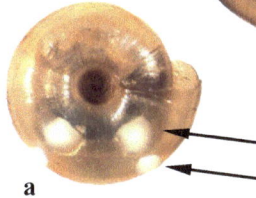

a

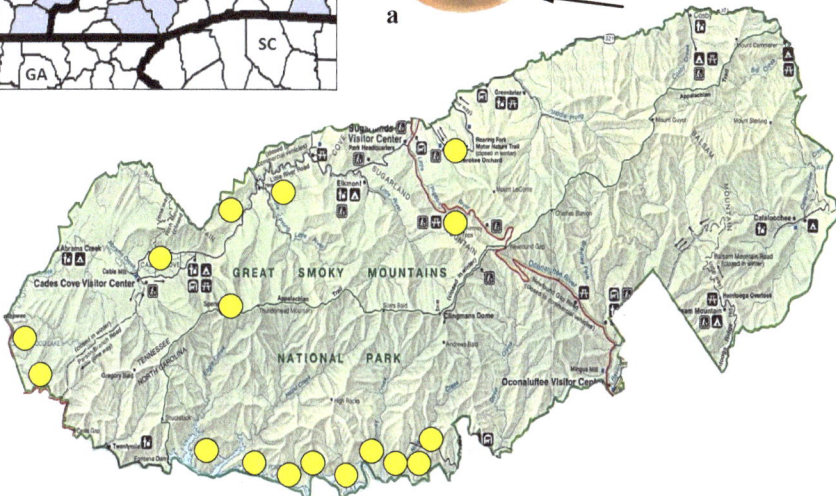

Variable supercoil

Pristilomatidae

Paravitrea variabilis (H.B. Baker, 1929)

Diameter: 3.5-3.6 mm

Description: Depressed heliciform; lip simple; shell with 5-6 tightly coiled whorls; umbilicate; shell thin, glossy and translucent in fresh and live shells; indented transverse striae are faint and closely-spaced and weak spiral striae; in half grown shells there are 1 to 3 rather large lamellae seen through the bottom of fresh and live shells (figure a), that usually exhibit weak and irregular subdivisions into five or more rounded points; adults shells usually without any notable lamellae.

Similar Species: *Paravitrea capsella*, *P. petrophila* and *P. clappi* are without lamellae in both juvenile and adult shells; *P. lamellidens, P. multidentata* and *P. umbilicaris* are with adult lamellae.

Habitat: Seeps, wooded hillsides and in deep leaf litter of mixed hardwood.

Status: G2/G3; Locally Rare; from west end of TN side of the park and White Oak Sinks.

Specimen: Tennessee, Blount County, Calderwood area, GSMNP (GSMNP collection).

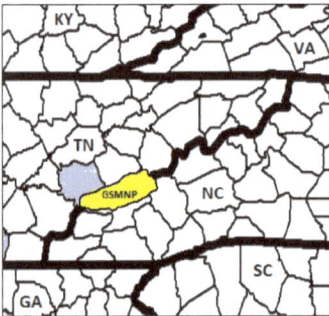

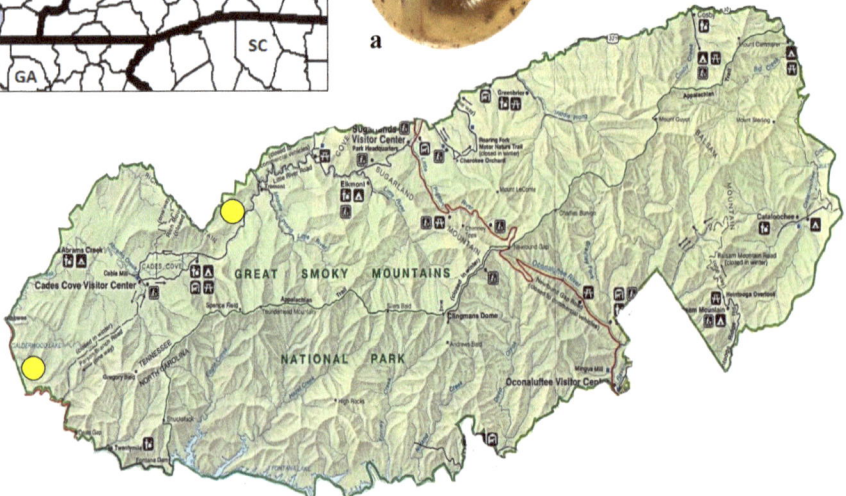

a

High mountain supercoil

Pristilomatidae

Paravitrea andrewsae (W. G. Binney, 1879)

Diameter: 6.5-8 mm

Description: Depressed heliciform; lip simple; 6-8 tightly coiled whorls; perforate; shells are corneous, translucent and glossy; numerous irregularly spaced transverse striae; several radial rows of 3-5 rather large teeth (b & e) that can be seen through the bottom of fresh shells although the number of teeth may vary down to 0 in the adult stage or less frequently at any stage of growth; teeth sometimes fused together; Figure b shows 2 sets of teeth (aperture and through shell); figure f is a fully adult shell.

Similar Species: The juvenile shells of *P. placentula* have paired teeth; *P. varidens* has a larger umbilicus and adults are without teeth.

Habitat: Found in leaf litter in higher elevation northern hardwood forests; also among leaf litters occurring in deep rock talus.

Status: G2; Locally Rare.

Specimen: Shells (b, c & d) from North Carolina, Mitchell County, Roan Mountain (authors collection); shell (e & f) from Roan Mountain (FMNH 248906).

b

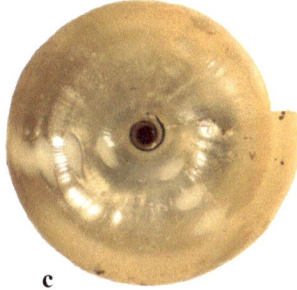

c

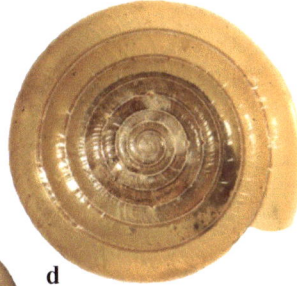

d

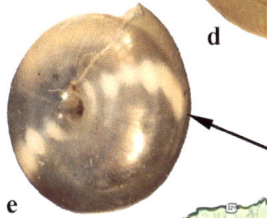

e

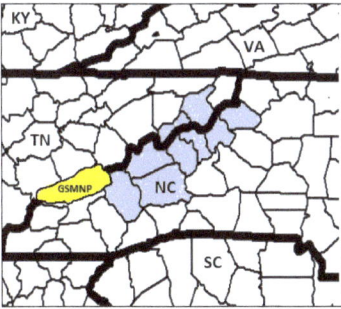

f

Lamellate supercoil

Pristilomatidae

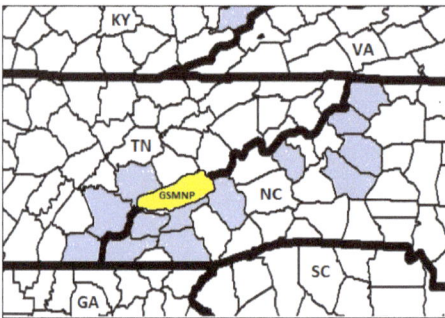

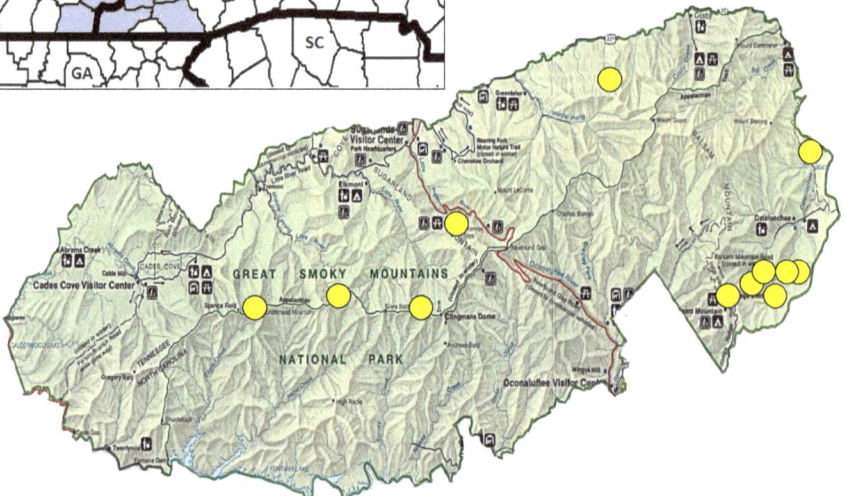

Paravitrea lamellidens (Pilsbry, 1898)

Diameter: 3.5-3.8 mm

Description: Depressed heliciform; lip simple; aperture half-moon shaped; shell with 6.5 tightly coiled whorls, last whorl not expanding; perforate; dark reddish-brown to cinnamon buff; shell thin and translucent in fresh and live shells; transverse striae are poorly developed but closely and evenly spaced; two to three rows of obliquely radial lamellae that are closely placed and well-angled, can be seen through bottom of both juveniles and adults.

Similar Species: *Paravitrea multidentata* is similar in form but is slightly smaller, has teeth instead of lamellae and a wider umbilicus.

Habitat: Found in deep pockets of moist leaf litter, especially in rock talus located in higher elevation mixed hardwood forests.

Status: G2; Uncommon; reported from Thunderhead Mountain, Clingman's Dome, Purchase Knob and other scattered locations.

Specimen: North Carolina, Swain County, Nantahala National Forest (author's collection).

Dentate supercoil

Pristilomatidae

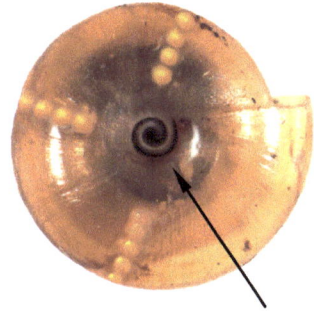

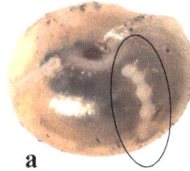

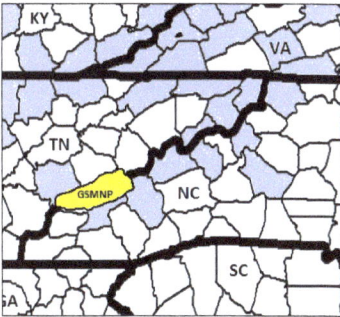

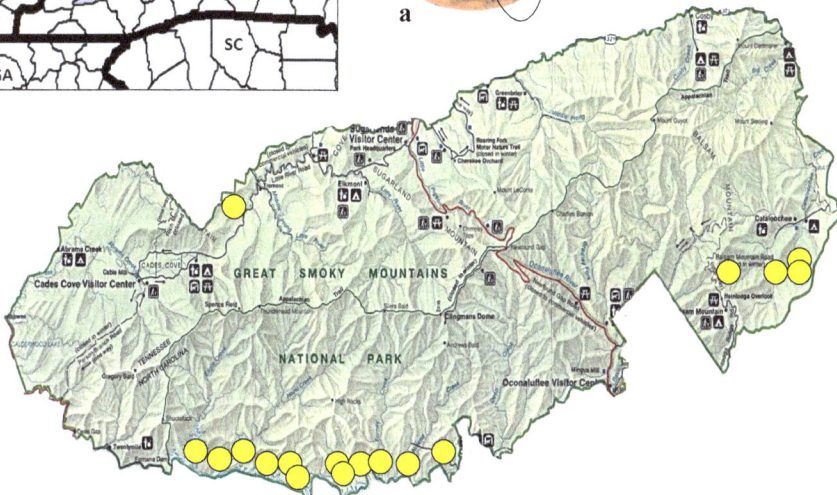

Paravitrea multidentata (A. Binney, 1840)

Diameter: 2.5-3 mm

Description: Depressed heliciform; lip simple, aperture half-moon shaped; shell with 6 tightly coiled whorls, perforate; last whorl not expanding; shell thin, glossy and translucent; young shells are perforate, adult becoming umbilicate; transverse striae are poorly developed but closely and evenly spaced; two to four radial rows of usually five teeth that can be seen through the bottom of fresh shells of juveniles and adults (a); on occasion, *P. multidentata* will have rows of lamellae.

Similar Species: *P. lamellidens* is similar in form but is slightly larger, has lamellae instead of teeth and has a notably smaller umbilicus.

Habitat: Found in pockets of moist leaf litter located on hillsides in mixed hardwood forests; especially common in wet leaf litter surrounding small seeps.

Status: G5; Common to Uncommon in scattered locations across the park.

Specimen: North Carolina, Swain County, Nantahala National Forest (author's collection).

a

116

Open supercoil

Pristilomatidae

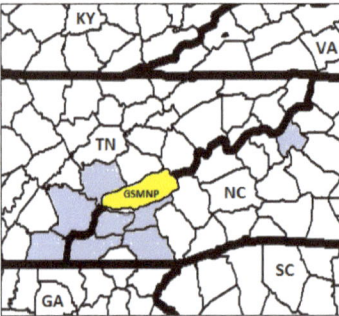

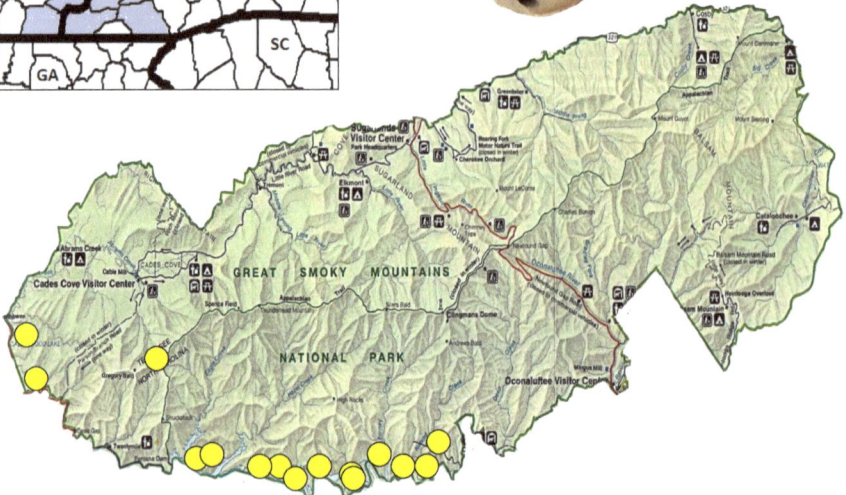

Paravitrea umbilicaris (Ancey, 1887)

Diameter: 2.9-3.6 mm

Description: Depressed heliciform; lip simple; shell with 6 tightly coiled whorls, umbilicate; the last whorl slightly expanding; shell thin, corneous with a cinnamon tinge, glossy and translucent; transverse striae are closely and regularly spaced with crossing (spiral) minute beads (page 119); two to four rows of radial lamellae (a) that can be seen through the bottom of fresh shells of juveniles and sometimes adults (page 119), these lamellae sometimes with finely denticulate edges (a); a single curving lamellae near the umbilicus (pg. 119).

Similar Species: *Paravitrea lamellidens* is similar in form and size but does not display the minute beading sculpture.

Habitat: Found in pockets of moist leaf litter located on hillsides in mixed hardwood forests and around small seeps.

Status: G3; Locally Rare; reported from south western and western portions of the GSMNP.

Specimen: Tennessee, Blount County, Calderwood area, GSMNP (GSMNP collection).

a

117

Toothed open supercoil

Pristilomatidae

Paravitrea umbilicaris form *dentata*
(Pilsbry, 1946)

Diameter: 3.5 mm

Description: Depressed heliciform; lip simple; shell with 6 tightly coiled whorls, the last whorl expanding only slightly; umbilicate; shell thin, glossy and translucent; transverse striae are closely and regularly spaced with crossing (spiral) minute beads (page 119); two to four rows of usually five or six radial teeth that can be seen through the bottom of fresh shells of both juveniles and adults (b), also page 119.

Similar Species: Differs primarily from *Paravitrea umbilicaris* by having radial rows of teeth instead of lamellae and by having a notably wider umbilicus.

Habitat: Found in pockets of moist leaf litter located on hillsides in mixed hardwood forests and around small seeps.

Status: G1; Globally Rare; a species Endemic to the GSMNP region; in need of further DNA investigation.

Specimen: Tennessee, Blount County, Calderwood area, GSMNP (GSMNP collection).

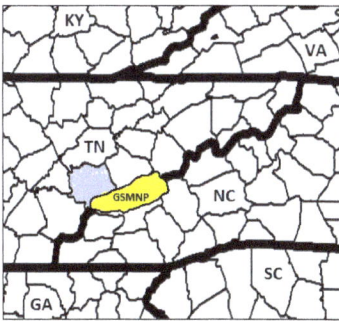

b

Comparing the internal armature of *Paravitrea umbilicaris* (lamellae) and *Paravitrea umbilicaris* form *dentata* (teeth)

Note the curving lamellae near the umbilicus, a feature first described by Hubricht (1985)

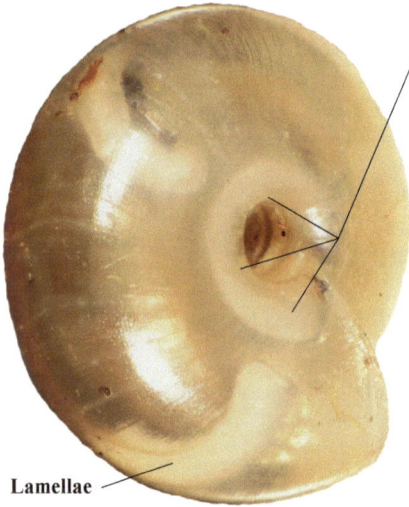

Lamellae

Teeth

P. umbilicaris (lamellae)

P. umbilicaris form *dentata* (teeth)

Top view of the shell of *Paravitrea umbilicaris* under high magnification showing the closely and regularly spaced beading (papillae).

Shells 3.2-8 mm, simple lip, notable raised shell sculpture under a lens of 10X, defense mucus fluorescent under UV (*Discus & Helicodiscus*),

Discus species are depressed heliciform and are notably larger than *Helicodiscus*. *Discus* are small (under 10 mm), less tightly coiled than *Helicodiscus* and do not feature color bands or blotches on shells like *Anguispira* species. *Discus* snails are usually without internal teeth, although *D. patulus* has a small callus tooth a short distance within its shell. *Discus* species have an entirely different shell sculpture than *Helicodiscus*, having transverse riblets (a), instead of spiral sculpture or fringed ornamentation (b). Under high magnification, *Helicodiscus* species are usually with teeth (c) and have some of the most extraordinary micro-sculpture of any land snails found in North America. Like that of *Anguispira*, the slime of *Discus* and *Helicodiscus* snails also glows fluorescent under a black light (Rawls and Yates 1971). Both *Discus* and *Helicodiscus* are generally found associated with rotting hardwood in advanced stages of decay and are sometimes found in large numbers.

Grouping 1: shells 6-8 mm, ribbed usually without teeth, pg 121

*D. patulus
*D. nigrimontanus
*D. bryanti

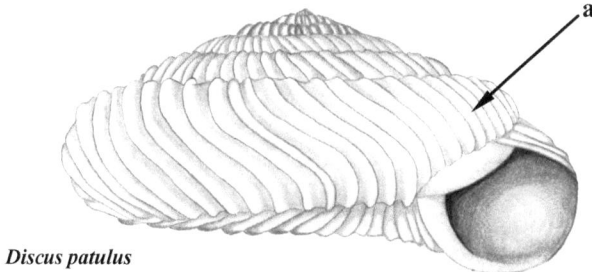

Discus patulus

Grouping 2: shells 3.3-5 mm, spiral striae strongly developed, with teeth, pg 124

*H. hexodon
*H. multidens
*H. notius
*H. parallelus

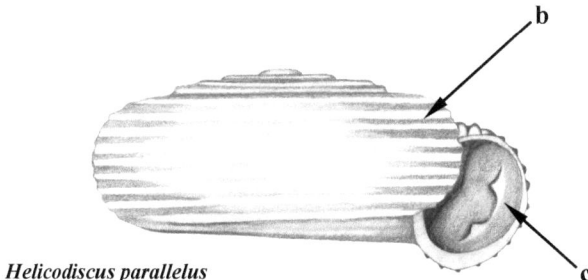

Helicodiscus parallelus

120

Domed disc

Discidae

Discus patulus (Deshayes, 1830)

Diameter: 7-8 mm

Description: Depressed heliciform; lip simple; aperture roundish to oval shape; shell with 5.5 whorls; widely umbilicate; there is a small rounded tubercle or callous tooth a short distance within (a) but this maybe a wanting feature in some specimens; transverse striae are well-developed, rib-like (3-4 ribs per mm) and are distinct on top, sides and base of shell; periphery rounded.

Similar Species: Other *Discus* species in the GSMNP have finer rib striae, a notably wider umbilicus and are without an internal parietal tooth.

Habitat: A species of upper slopes in mixed hardwood under leaf litter; it is most common on or under the exfoliating bark of rotting hardwood logs in advanced stages of decay.

Status: G5; Common; the most common *Discus* species in the GSMNP and eastern North America.

Specimen: North Carolina, Macon County, Nantahala National Forest (author's collection).

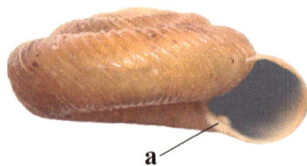
a

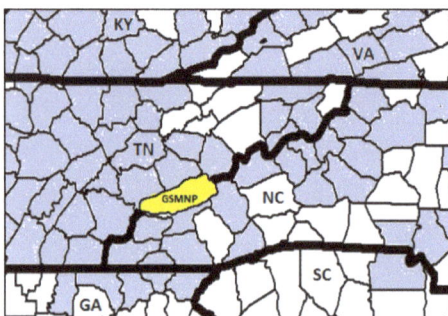

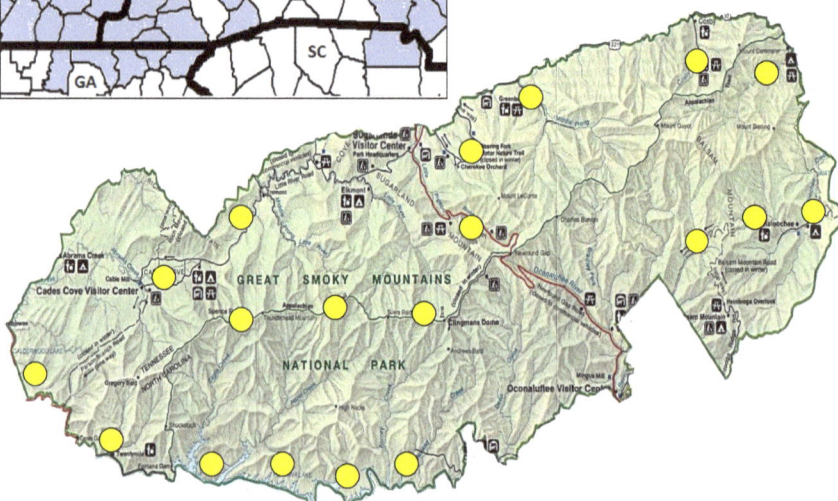

Black Mountain disc

Discidae

Discus nigrimontanus (Pilsbry, 1924)

Diameter: 7.4 mm

Description: Depressed heliciform; lip simple; aperture roundish to oval; shell with 5.25 whorls; widely umbilicate; striae are well-developed and rib-like; periphery sub-angular; the microscopic crisscross sculpture of the embryonic whorl is weak and wanting; no internal columellar tubercle as seen in *D. patulus*.

Similar Species: *D. patulus* has a rounded periphery; a small parietal tooth; coarser riblets and a smaller more well-like umbilicus.

Habitat: A species of upland rocky talus slopes in mixed hardwood forests where leaf litter is sparse; common in the lower layers of deep limestone talus along the base of outcropping limestone.

Status: G4; Locally Rare; reported from the Whites Oak Sinks area only, but the species is expected to occur elsewhere in the park.

Specimen: Tennessee, Blount County, White Oak Sinks, GSMNP (GSMNP collection).

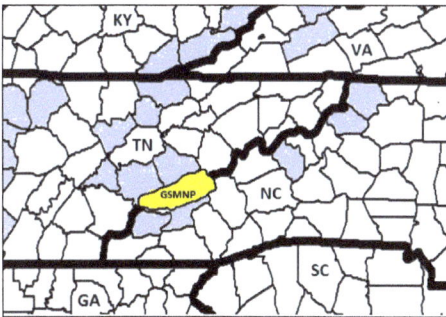

Sawtooth disc

Discidae

Discus bryanti (Harper, 1881)

Diameter: 6-7.5 mm

Description: Depressed heliciform; lip simple, aperture roundish, the outer portion pinched (a); shell with 5.25 whorls; widely umbilicate; striae are well-developed and rib-like; periphery strongly carinate; no internal columellar tubercle as seen in *D. patulus*.

Similar Species: *D. patulus* has a rounded periphery, a small parietal tooth, coarser riblets and a smaller more well-like umbilicus; *D. nigrimontanus* is without the pinched aperture, less angular periphery and smaller umbilicus.

Habitat: A species of mixed hardwood forested slopes found around logs and stumps in advanced stages of decay at higher elevations.

Status: G3; Locally Rare; currently reported from around the Purchase Knob area only; the species is most common in Virginia becoming more scarce in the southern mountains.

Specimen: North Carolina, Mitchell County, Roan Mountain around Carvers Gap (author's collection).

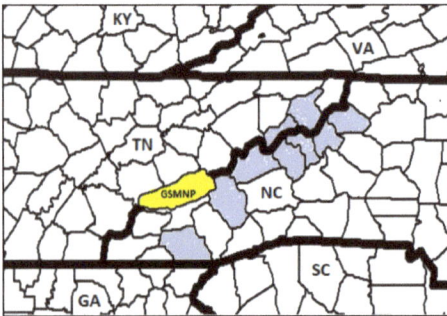

Toothy coil

Helicodiscidae

Helicodiscus hexodon (Hubricht, 1966)

Diameter: 4.9 mm

Description: Depressed heliciform; lip simple, aperture oval-distorted; shell with 5-5.5 whorls; widely umbilicate; shell yellowish; in the last whorl there are usually 3 pairs of teeth on the outer and basal wall (b & c), teeth transversely elongate and angled, with basal teeth a little in front of their respective outer teeth (may not be seen in aperture view); spiral striae and fringes remain uniform in length on all whorls.

Similar Species: *Helicodiscus fimbriatus* has both short and longer fringes, having at least three that are more prominent than the rest of the fringes.

Habitat: A calciphile of mixed hardwood forests under leaf litter and among rock slides.

Status: G1; Globally Rare; Until recently this species was known only from the type locality (listed below), but a recent (2010) survey around Calderwood found specimens of this rare *Helicodiscus*.

Specimen: Tennessee, Bledsoe County, Pikeville, base of Walden Ridge (FMNH 147041).

b

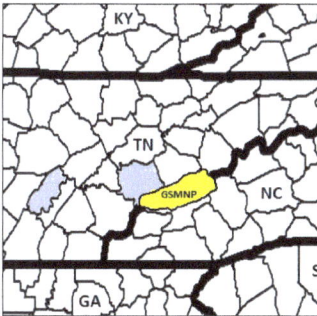

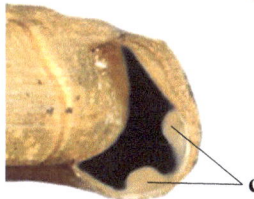

c

Aperture

124

Twilight coil

Helicodiscidae

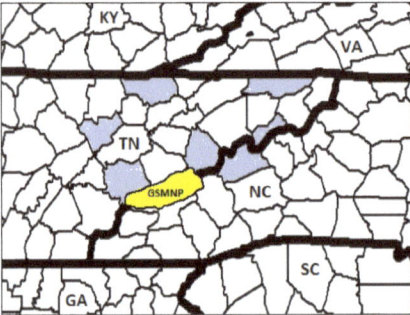

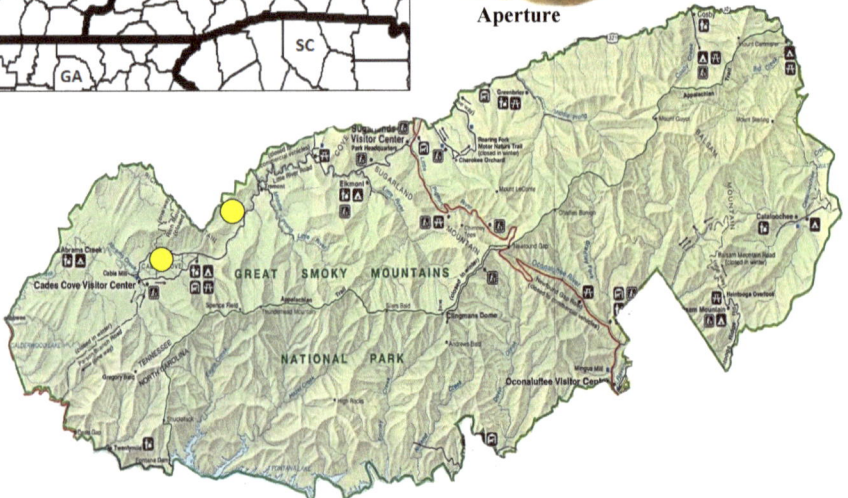

Helicodiscus multidens Hubricht, 1962

Diameter: 4.75 mm

Description: Depressed heliciform; lip simple, aperture roundish; shell with 4.5-5 whorls; shell pale greenish-yellow, dull, opaque; widely umbilicate; in the last whorl there are usually three pairs of teeth on the outer and basal wall (a); these teeth are radially elongate and are about twice as broad as high (unlike the teeth of other *Helicodiscus* species which are more knob-like), alternating with these teeth are usually three teeth on the parietal wall (b); under a strong lens the spiral striae are fringed (figured on opposite page) and present on all whorls but this feature is typically lost in aging shells.

Similar Species: *Helicodiscus parallelus* and *H. notius* have knob-like teeth.

Habitat: A calciphile found under rocks and leaf litter on river bluffs and in caves.

Status: G3; Locally Rare; known from Cades Cove and White Oak Sinks only.

Specimen: Tennessee, Smith County, Elmwood (FMNH 239082).

Aperture

Above illustrating the extraordinary micro-ornamentation found on the shell surface of *Helicodiscus multidens* magnified 1000X; not an uncommon feature on other species of land snails under five millimeters. SEM image by R. Wayne Van Devender.

Tight coil

Helicodiscidae

Helicodiscus notius Hubricht, 1962

Diameter: 3.66 mm

Description: Depressed heliciform; lip simple; shell with 5-5.5 whorls; widely umbilicate; within the last whorl of juvenile and adult shells there are 2 to 3 pairs of conical teeth which can be seen through the bottom of fresh shells; spiral, raised striae are well-developed including the embryonic whorl (a).

Similar Species: *Helicodiscus parallelus* has poorly developed spiral striae on the embryonic whorl; *H. notius specus,* a troglobitic land snail is reported from Kentucky and several eastern Tennessee cave systems within a reasonable distance of the GSMNP; adult *H. n. specus* shells are without internal armature.

Habitat: A species found on forested hillsides, in ravines and occasionally in caves; usually found in drier and higher habitats than *H. parallelus*.

Status: G5; Uncommon but likely more common than current records indicate.

Specimen: Kentucky, Letcher County, Bad Branch (author's collection).

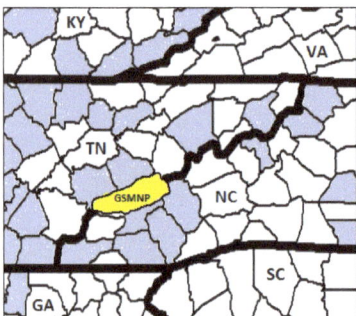

a

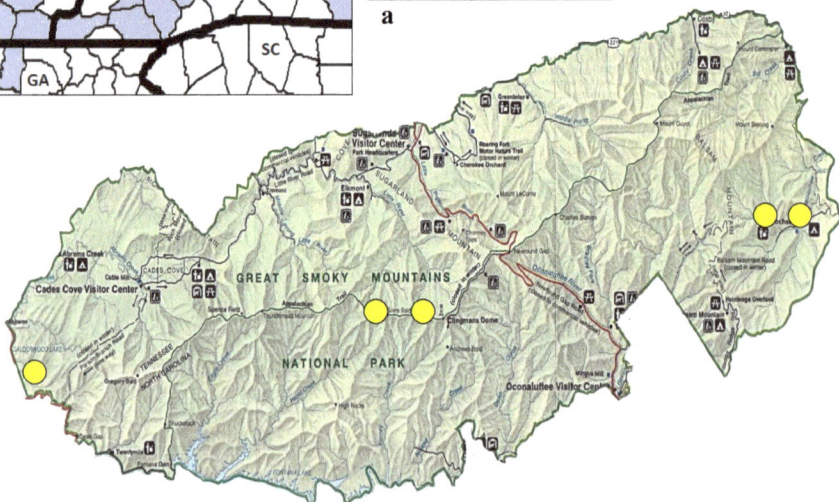

Compound coil

Helicodiscus parallelus (Say, 1817)

Diameter: 3.2-3.5 mm

Description: Depressed heliciform; lip simple; shell with 4-4.5 whorls; widely umbilicate; two teeth present usually deep within the aperture; spiral striae are well-developed except on the embryonic whorl where there is only a hint of spiral striae (b); periphery rounded.

Similar Species: *Helicodiscus notius* is larger by one whorl, has a broader umbilicus and most importantly, the spiral striae on the embryonic whorl is more strongly developed.

Habitat: Found in floodplains and upland mixed hardwoods under leaf litter and around rock structure; also in urban areas under wood debris and around vacant lots.

Status: G5; Uncommon; the species is likely more common than records currently indicate and leaf litter surveys will most certainly reveal additional locations for the species.

Specimen: Kentucky, Jefferson County, near Harrods Creek, at US Route 42, S. of Prospect (FMNH 239415).

Helicodiscidae

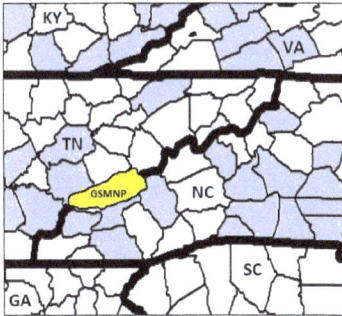

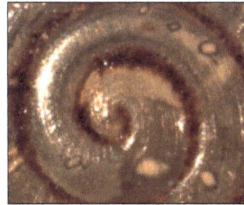

b

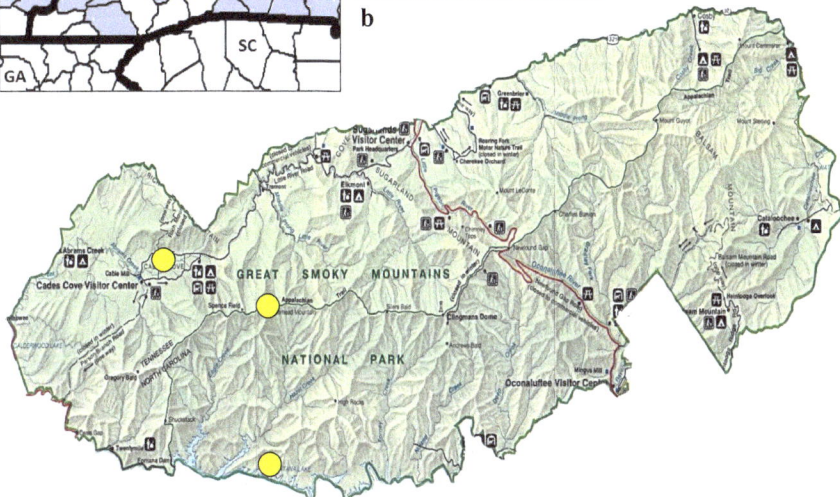

Fringed coil

Helicodiscus fimbriatus Wetherby, 1881

Diameter: 5 mm

Description: Depressed heliciform; lip simple; shell with 5 whorls; widely umbilicate; paired teeth present (sometimes up to six pairs) can be seen through the bottom of fresh and live shells (a), located on the outer and basal walls, these teeth opposite of each other (may not be seen in aperture view); spiral striae are unequal, several developed into long conspicuous fringes (not hairs); these fringe augmentations are frequently lost in older shells.

Similar Species: *Helicodiscus bonamicus* has hairs not fringes; *H. hexodon* fringes are short and uniform in length, not containing the longer fringes seen in *H. fimbriatus*.

Habitat: Found in leaf litter, under flat rocks on rocky hillsides and in mixed hardwoods in ravines.

Status: G4; Locally Rare but likely more common than current records show.

Specimen: Tennessee, Monroe County, Cherokee National Forest (author's collection).

Helicodiscidae

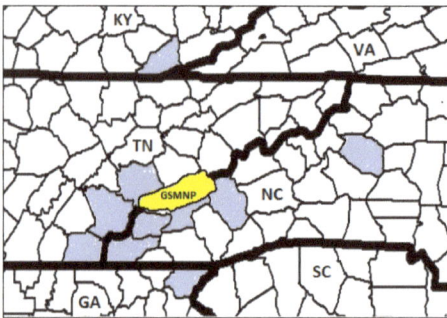

a

Crawling by a Thread

With perfect balance a five millimeter wide juvenile flat bladetooth, *Patera appressa* crawls across a single strand of silk, demonstrating its remarkable adaptable foot. Red River Gorge, Powell County, Kentucky.

Shells 6-20 mm (most species are between 7-12 mm), with or without teeth, simple lip (*Ventridens*)

Superficially, *Ventridens* look like miniature *Mesomphix*, but a closer look reveals that *Ventridens* have tightly-coiled shells (a), the last whorl not expanded like seen in *Mesomphix*. *Ventridens* are also more compact in form and average 2-3 more whorls than *Mesomphix*. *Ventridens* often have teeth or lamellae (b) while *Mesomphix* are without these defensive features. Like *Mesomphix* however, *Ventridens* have a thickening or thin callus usually a whitish color just inside the aperture's bottom (c). The shell is perforate (never completely closed) to umbilicate and in many *Ventridens* the umbilicus is proportionately wider in the youngest shells. The bidentate species of the genus *Ventridens* are characterized by the presence of 2 apertural lamellae in the pre-adult shells. *Ventridens* in this group can be divided readily into two basic groups by the color of the live animal and the presence or absence of internal armature. In the *V. gularis* group, the live animal is dark, olive or bluish-gray (opposite page). In the *V. pilsbryi* group, the live animal is pale, yellow with perhaps some grayish flecking along the back (opposite page). Interestingly, species that fall in each of the groups are not found living together while it is common to find species of different groups in the same vicinity. One exception is *V. lasmodon* which has been documented with other species in its group at three locations (Hubricht, 1964). Not only do the different species occur in separate colonies, they are often geographically exclusive creating a crazy-quilt pattern of distribution that appears to have no relation to underlying geological formations, exposure or other ecological factors (Hubricht, 1964). The third group of *Ventridens*, are identified by adult shells are without teeth. The color of live animals should be noted in the field (before preservation) when possible as well as the presence of any teeth in shells collected.

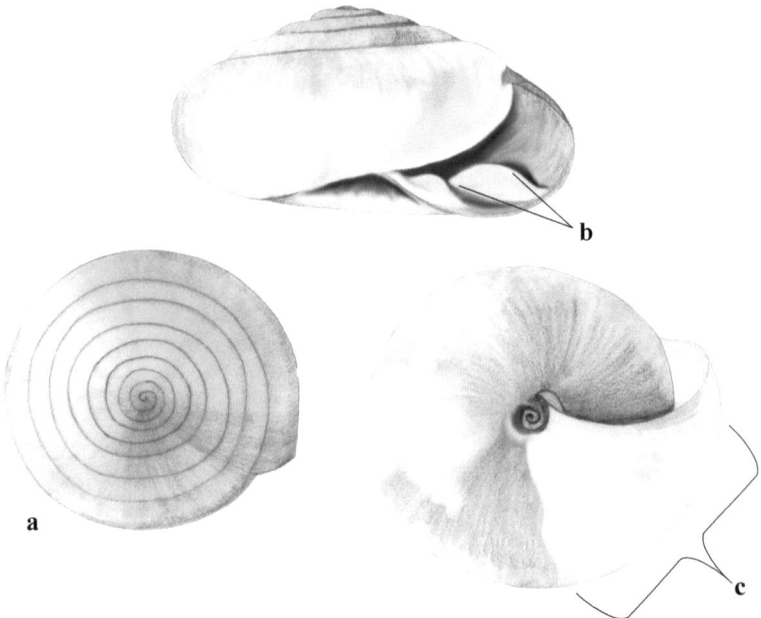

Grouping 1: *Ventridens gularis* group: Live animal dark, olive or bluish-gray. Two or more apertural lamellae (d) in the juvenile shell-stage. One, both or all of these lamellae may be reduced or wanting in mature shells of this group and shells rarely have more than 6.5 whorls. The aperture is proportionally larger than in the *V. pilsbryi* group of shells, starting on pg. 133.

*V. gularis
*V. suppressus magnidens

d

Grouping 2: *Ventridens pilsbryi* group: Animal pale, (may have a hint of yellow) with some grayish flecking along the back and sides. Like the *V. gularis* group, there are two or more apertural lamellae (d) in the juvenile shell stage and one, both or all of these lamellae may be reduced or wanting in mature shells of this group. The diameter of the whorls usually increases more slowly and the whorls are more flattened than seen in the *V. gularis* group. *V. pilsbryi* shells may have as many as 9 whorls, starting on pg. 135.

*V. pilsbryi
*V. collisella
*V. theloides
*V. lawae
*V. lasmodon
*V. decussatus

Grouping 3: Other *Ventridens*, no teeth in the adult stage, pg 141

*V. demissus
*V. intertextus
*V. ligera
*V. arcellus
*V. acerra

No internal armature in adult shells

132

Throaty dome

Gastrodontidae

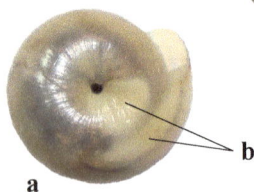

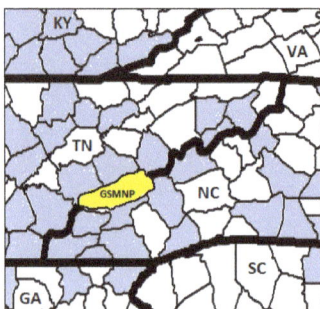

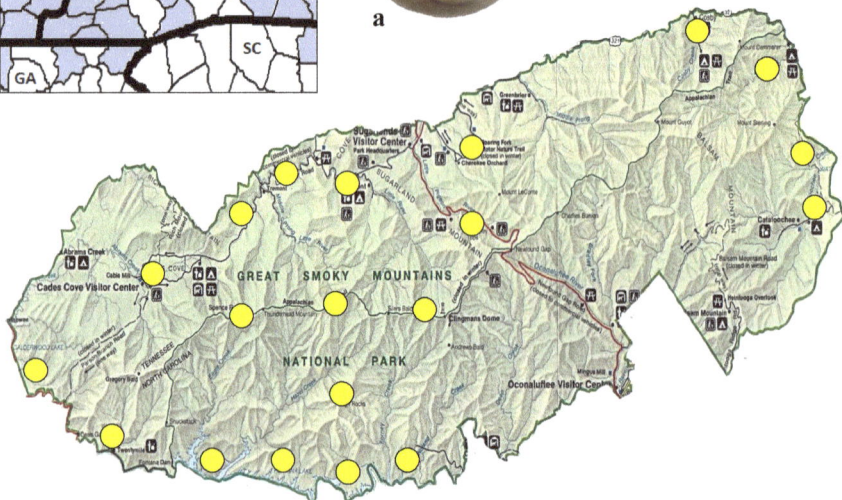

Ventridens gularis (Say, 1822)

Diameter: 7.5-8 mm

Description: Depressed heliciform to helici-form, young shells are more depressed; lip simple; shell with 6-7 tightly coiled whorls; perforate, usually widest in young shells (a) but in some adults, the perforation is as wide as half a millimeter; live animal nearly black (an important feature); shell translucent and glossy; transverse striae are distinct on top, but weaker on the sides and bottom; spiral striae are indistinct; within the aperture of both young and adult shells there are two lamellae which can be seen through the bottom of fresh shells (b); a member of the *V. gularis* group.

Similar Species: *Ventridens pilsbryi* has a lower shell profile, 7-8 whorls and the live animal is pale yellow.

Habitat: Found on wooded hillsides, ravines, under leaf litter and logs.

Status: G5; Common; the most common low elevation *Ventridens* in the park.

Specimen: North Carolina, Haywood County, Purchase Knob, GSMNP (GSMNP collection).

Corneous dome

Gastrodontidae

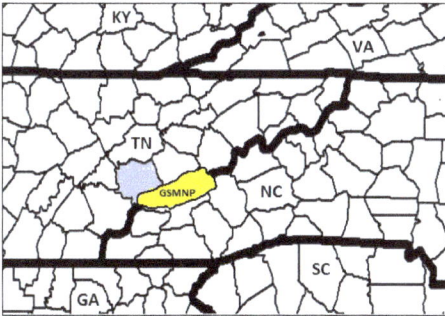

Ventridens suppressus magnidens
(Pilsbry, 1946)

Diameter: 6-7 mm

Description: Depressed heliciform; lip simple; shell with 6 tightly coiled whorls; perforate; shell translucent and very glossy; thin callus just inside the aperture's bottom (c); transverse striae are only a faint feature on top, weaker on sides and bottom; spiral striae a diminished feature; live animal dark, olive or bluish color; within the aperture there are two lamellae or teeth that are present at any stage of growth and can be seen through the bottom of live and fresh shells; a member of the *V. gularis* group.

Similar Species: *Ventridens pilsbryi* and *V. gularis* are larger, have thicker shells and are generally found in drier sites.

Habitat: Found in low wet grassy places and around the edges of wetlands.

Status: G1; Locally Rare; reported from the Cades Cove area only; a subspecies in need of further DNA investigation.

Specimen: Tennessee, Blount County, Cades Cove, GSMNP (GSMNP collection).

c

134

Yellow dome

a

Gastrodontidae

Ventridens pilsbryi Hubricht, 1964

Diameter: 8.7-9.6 mm

Description: Depressed heliciform; lip simple; shell with 7-8 tightly coiled whorls; perforate, widest in young shells (a); live animal pale; shell translucent; transverse striae are distinct on top and somewhat weaker on the bottom; spiral striae are indistinct; within the aperture of both young and adult shells there are two lamellae (b) which can be seen through the bottom of fresh shells (c); a member of the *V. pilsbryi* group (Hubricht 1964).

Similar Species: *V. pilsbryi* has been confused with *V. gularis,* which has a smaller more globose shell, with fewer whorls and the live animal is nearly black.

Habitat: Found on wooded hillsides and in ravines under leaf litter and on logs, more common on limestone but also occurs on acidic sandstone sites.

Status: G5; Uncommon; reported from scattered locations across the park.

Specimen: Tennessee, Monroe County, Cherokee NF (authors collection).

c

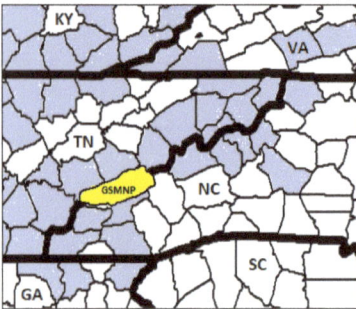

b

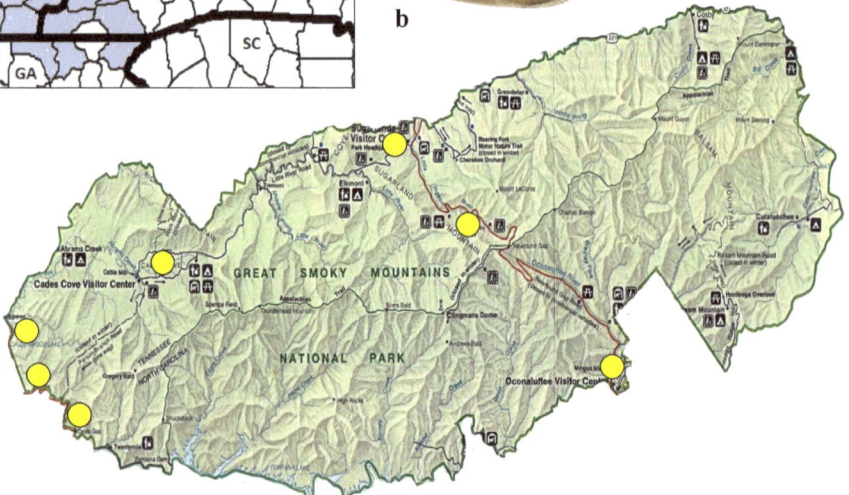

135

Crossed dome

Gastrodontidae

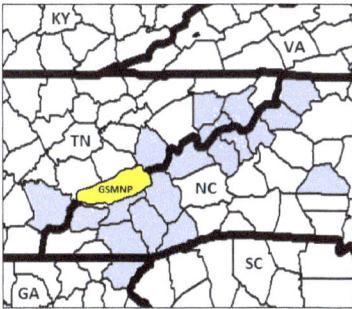

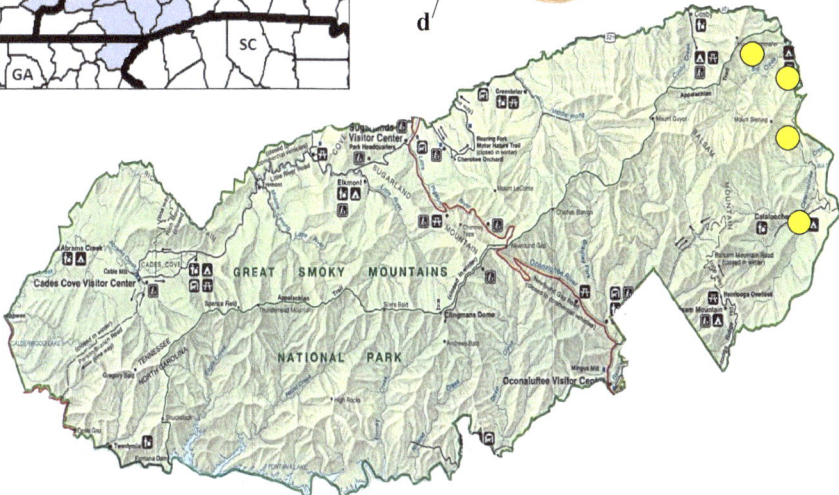

Ventridens decussatus (Walker & Pilsbry, 1902)

Diameter: 7.8-8.2 mm,

Description: Heliciform, young shells are flatter and have an angular periphery (d); lip simple; shell with 8 tightly coiled whorls; perforate; transverse striae are well-developed and deeply cut into the shell surface, continuing to the base; there are well-developed basal lamellae in both juvenile and adult shells.

Similar Species: Other *Ventridens* species have only weakly developed transverse striae on their base; *Ventridens ligera* is larger, has a smaller umbilicus, no teeth in juvenile shells and has a higher shell than *V. decussatus*; *V. gularis* and *V. pilsbryi* have smoother top surfaces and in general a tighter umbilicus.

Habitat: Found on wooded hillsides, in ravines under leaf litter.

Status: G3; Uncommon; currently reported from the eastern portions of the park in Big Creek and Purchase Knob areas only.

Specimen: North Carolina, Haywood County, Big Creek, GSMNP (GSMNP collection).

d

Sculptured dome

Gastrodontidae

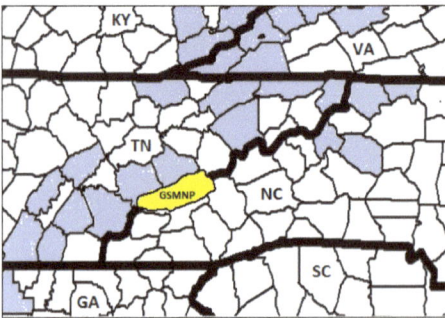

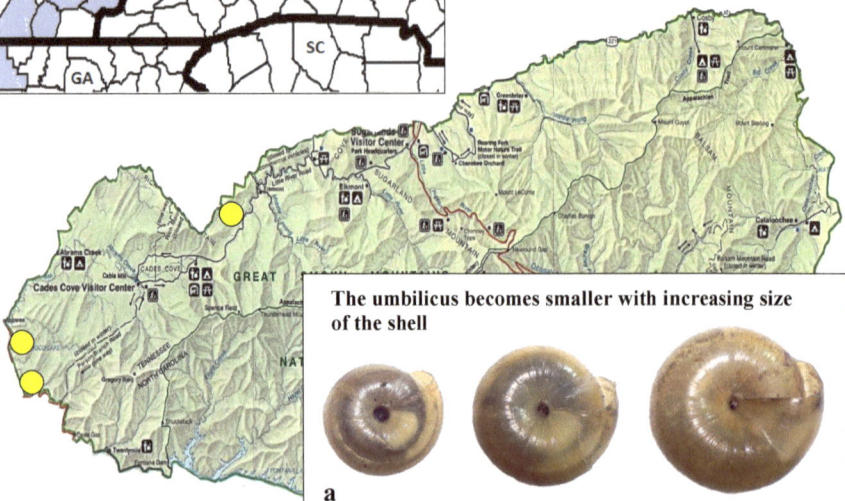

Ventridens collisella (Pilsbry, 1896)

Diameter: 8.4-9.6 mm

Description: Heliciform; lip simple; shell with 7.5-8 tightly coiled whorls; perforate, young shells are umbilicate, the umbilicus becoming smaller as shells mature (figure a below); live animal pale; shell pale yellowish horn color and dull-glossy; transverse striae are well-defined and boldest near the sutures; without any notable spiral striae; usually with two entering lamella, in adults shells the lamellae becoming lower in stature; although some specimens will be without this armature; a member of the *V. pilsbryi* group (Hubricht 1964).

Similar Species: *Ventridens ligera* is larger and does not contain teeth at any stage of growth.

Habitat: A calciphile species occurring on wooded hillsides in hardwood forests; usually below 300 m.

Status: G4; Uncommon; found in around White Oak Sinks and the west end of TN side around the Calderwood area.

Specimen: Virginia, Tazewell County, (author's collection).

The umbilicus becomes smaller with increasing size of the shell

a

137

Copper dome

Gastrodontidae

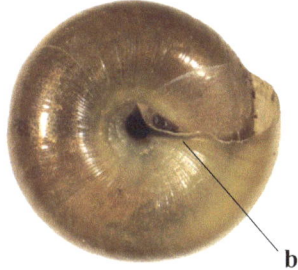

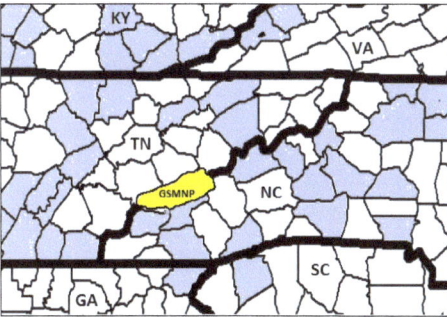

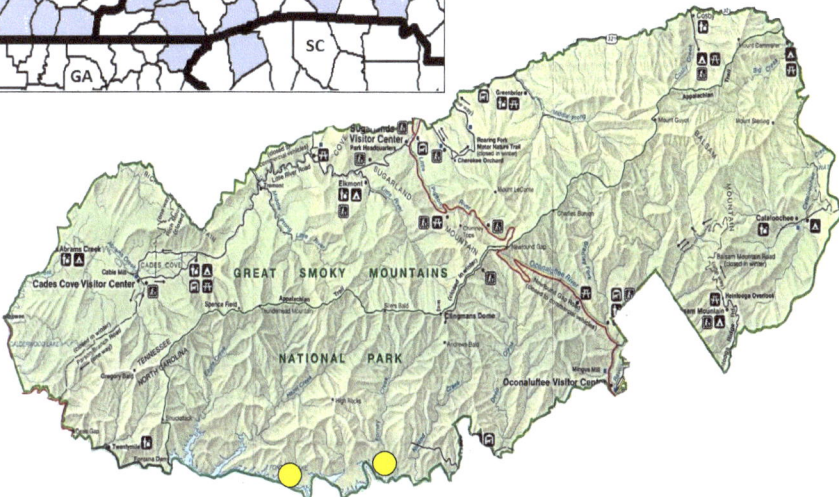

Ventridens theloides (Walker & Pilsbry 1902)

Diameter: 7.5-8 mm

Description: Heliciform and dome shaped; lip simple or with a slight basal deflection (b); 7.5-9 tightly coiled whorls; perforate but open, the area around umbilicus well excavated and funnel-like; live animal pale; shell yellowish or coppery, the base very glossy; transverse striae are moderately well-developed on top but are diminished on the base; young shells are with basal lamellae, adult shells typically without; a member of the *V. pilsbryi* group.

Similar Species: *Ventridens gularis* has well developed basal lamellae and a smaller umbilicus; *V. lawae* has a much wider umbilicus and is more compressed.

Habitat: Found in upland hardwood forests, in ravines, under leaf litter and around logs; generally a species of limestone but also occurs on sandstone sites.

Status: G5; Locally Rare; in the park currently known only from areas above Fontana Lake.

Specimen: North Carolina, Swain County, Nantahala NF (author's collection).

b

138

Rounded dome

Gastrodontidae

Ventridens lawae (W. G. Binney, 1892)

Diameter: 7.8-9 mm

Description: Depressed heliciform; lip simple; shell with 7.5-8.5 tightly coiled whorls; umbilicate, the umbilicus is deep with nearly parallel sides (well-like); in adults the umbilicus may vary from 0.4 mm to 2 mm in size (Hubricht 1964) ; live animal pale; shell light horn to yellow and glossy; finely developed transverse striae are distinct on top but dwindle on the sides and base of shell; young shells with lamellae (a); adults with or without a basal lamella; a member of the *V. pilsbryi* group.

Similar Species: *Ventridens lasmodon* is slightly smaller, has a notably wider umbilicus and has a thinner shell.

Habitat: Found in mixed hardwood forests on hillsides and in ravines under leaf litter and on logs, more common on limestone but also occurs on sandstone sites (Hubricht 1985).

Status: G5; Uncommon; reported from Cades Cove area, Twin Creeks and White Oaks Sinks.

Specimen: North Carolina, Macon County, Nantahala National Forest (author's collection).

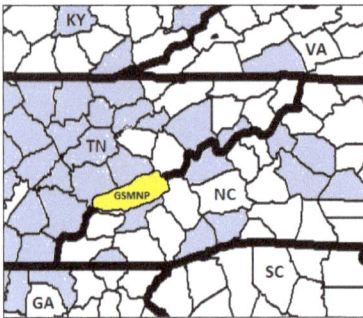

a

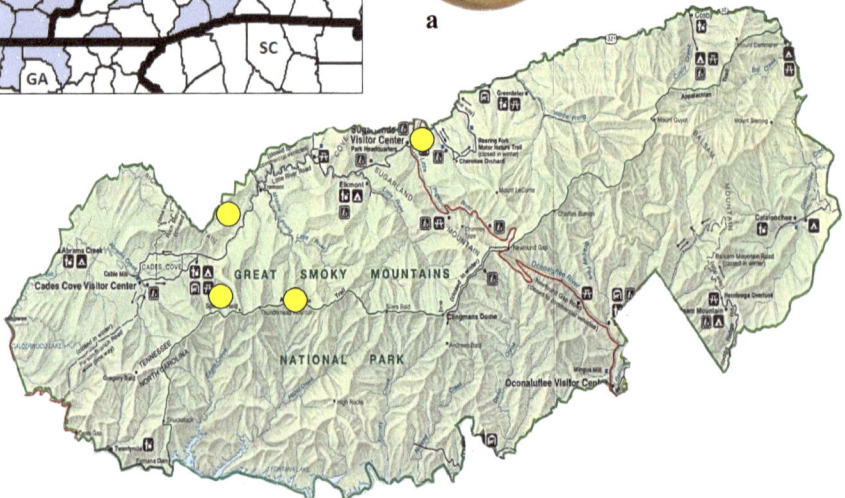

139

Hollow dome

Gastrodontidae

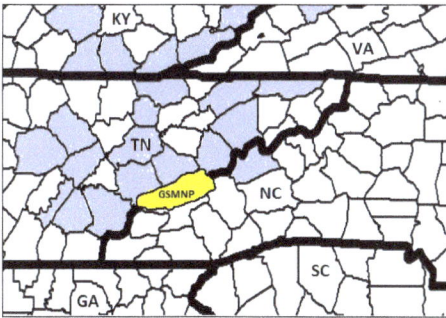

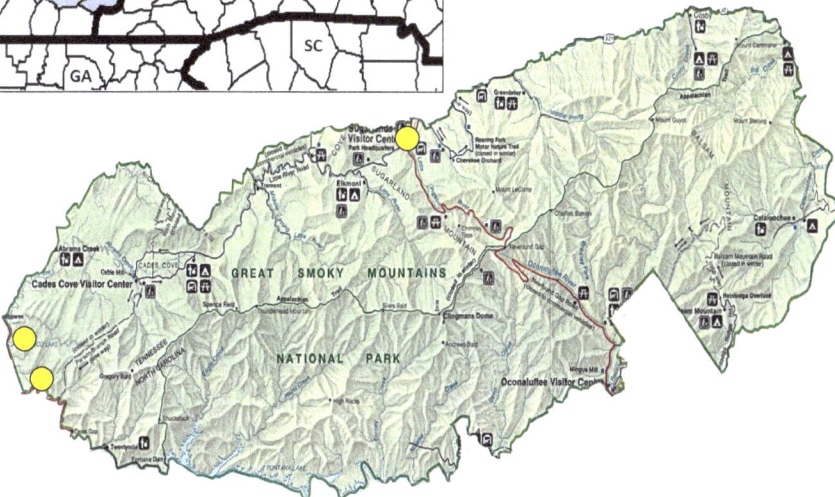

Ventridens lasmodon (Phillips, 1841)

Diameter: 7.5-7.8 mm

Description: Depressed heliciform; lip simple; shell with 7.5-8 tightly coiled whorls; umbilicate (the widest of any *Ventridens*), the umbilicus is funnel-shaped and nearly one third the diameter of the shell; live animal pale; shell translucent, thin and glossy; finely developed transverse striae (growth wrinkles) are distinct on top but grow weaker on the sides and base of shell; spiral striae not well-defined; within the aperture are two elongated lamellae (b) or teeth that are present at every stage of growth, member of the *V. pilsbryi* group.

Similar Species: *Ventridens lawae* has a higher shell profile, a thicker shell and a notably smaller umbilicus.

Habitat: Found on wooded hillsides and in ravines under leaf litter on both limestone and sandstone sites.

Status: G4; Locally Rare; in the GSMNP currently reported from only three locations.

Specimen: Tennessee, Monroe County, Cherokee National Forest (author's collection).

b

Perforate dome

Gastrodontidae

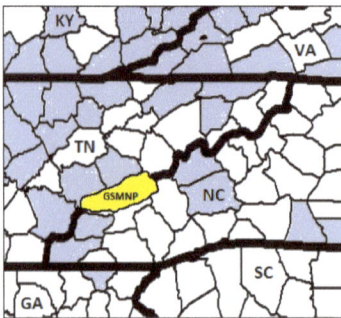

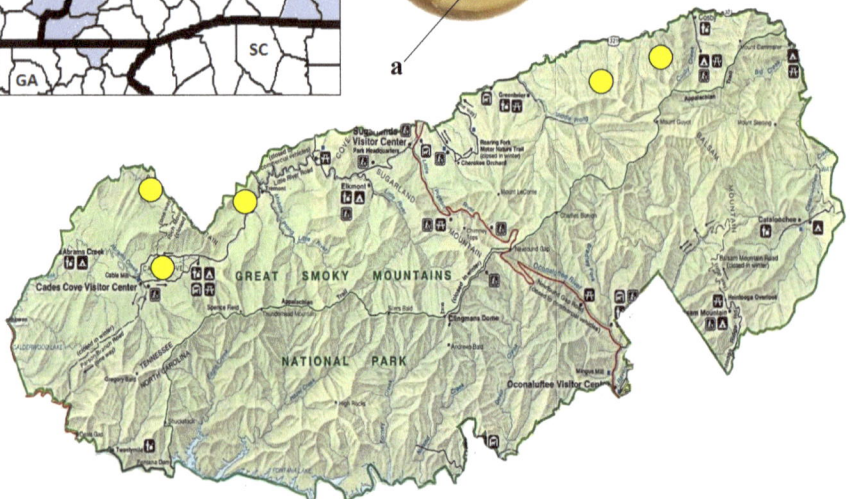

Ventridens demissus (A. Binney, 1843)

Diameter: 7.5-11 mm,

Description: Heliciform; lip simple; shell with 6-7.5 tightly coiled whorls; perforate; shell horn-yellow and glossy; transverse striae (growth wrinkles) are moderately developed on top of shell but less so on the sides and bottom; aperture typically without basal lamella in adult shells, but young shells (a) are with this feature.

Similar Species: *Ventridens ligera* is larger, has a slightly smaller umbilicus, no teeth in juvenile shells and has a higher shell than *V. demissus*; *V. collisella* is smaller, has better developed transverse striae and teeth in the adult stage.

Habitat: Found on wooded hillsides, in ravines and floodplains under leaf litter; also a snail found frequently in urban areas.

Status: G5; Uncommon; several sites reported including Rich Mountain, White Oak Sinks and the Albright Grove area.

Specimen: Kentucky, Letcher County, Bad Branch (author's collection).

a

Pyramid dome

Gastrodontidae

Ventridens intertextus (A. Binney, 1841)

Diameter: 8-20 mm

Description: Heliciform; lip simple; shell with 5.5-6 whorls; perforate; shell yellowish-horn to olive-buff and dull; broken transverse striae (b) are well-developed and are the single best feature for identifying this species; spiral striae are well-developed; without teeth or basal lamellae in all stages of growth; shell periphery can be round or angular, rarely with a faint color band.

Similar Species: *V. ligera* has a similar form but does not have the unique broken striae found in *V. intertextus*.

Habitat: Found in mixed hardwood forests on hillsides, ravines and acidic ridgetops, under leaf litter and in discarded bottles with their openings pointing uphill.

Status: G5; Uncommon; found in the southwestern portion of the park along the north shore of Fontana Lake, and near the Calderwood area of the park.

Specimen: Kentucky, Letcher County, Bad Branch (author's collection).

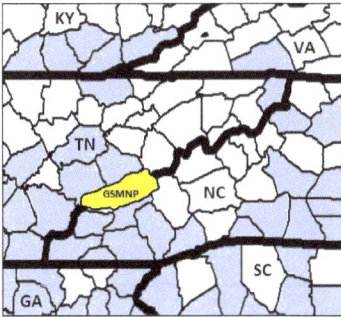

b

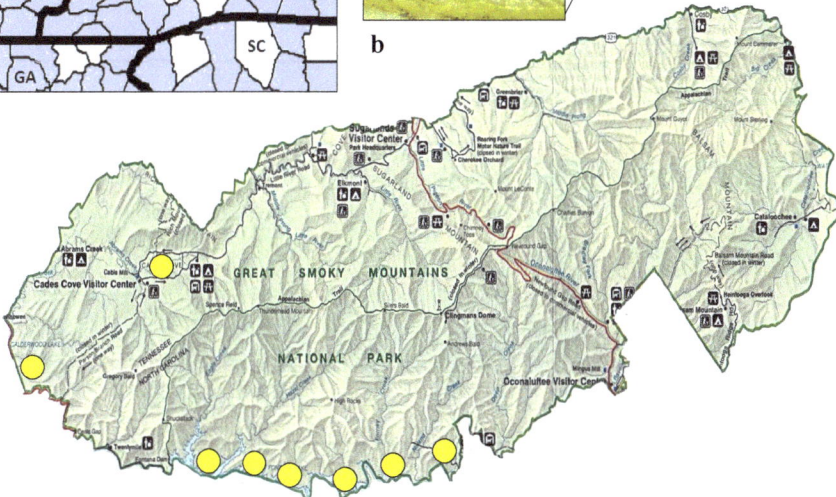

142

Glossy dome

Gastrodontidae

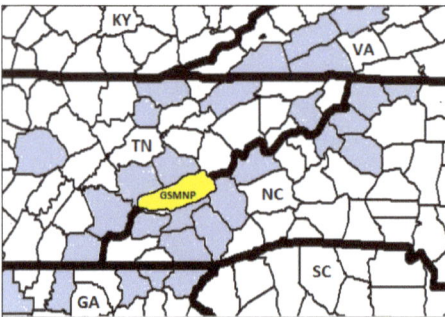

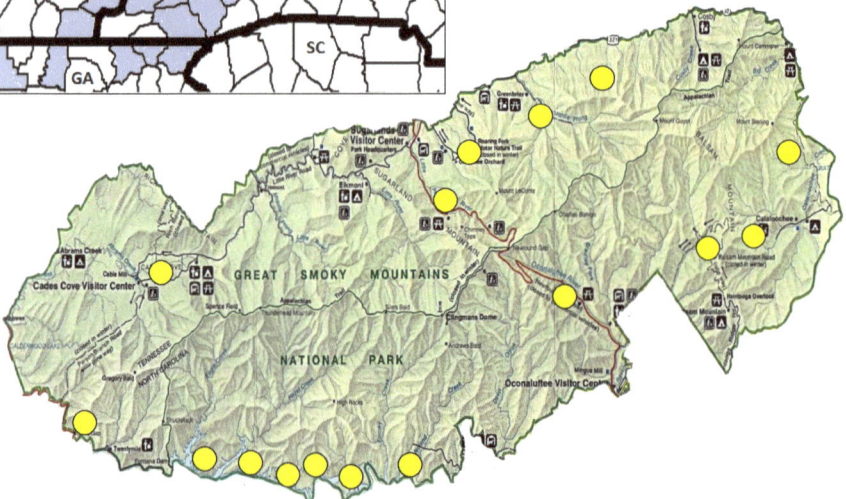

Ventridens acerra (J. Lewis, 1870)

Diameter: 12.6-19 mm

Description: Heliciform; lip simple; shell with 7-8 whorls, the last whorl only slightly expanded; perforate; shell light yellowish olive and very glossy; transverse striae are fairly well developed, widest on the last whorl, these striae strongest on the top but diminished on the sides and base; without notable spiral striae; no teeth at any stage of growth.

Similar Species: *Ventridens arcellus* is smaller, has a larger umbilicus, is more tightly wound (the last whorl not expanding greatly) and is usually found above 1200 m; *V. ligera* is less glossy, has a higher shell, is less excavated around the perforation and is a species of low places.

Habitat: A habitat generalist found in a variety of mixed hardwood forests on hillsides under leaf litter, generally under 1200 m in elevation.

Status: G5; Relatively Common; mostly found throughout the park at lower elevations.

Specimen: North Carolina, Swain County, Nantahala National Forest (author's collection).

Golden dome

Ventridens arcellus Hubricht, 1976

Diameter: 12.9-13.6 mm

Description: Heliciform; lip simple; shell with 6-7.5 whorls, the last whorl only slightly expanded; perforate; shell light yellowish olive and glossy; transverse striae are poorly developed, strongest on the top but fading away on the sides and base; without notable spiral striae; no teeth at any stage of growth.

Similar Species: *Ventridens acerra* is larger, slightly more compressed, has a smaller umbilicus, is less tightly coiled and is generally found under 1200 m; *V. ligera* is less glossy and found in low places.

Habitat: A habitat generalist found in a variety of mixed hardwood and northern hardwood forests on hillsides and mountaintops under leaf litter, above 1200 m elevation.

Status: G4; Locally Rare; in the park restricted to mid and higher elevations.

Specimen: Tennessee, Swain County, Clingman's Dome, GSMNP (GSMNP collection).

Gastrodontidae

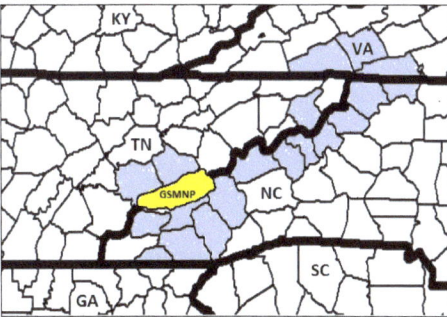

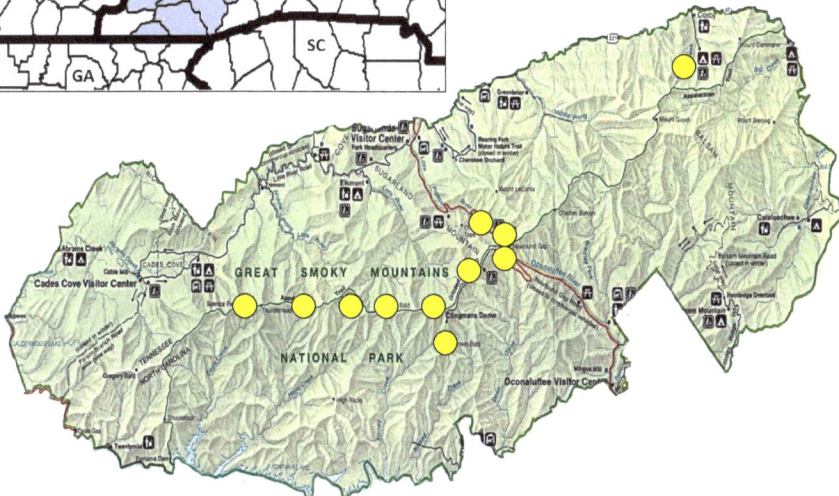

Globose dome

Gastrodontidae

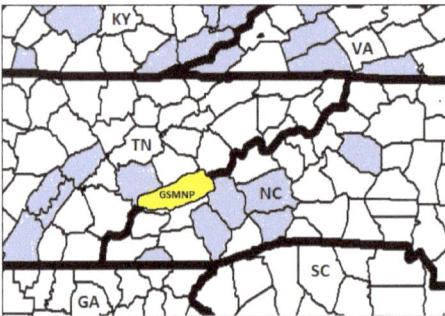

Ventridens ligera (Say, 1821)

Diameter: 11-15.6 mm

Description: Heliciform; lip simple; shell with 6-7 whorls; perforate; shell pale-yellowish-horn and somewhat glossy; transverse striae (growth wrinkles) are relatively well-developed on top of shell but much less defined on the sides and bottom; aperture without a basal lamella in any stages of growth.

Similar Species: *Ventridens intertextus* has a similar form but has better developed transverse striae, a slightly larger umbilicus and often a weakly angular periphery, occasionally displaying a light a color band on its periphery; *V. demissus* is smaller and young shell are with internal armature.

Habitat: Found in a variety of open, weedy and mixed hardwood forests in floodplains and other wet low-lying areas; also along roadsides.

Status: G5; Locally Rare; currently known only from the Foothills Parkway.

Specimen: Kentucky, Powell County, Furnace Mountain (author's collection).

Ventridens Shells Compared X 3 (proportionate)

V. gularis

V. pilsbryi

V. collisella

V. lawae

V. lasmodon

V. decussatus

V. s. magnidens

Adult shells with teeth

Adult shells without teeth

V. acerra

V. arcellus

V. ligera

V. intertextus

V. theloides

V. demissus

Shells 15-36 mm, without teeth, simple lip (*Mesomphix*)

Mesomphix species have medium to large shells that are either depressed heli-ciform or heliciform. The shells are loosely coiled, the last whorl greatly expanded (a); in contrast, *Ventridens* species have tightly coiled whorls (b) and are smaller. The shell surface can be dull to glossy (some species with a glass-like surface). Lips are simple, not reflected and they are without teeth in all stages of development. The transverse striae vary from closely-spaced to indistinguishable. Several species of this genus exhibit spirally arranged papillae. The shell is umbilicate to perforate; the umbilicus never completely closed. Like *Ventridens* species, most *Mesomphix* have a thickening or thin whitish callus usually just inside the aperture's bottom of the last whorl. Two or three fresh shells of *M. cupreus,* shaken in one loosely closed hand will sound remarkably like glass. Shaking other species of snails like *Mesodon* will have a duller sound. Immature *Mesodon* species are easily confused with mature *Mesomphix* snails (see below comparison); the primary difference is that mature *Mesomphix* species will have an aperture that is more egg-shaped and a non-reflected lip. The following *Mesomphix* species will be arranged according to some of these features and are as follows. **Group 1** shells that have spiral papillae and **Group 2** shells that are generally without any spiral papillae, although some shells in this group may occasionally have scant traces of this micro-feature (see next page)

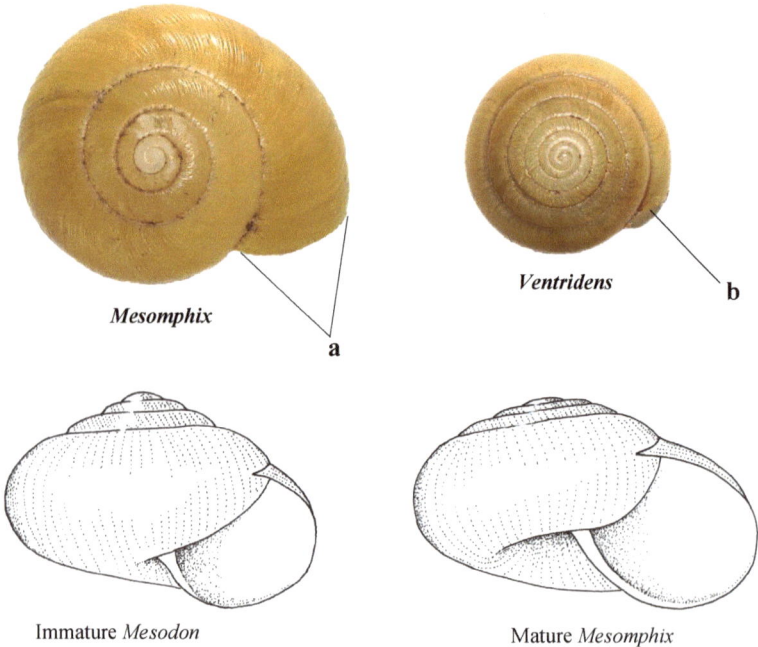

Mesomphix

a

Ventridens

b

Immature *Mesodon*

Mature *Mesomphix*

Important Note: Immature *Mesodon* shells are often confused with immature and mature *Mesomphix* species (see above illustrations). An immature **Mesodon** has a more rounded lip, while the mature **Mesomphix** lip is more oval in shape. **Mesodon** species will have a reflected lip only when fully mature.

Grouping 1: typically with spiral rows of papillae, pg 149

M. perlaevis
M. vulgatus
M. latior
M. latior, form *monticola*
M. capnodes

spiral rows of papillae

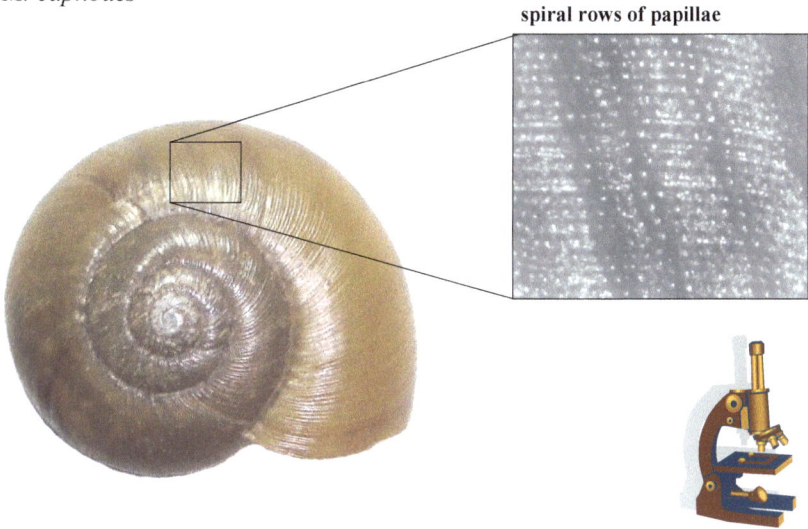

Grouping 2: typically without spiral rows of papillae, pg 154

M. andrewsae
M. cupreus
M. cupreus form *politus* (with spiral rows of papillae)
M. rugeli
M. subplanus

Shell generally smooth without spiral papillae

Smooth button

Gastrodontidae

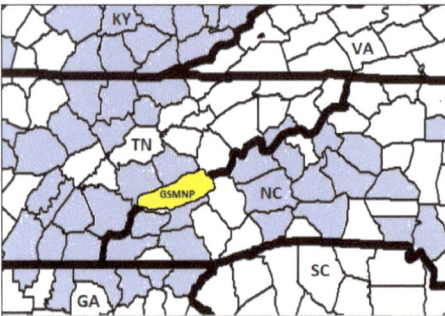

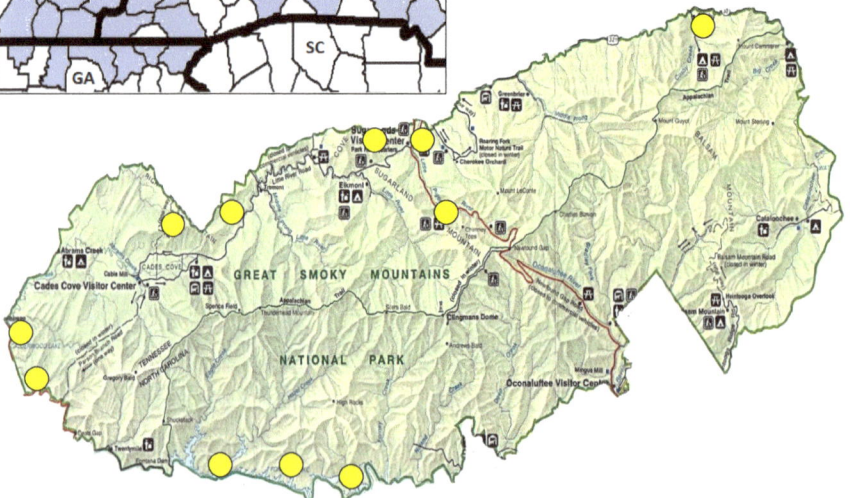

Mesomphix perlaevis (Pilsbry, 1900)

Diameter: 17-21 mm

Description: Heliciform; lip simple; shell with 5 whorls; perforate; shell thin but not fragile, light-olive, glossy, embryonic whorl with notable transverse striae if not worn (can be seen with a hand lens of 10X); with or without a thin whitish callus just inside the aperture's bottom of the last whorl; transverse striae are strongest on the third whorl, the striations of the last whorl become notably weaker; fine spiral rows of papillae are not well-developed (wanting in most places) but usually present on at least some portions of the upper shell surface.

Similar Species: *Mesomphix vulgatus* is generally larger and has a lower shell profile.

Habitat: A snail of rich upland mixed hardwood forests usually found under and among moist leaf litters and detritus.

Status: G5; Common; throughout the GSMNP in low to mid elevation.

Specimen: Tennessee, Cocke County, Cosby Campground, GSMNP (GSMNP collection).

Common button

Gastrodontidae

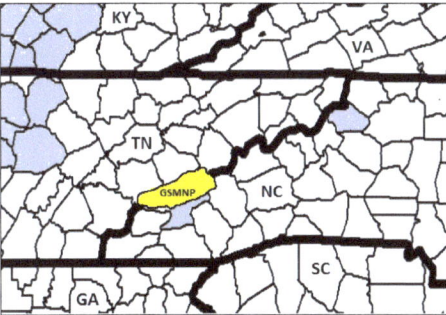

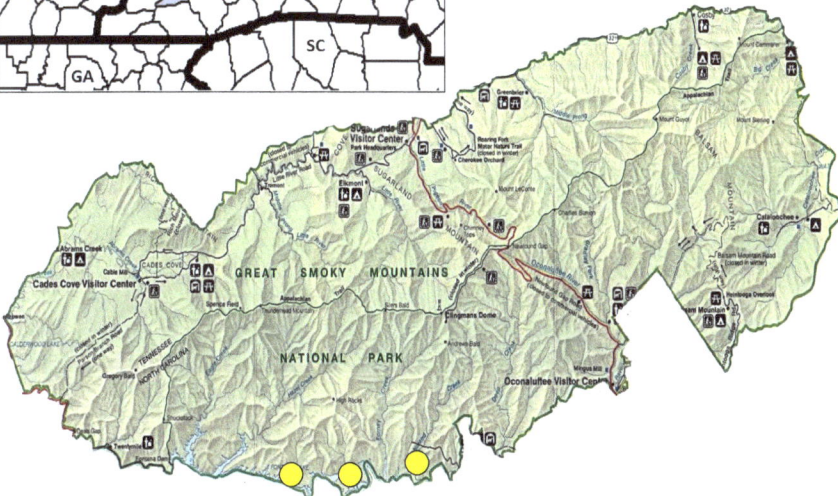

Mesomphix vulgatus H. B. Baker, 1933

Diameter: 20-27.8 mm

Description: Depressed heliciform; lip simple; shell with 4.5-5.5 whorls; perforate; shell thin but not fragile; periostracum dark, (the chitinous protein-based external outer layer of the shell); darker and lighter streaks; top surface a dull sheen; transverse striae (growth wrinkles) are well-developed and crowded; spiral rows of papillae are well-developed continuing to the lower surface.

Similar Species: *Mesomphix latior* is around the same size and shell form but glossier and has less well-developed rows of spiral papillae.

Habitat: In the GSMNP, found in dry habitats on hillside and ravines in hardwood forests.

Status: G5; Locally Rare; currently reported from the south of the park above Fontana Lake; this population is a considerable distance from known *M. vulgatus* reported Hubricht (1985);it may in fact represent a new species that warrants further investigation.

Specimen: North Carolina, Swain County, Forney Creek, GSMNP (GSMNP collection).

Broad button

Gastrodontidae

Mesomphix latior (Pilsbry, 1900)

Diameter: 20-28 mm

Description: Depressed heliciform; lip simple; shell with 4.5-5 whorls; perforate; shell thin but not fragile, olive-tan with darker streaks, base glossy; top with a more dullish sheen; with or without a thickening or thin whitish callus; transverse striae are smooth and the shell is usually with weakly developed spiral rows of papillae.

Similar Species: *Mesomphix perlaevis* has better developed transverse striae, a higher shell profile, but the spiral rows of papillae are more or less the same.

Habitat: A snail of lower elevation mixed hardwood forests usually found under and among moist leaf litter and detritus.

Status: G4; Uncommon in the southern and western portions of the park and around Sugarlands Visitor Center.

Specimen: North Carolina, Swain County, Nantahala National Forest (author's collection).

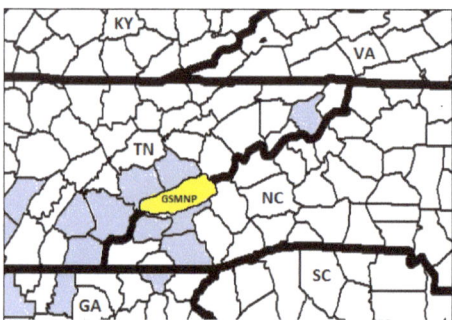

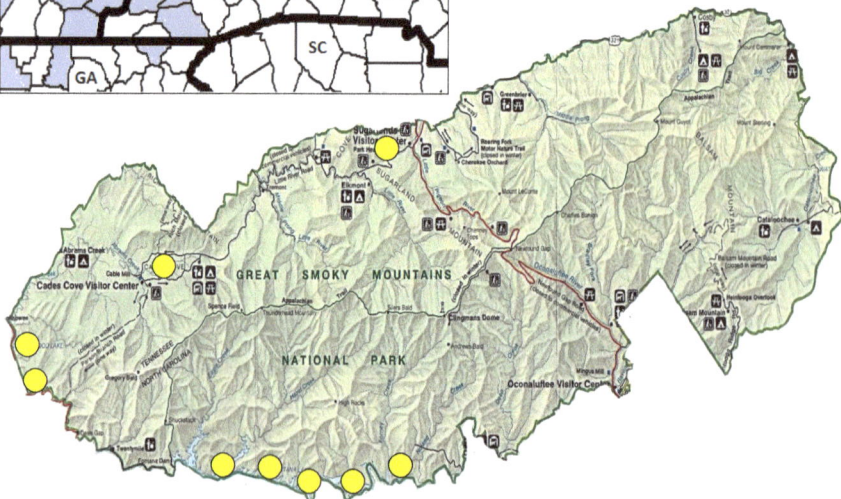

Globose broad button

Mesomphix latior form *monticola*
(Pilsbry, 1900)

Diameter: 20-24 mm

Description: Heliciform; lip simple; shell with 4.5-5 whorls; perforate; shell thin but not fragile, olive-tan with darker streaks, base glossy; top more dull; with or without a thickening or thin whitish callus; transverse striae are smooth and the shell is usually with at least some traces of spiral rows of papillae.

Similar Species: Differs from typical *Mesomphix latior* by its more globose, elevated shell and more compact form.

Habitat: A snail of lower elevation mixed hardwood forests usually found under and among moist leaf litter and detritus.

Status: G1; Endemic to the Swain County region; Uncommon in lower elevations of southern portions of the park; in general not usually found with *M. latior*; little is known of this interesting form and further DNA investigation seems warranted.

Specimen: North Carolina, Swain County, Nantahala National Forest (author's collection).

Gastrodontidae

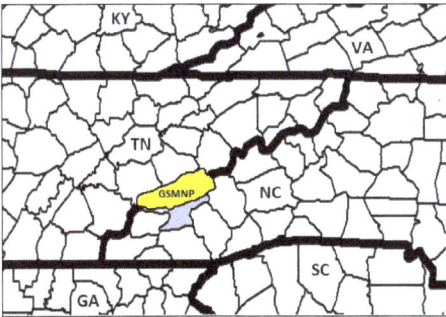

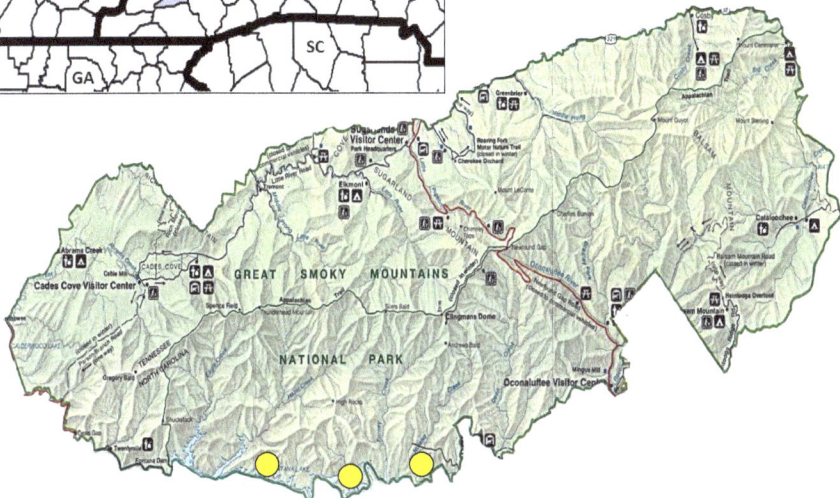

Dusky button

Mesomphix capnodes (W. G. Binney, 1857)

Diameter: 29.5-35.5 mm

Description: Depressed heliciform; lip simple; shell with 5 whorls; perforate to umbilicate; shell thin but not fragile; periostracum dark; somewhat glossy; embryonic whorl usually worn; no teeth present; with or without a thickening or thin whitish callus that is just inside the aperture's bottom of the last whorl; transverse striae are poorly developed and mostly smooth; spiral rows of papillae are present but not well-developed.

Similar Species: *Mesomphix cupreus* is smaller, has a slightly higher shell profile, a wider umbilicus and very rarely displays any spiral papillae.

Habitat: A calciphile snail of upland mixed hardwood forests usually found in leaf litter.

Status: G5; Locally Rare; known only from two sites at the west end of TN side near Calderwood.

Specimen: Tennessee, Monroe County, Cherokee National Forest (author's collection).

Gastrodontidae

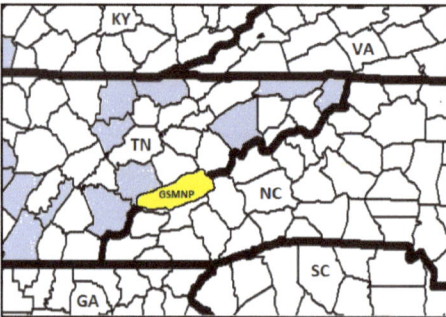

Mountain button

Gastrodontidae

Mesomphix andrewsae (Pilsbry, 1895)

Diameter: 16-21 mm

Description: Depressed heliciform; lip simple; shell with 5 whorls; perforate; shell thin and fragile; aperture proportionately wide; shell surface very glossy with a yellowish cast; no teeth present; with or without a thickening or thin whitish callus that is just inside the aperture's bottom of the last whorl; transverse striae are poorly developed and mostly smooth; without spiral rows of papillae, or with only very weakened traces under a strong lens.

Similar Species: *Mesomphix subplanus* is flatter and is without any spiral papillae; other *Mesomphix* species are taller in form.

Habitat: A snail of acidic upland mixed hardwood forests usually found in leaf litter or under log structure.

Status: G4; Common; one of the most common *Mesomphix* species in the park and reported from nearly all elevations.

Specimen: North Carolina, Swain County, Forney Creek, GSMNP (GSMNP collection).

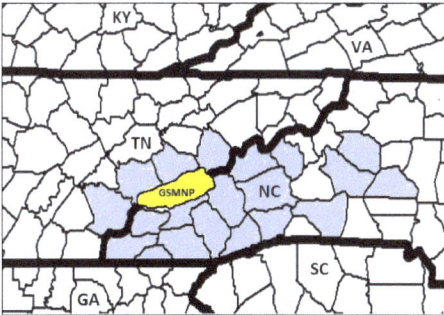

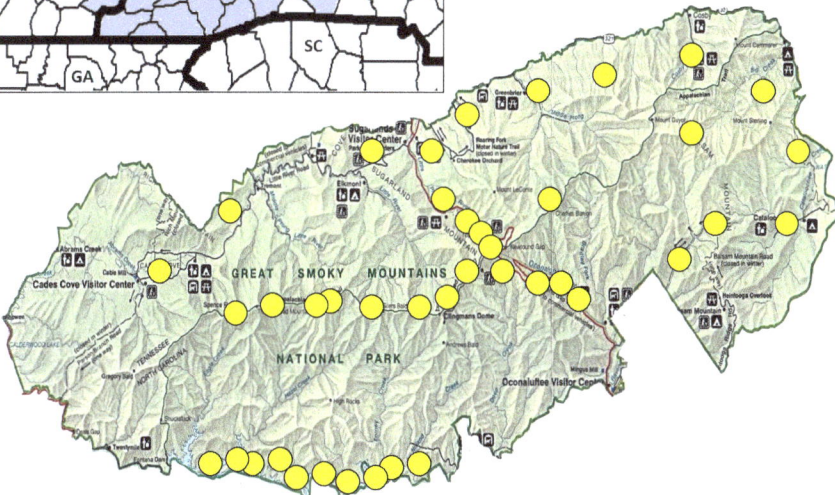

Copper button

Gastrodontidae

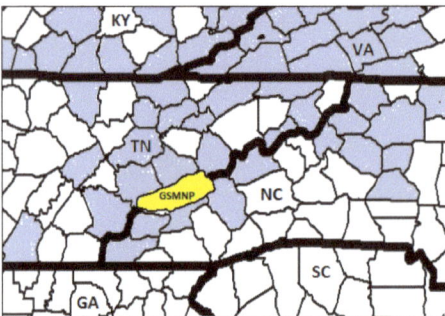

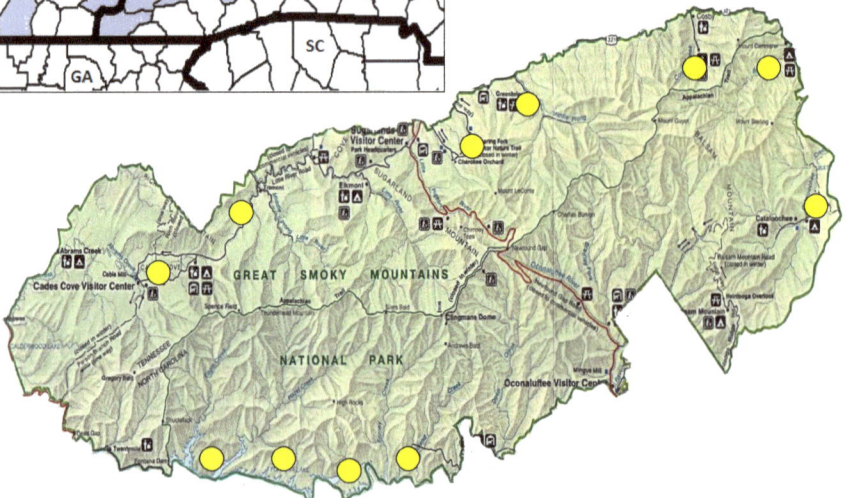

Mesomphix cupreus (Rafinesque, 1831)

Diameter: 22-29.5 mm

Description: Heliciform; lip simple; shell with 4.5-5 whorls; umbilicate, has the widest umbilicus of any *Mesomphix* species found in the park; shell thin but not fragile; periostracum dark; shell glossy; embryonic whorl usually worn; a thickening or thin whitish callus just inside the aperture; transverse striae are poorly developed and smooth; without detectable rows of papillae.

Similar Species: *Mesomphix capnodes* is larger, has a slightly lower shell profile, a smaller umbilicus and weak but distinct spiral rows of papillae.

Habitat: A snail of acidic upland mixed hardwood forests usually found under and among moist leaf litters and detritus.

Status: G5; Relatively Common; one of the most common large *Mesomphix* in the park; found at all but the highest elevations.

Specimen: Tennessee, Cocke County, Cosby Campground, GSMNP (GSMNP collection).

Varnish button

Gastrodontidae

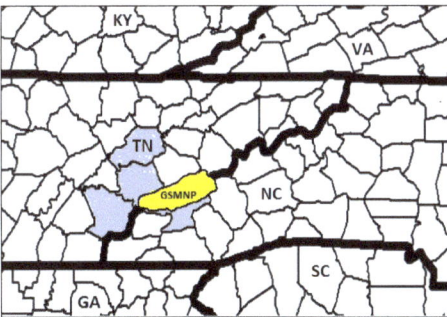

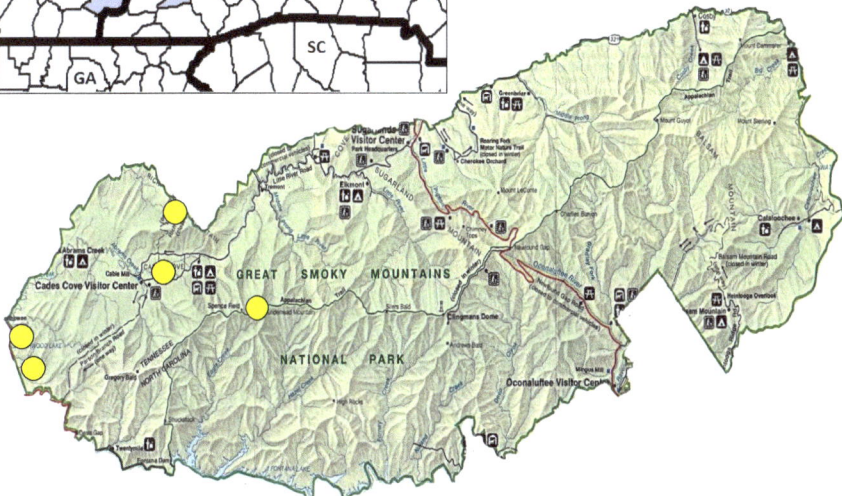

Mesomphix cupreus form *politus*
(Pilsbry, 1946)

Diameter: 25-30 mm

Description: Depressed heliciform; lip simple; shell with 4.5-5 whorls; umbilicate; shell thin but not fragile; shell surface brilliantly glossy, as though painted in a dark varnish; embryonic whorl usually worn; with a thickening or thin whitish callus; transverse striae are poorly developed and smooth but rows of papillae are usually present.

Similar Species: Differs from typical *Mesomphix cupreus* by its brilliantly glossy surface, more compressed shell and by having distinct rows of notable spiral papillae.

Habitat: Upland mixed hardwood forests (generally where limestone occurs) usually found under and among moist leaf litter.

Status: G1; Globally Rare; Endemic to the GSMNP; a range-restricted form of *M. cupreus* currently known from Rich Mountain, Cades Cove and Calderwood region of the GSMNP.

Specimen: Tennessee, Blount County, Calderwood area, GSMNP (GSMNP collection).

Wrinkled button

Gastrodontidae

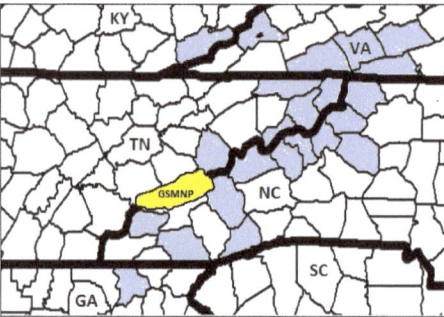

Mesomphix rugeli (W. G. Binney, 1879)

Diameter: 16-22.5 mm

Description: Heliciform; lip simple; shell with 5.5-6 whorls; perforate; shell thin but not fragile, greenish-horn, glossy; with or without a thin whitish callus just inside the aperture's bottom (can be seen in frontal view); transverse striae are smooth and the shell is usually without discernible spiral rows of papillae; cannibalistic, will feed on the fresh dead of its own kind; also feeds on bird droppings (authors personal observation).

Similar Species: *Mesomphix perlaevis* has better developed transverse striae and spiral rows of papillae; *M. rugeli* form *oxycoccus* (Pilsbry 1946) differs from typical *M. rugeli* by having well-developed and crowded spiral papillae.

Habitat: Higher elevation mixed hardwood and spruce/fir forests under moist leaf litters.

Status: G4; Locally Rare; Thompson (1981) reported this species in two upper elevation locations and the Albright Grove.

Specimen: North Carolina, Mitchell County, Roan Mountain (author's collection).

Flat button

Gastrodontidae

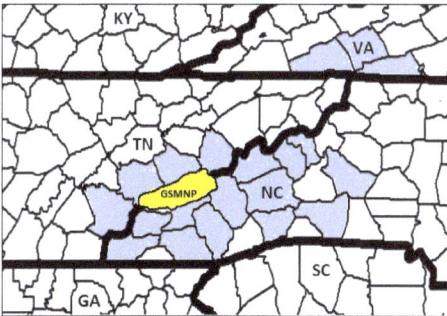

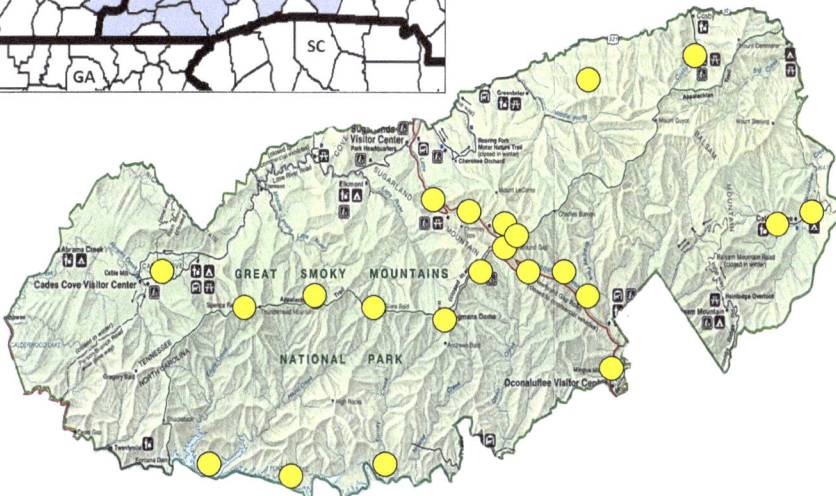

Mesomphix subplanus (A. Binney, 1842)

Diameter: 16-23 mm

Description: Exceedingly depressed heliciform; lip simple; shell with 5.5-6 whorls; perforate; shell thin but not fragile; shell glossy;
like most *Mesomphix* species the embryonic
whorl is usually worn; with a thickening or thin
whitish callus; transverse striae (growth wrinkles) are poorly developed and smooth; without
rows of papillae.

Similar Species: This is the flattest *Mesomphix*
species recorded from the GSMNP and the
Southern Appalachians mountains, although
specimens of *M. subplanus* from Mt Mitchell,
North Carolina are even flatter.

Habitat: A snail of upland (up to 2000 m)
mixed and northern hardwood forests usually
found under and among moist leaf litters or
sometimes living among well-developed talus.

Status: G4; Common to uncommon at all elevations, but appears to be most frequent in upper altitudes.

Specimen: Tennessee, Swain County, Nantahala National Forest (author's collection).

Mesomphix **Shells of the GSMNP Compared** (shells are proportionate to each other)

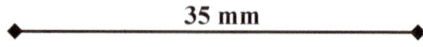
35 mm

M. capnodes

M. vulgatus

M. cupreus

M. cupreus politus

M. latior

M. latior monticola

M. rugeli

M. subplanus

M. perlaevis

M. andrewsae

159

Shells 15-31 mm, without teeth, simple lip, shell typically with bands and streaks of color, defense mucus fluorescent under UV light (*Anguispira*)

Anguispira species are the most colorful native land snails in the GSMNP and interestingly, the slime from this genus glows fluorescent under black light (Rawls & Yates, 1971). In general, these medium-size gastropods are most common around outcroppings of limestone and are sometimes found in significant numbers, but are also common in acidic conditions as well. Shells of *Anguispira* are usually depressed heliciform. There are crisscrossing striae on the embryonic whorls of most species of *Anguispira* (often a lost feature in worn shells), but not all specimens within the same population have this feature. *Anguispira* lips are simple in all stages of growth and there are never any teeth present in the aperture. The transverse striae or ribbed surfaces are easily observed with a hand lens of 10X magnification. *Anguispira* shells are always umbilicate. The periphery of the body whorl is either carinate, bluntly angular or rounded. The color features are always present but this character may be poorly defined in some species and in shells that are badly weathered. The following *Anguispira* are arranged according to rib-counts and are as follows: **Group 1** shells that have more than 75 ribs on the last whorl and **Group 2** shells that have less than 65 ribs on the last whorl.

<u>**Grouping 1: more than 75 ribs on the last whorl of adult shells, page 161**</u>

*A. alternata
*A. knoxensis

<u>**Grouping 2: less than 65 ribs on the last whorl of adult shells, page 164**</u>

*A. mordax
*A. jessica

Counting Ribs

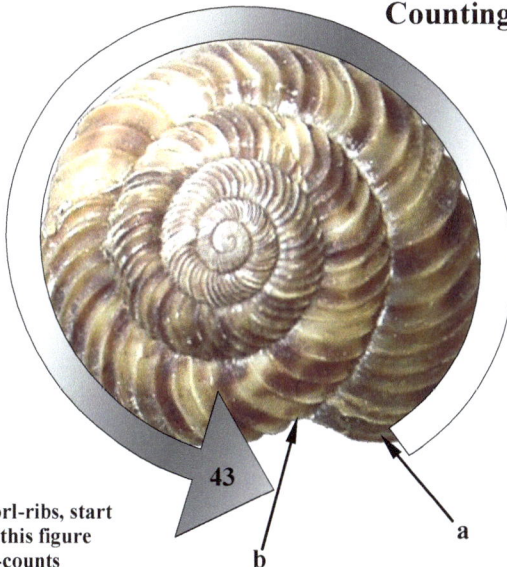

43

Counting the last whorl-ribs, start at (a) and stop at (b); this figure having around 43 rib-counts

a

b

Flamed tigersnail

Discidae

Anguispira alternata (Say, 1816)

Diameter: 15-30 mm

Description: Depressed heliciform; lip simple; shell with 5-6 whorls; umbilicate; color features usually strong on entire shell surface; in some population there are crisscrossing striae on the embryonic whorl (lost in worn shells); no teeth present; transverse striae are closely-spaced and well-developed, adult shells with at least 5 whorls have a count of over 80 ribs in the last whorl; periphery can be carinate, angular (a) or rounded (b).

Similar Species: *Anguispira mordax* is smaller having fewer than 50 ribs on the last whorl; *A. jessica* has fewer than 70 rib counts on the last whorl and a smaller umbilicus.

Habitat: A species found in a wide range of lower elevation wooded and open habitats including rocky limestone regions, glades, along forested hillsides adjacent to rivers.

Status: G5; Uncommon; in the park reported from scattered locations.

Specimen: Kentucky, Powell County, Red River Gorge (author's collection).

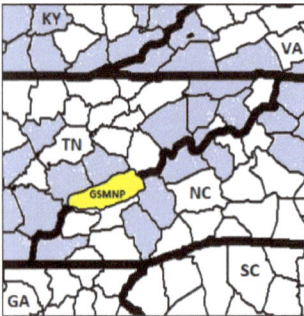

b

Preston Co., WV

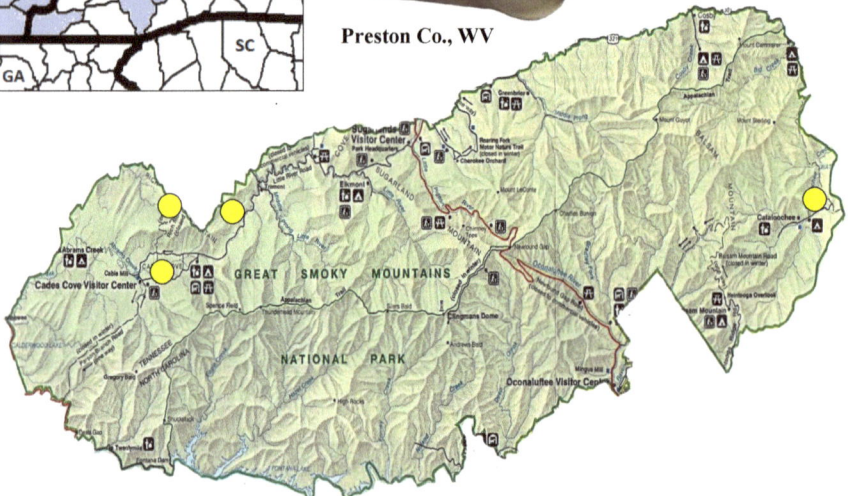

161

Above image of the flamed tigersnail, *Anguispira alternata* in typical defense posture; the live animal retracted inside its shell and its aperture blocked by a bubbly orange mucus. Below the same specimen under UV light. The function of fluorescence in land snails remains a mystery and some scientists speculate it is simply a random act of evolution. But the glow might serve as an early warning sign of distasteful compounds found in the mucus, which may repel predators and help conserve body moisture. Red River Gorge, Kentucky.

Rustic tigersnail

Discidae

Anguispira knoxensis Hubricht, 1938

Diameter: 20-25 mm

Description: Depressed heliciform; lip simple; shell with 5.5-6 whorls; widely umbilicate, widest of the *Anguispira*; color features faint but always present; no teeth; transverse striae are closely-spaced and well-developed, in adult shells having a count of over 80 ribs in the last whorl; periphery rounded.

Similar Species: *Anguispira alternata* has finer sculpture of transverse striae, more prominent color features and a proportionately smaller umbilicus; *A. mordax* is smaller, with stronger color features and more widely spaced ribs; *A jessica* is more colorful and has a smaller umbilicus.

Habitat: A species found around large rotting hardwood logs in advanced stages of decay, close to outcrops of limestone.

Status: G2; Rare; Endemic to regions surrounding the GSMNP; historic records include Cades Cove and Hazel Creek.

Specimen: Tennessee, Blount County, around Townsend (author's collection).

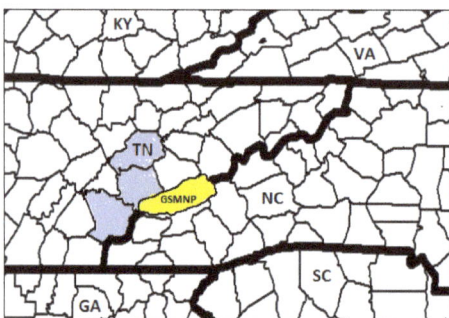

Appalachian tigersnail

Discidae

Anguispira mordax (Shuttleworth, 1852)

Diameter: 15-20 mm

Description: Depressed heliciform; lip simple; shell with 5-6 whorls; umbilicate; color streaks can be either a strong or weakly developed feature but is always present; transverse striae are modified into distinct and widely spaced ribs, having a count of 40 to 45 ribs in the last whorl; periphery either rounded or bluntly angular.

Similar Species: *Anguispira alternata* is larger in size and is more finely sculptured, having more than 80 ribs in the last whorl; *A. strongylodes* (page 281) has a higher number of ribs (56 to 62) in the last whorl.

Habitat: Found in a variety of forested habitats, especially around limestone outcrops where it can reach large numbers.

Status: G5; Locally Rare; common only around Cades Cove and White Oaks Sinks regions of the park.

Specimen: Tennessee, Blount County, White Oak Sinks, GSMNP (GSMNP collection).

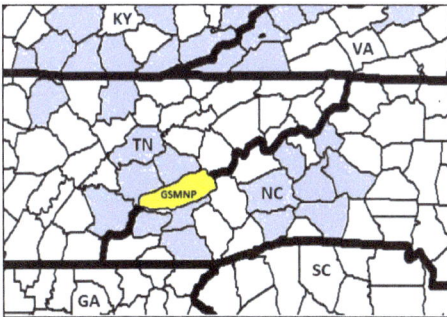

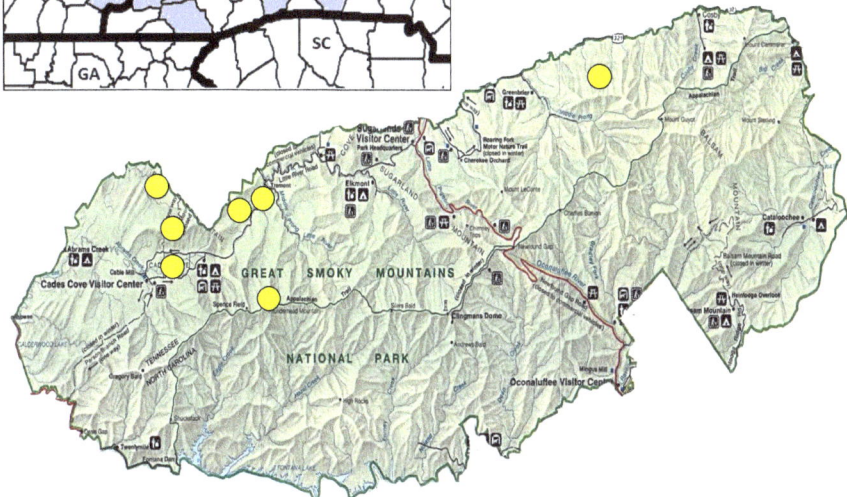

The Tree Climber

A mountain tigersnail, *Anguispira jessica* climbing a tree trunk at night. The species has been observed in excess of 20 meters high in tree tops where it is suspected to feed on slime molds (Keller & Snell 2002). Several theories have emerged explaining why the shell parades a range of contrasting color streaks. They may act as disruptive patterns, breaking up the round form, in turn making it harder to spot by predators such as birds, or, like the black and white warning colors of the skunk's bad smell, warn of distasteful defense-mucus produced by the snail under attack. Many tropical snails found living in trees are also colorfully marked (Dourson, 2009). This image was captured by back-lighting the snail's translucent shell at night, which shows the colorful alternating markings. Below the same species feeds on the sap of sugar maple from holes created by the yellow-bellied sapsucker. Both images taken on Roan Mountain, Mitchell County, North Carolina.

Mountain tigersnail

Discidae

Anguispira jessica Kutchka, 1938

Diameter: 15-20 mm

Description: Depressed heliciform; lip simple; shell with 5-6 whorls; umbilicate; color features usually boldly developed on top and side but usually absent on the base; crisscrossing sculpture on embryonic whorl (a) which is lost in worn shells; transverse striae are closely-spaced and well-developed, in adult shells having a count of 55 to 60 ribs in the last whorl; periphery slightly or bluntly angular.

Similar Species: Size of *Anguispira mordax* but having more than 55 ribs on the last whorl.

Habitat: A species usually associated with rotting hardwood logs in advanced stages of decay in upland mixed hardwood forests (above 800 m); the snail has also been documented in trees over 20 meters high (Keller & Snell 2002).

Status: G4; Common; the most common *Anguispira* species in the GSMNP.

Specimen: Tennessee, Haywood County, Big Creek Picnic area, GSMNP (GSMNP collection).

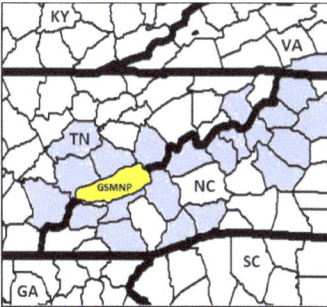

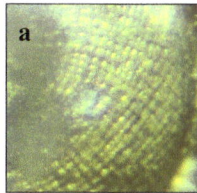

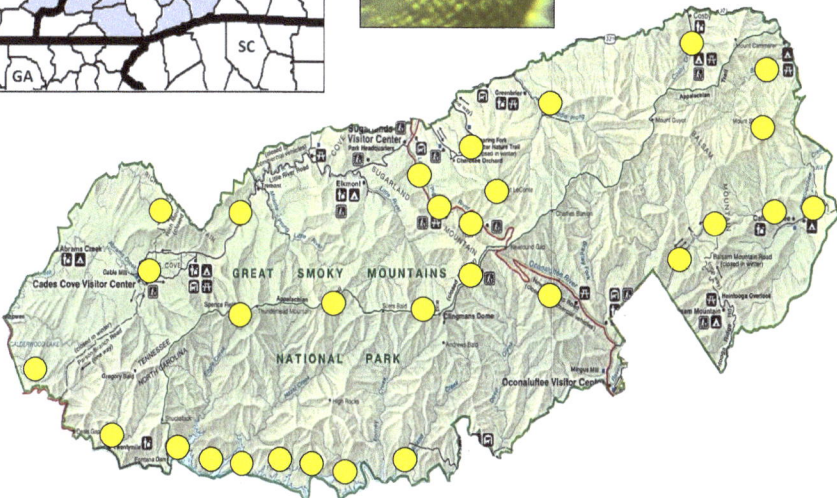

Blue Snails of the Southern Appalachians

R. Wayne Van Devender

Above a Blue-foot lancetooth and below a glassy grapeskin, illustrating the atypical deep blue color of the live animals. Bottom photo of three glassy grapeskins and a black mantleslug, *Pallifera hemphilli* feeding on a fresh dead balsam globe, *Mesodon andrewsae*, Roan Mountain, North Carolina.

Two Curious Gastropods, *Haplotrema* and *Vitrinizonites*

Haplotrema are the wolves of land snails, eating a variety of terrestrial gastropods including *Novisuccinea ovalis*, *Zonitoides arboreus*, *Discus catskillensis* and *Strobilops labyrinthicus* (Pearce and Gaertner 1996). *Patera* and *Mesodon* species are also taken by this formidable hunter (author's personal observation). To access the snail flesh, the species will typically enter through the aperture of the prey's shell where it slowly eats the snail alive. If the aperture is too small for entry, *Haplotrema* will drill (actually grind) through the side of the shell by using its radula (Pearce and Gaertner 1996;). *Haplotrema* snails are medium size gastropods having one very distinctive feature in their shell, an extra wide umbilicus (one of widest of any eastern land snail, figure a). Two species occur in the GSMNP, *H. concavum* and *H. kendeighi,* the latter being the rarer of the two species.

Haplotrema concavum

a

Like *Haplotrema*, **Vitrinizonites** or the glassy grapeskin is not so ordinary and is a hunter of snail flesh. It is unlike any other land snail in the eastern US, possessing a flexible shell with a proteinaceous coating on its outer surface and a thin calcium layer on its interior. This works well for the species since it typically lives in highly acidic environments requiring less calcium carbonate for shell building. The live animal of *Vitrinizonites latissimus* is a rich bluish color (opposite page), an infrequent color for a land snail of temperate forests. If placed in one's hand, this species will make painless scrapes by using its radula (a defense technique thought to be used against attacking shrews). Lastly the shell of the glassy grapeskin has the silhouette and elegance of the famous nautilus found in deep ocean environments.

Glassy grapeskin

The nautilus

Gray-foot lancetooth

Haplotrematidae

Haplotrema concavum (Say, 1821)

Diameter: 16-25 mm

Description: Depressed heliciform; lip simple but in fully developed shells there is a slight reflection in the lower lip; shell with 4.5-5.5 whorls; widely umbilicate; shell light greenish-yellow and glossy; transverse striae poorly developed; spiral striae present but may be a wanting feature on some portion of the shell; live animal gray (figure a, Newfound Gap, GSMNP); a carnivorous species that feeds on other gastropods.

Similar Species: No other land snail of similar size has the wide umbilicus; the live animal of *H. kendeighi* is blue whereas the live animal of *H. concavum* is gray.

Habitat: Found in mixed hardwood forests at all elevations.

Status: G5; Common; one of the most common land snails in eastern North America and a frequent and widespread resident of the GSMNP.

Specimen: Shell images from Tennessee, Blount County, Chestnut Top Trail, GSMNP (GSMNP collection).

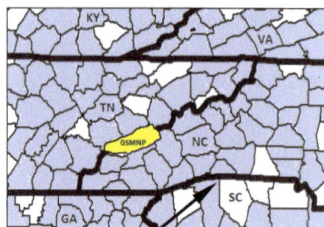

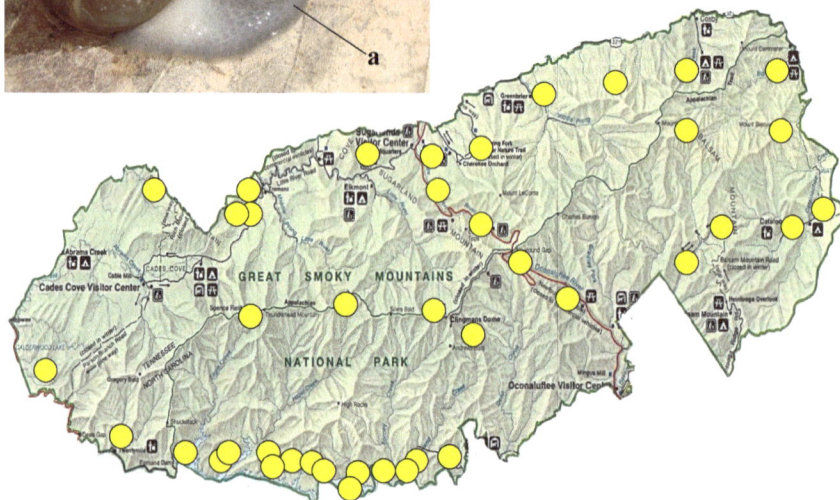

Blue-foot lancetooth

Haplotrematidae

Haplotrema kendeighi Webb,1951

Diameter: 16-22 mm

Description: Depressed heliciform; lip simple with a slight reflection in the lower lip; 4.5-5.5 whorls; widely umbilicate; shell light greenish-yellow and glossy; transverse striae poorly developed; live animal blue (figure b, Snowbird Mountains, Graham County, NC).

Similar Species: The live animal of *H. concavum* is gray; reproductive anatomy differs (the penis of *H. kendeighi* is longer); shells of *H. concavum* and *H. kendeighi* are reported to be indistinguishable (Hubricht 1985).

Habitat: Found in mixed hardwood forests, restricted to higher elevations in the park, but also found in lower elevations and cooler river gorges outside the park.

Status: G2; Globally Rare; Endemic to regions surrounding the GSMNP; the overall status of this species in the GSMNP and elsewhere remains unknown, a result of too few collections that paid close attention to the color of live animals.

Specimen: North Carolina, Swain County, Clingmans Dome (GSMNP collection).

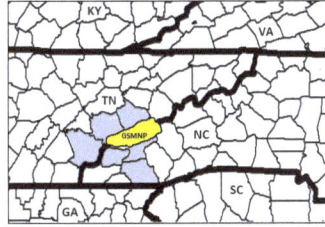

170

Glassy grapeskin   # Gastrodontidae

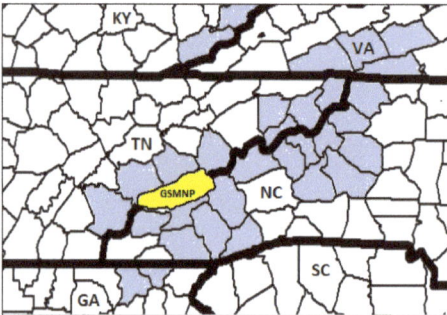

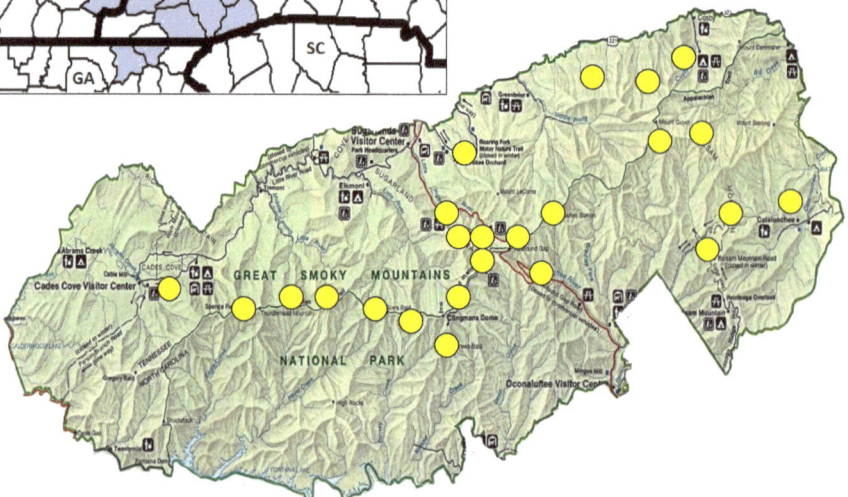

Vitrinizonites latissimus (J. Lewis, 1875)

Diameter: 16.2-19.5 mm

Description: Vitriniform; lip simple; shell with 2.5-3 whorls; imperforate; shell paper-thin, translucent and flexible with a proteinaceous coating on its outer surface and a thin calcium layer on its interior; brown to olive, yellow or sometimes green; shell smooth and glossy giving the shell a glass-like surface; live animal bluish to black; this snail will feed on the carcasses of *Mesodon* and *Mesomphix* species (pers. obs.).

Similar Species: No other land snail in the GSMNP has a flexible shell and nautilus-shape of *V. latissimus*.

Habitat: Usually at elevations above 700 m, under leaf litter located next to or under rotting logs in mixed hardwood, but also common in acidic soils thriving in such places as heath balds and spruce/fir forests.

Status: G4; Common; the most common land snail found at elevations above 1400 m.

Specimen: North Carolina, Mitchell County, Roan Mountain (author's collection).

171

Shells 2.4-8 mm, no teeth, simple lip (*Euconulus & Hendersonia*)

Euconulus species are small dome or bee-hive shaped land snails that have many tightly coiled whorls. They are generally uncommon in leaf litters of mixed hardwood forests. They differ from *Strobilops aeneus* by having a simple lip (not reflected), a smoother surface (*Strobilops* having a ribbed surface) and being more domed-shaped. *Euconulus* are typically without teeth, although *E. dentatus* has internal armature in young shells. Four species are reported from the park.

Grouping 1: shells 2.4-3.4 mm, dome or bee-hive shape, shell surface more or less smooth (not ribbed), pg. 173

Euconulus chersinus
Euconulus dentatus
Euconulus trochulus
Euconulus fulvus

Euconulus chersinus

Hendersonia species are similar in shape to *Euconulus* but are much larger with a thicker shell and in live specimens, containing an operculum (door). *Hendersonia* differ from *Strobilops aeneus* by having a simple lip (not reflected), a smoother surface (*Strobilops* having a ribbed surface, see below) and averages 5-6 mm larger.

Grouping 2: shells 6-8 mm, heliciform, shell surface more or less smooth (not ribbed), pg. 177

Hendersonia occulta

Shells are proportionate

Hendersonia occulta

Strobilops aeneus

Wild hive   # Euconulidae

Euconulus chersinus (Say, 1821)

Diameter: 2.4-3.4 mm, height 2.9-3.4 mm

Description: Dome-shape; lip simple; shell thin with 6-8 tightly coiled whorls; perforate; shell surface yellowish-white and rather dull; shells are translucent with a silky luster in live snails and fresh dead; transverse striae are poorly developed (microscope required); spiral striae not well-defined but always present (a really strong lens required here); without teeth; body whorl slightly angular especially in young shells.

Similar Species: *Euconulus fulvus* is less finely sculptured, more glossy and has two less whorls; *E. dentatus* has internal armature in young shells.

Habitat: Under moist leaf litter in mixed hardwood forests.

Status: G5; Uncommon; reported from the north shore of Fontana Lake, Cades Cove and Purchase Knob areas only.

Specimen: North Carolina, Swain County, Nantahala National Forest (author's collection).

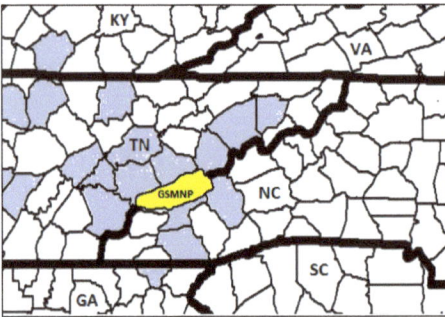

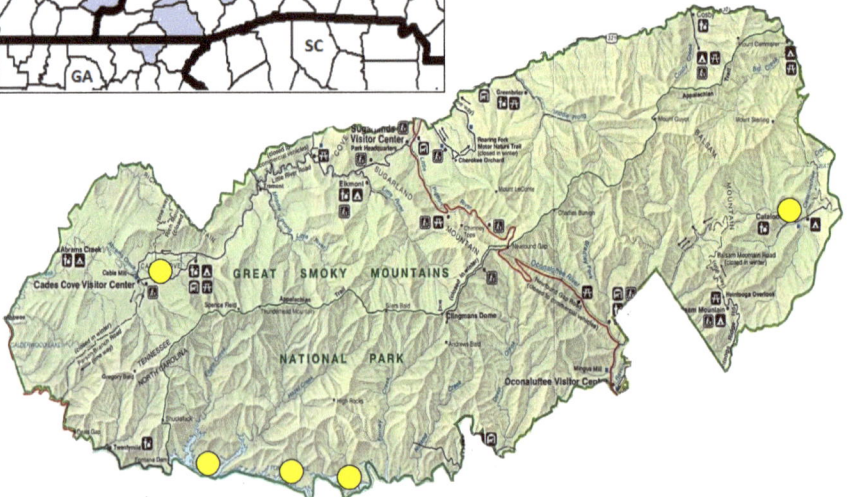

Toothed hive

Euconulidae

Euconulus dentatus (Sterki, 1893)

Diameter: 2.4 mm, height 2.3 mm

Description: Dome-shape; lip simple; shell with 6.5 tightly coiled whorls; perforate; yellowish-white and rather dull; shell thin, translucent in live snails and fresh dead; transverse striae are poorly developed (microscope required); spiral striae not well defined but always present; teeth or lamellae in young shells (a), adults usually without teeth.

Similar Species: Most closely related to *Euconulus chersinus* but differs in having armature in young shells and 2 to 3 less whorls.

Habitat: A species found in mixed hardwood forests on hillsides and ravines living under layers of leaf litter usually in drier sites than other species of *Euconulus*; like all *Euconulus* species, it is rarely found in numbers.

Status: G5; Uncommon; found in the southwestern portion of the park north of Fontana Lake, along the Foothills Parkway, Purchase Knob and White Oak Sinks area.

Specimen: Kentucky, Hardin County, Spurrier (FMNH 251488).

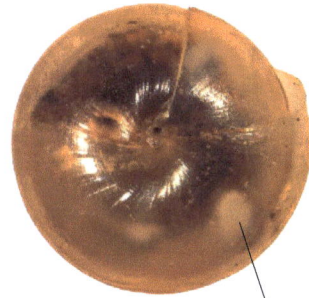

a

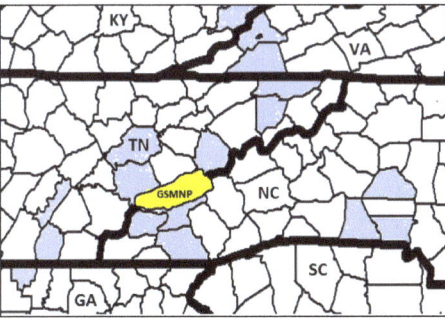

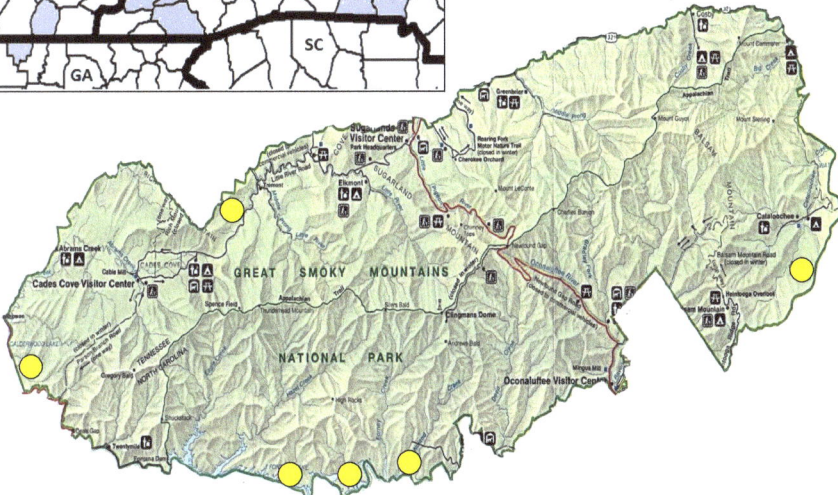

174

Silk hive

Euconulidae

Euconulus trochulus (Reinhardt, 1883)

Diameter: 2.9 mm, height 2.5 mm

Description: Dome-shape; lip simple; shell with 7 tightly coiled whorls; imperforate, the umbilicus completely covered by the reflected columellar margin; pale-horn with a silky luster; shell thin, translucent in live snails and fresh dead; transverse striae are poorly developed (microscope required); spiral striae not well-defined but always present (a strong lens required here); without teeth; in mature specimens, the periphery is slightly angular or sometimes almost rounded (Pilsbry 1946).

Similar Species: Shell much like *E. chersinus*, but it has one whorl more in shells of similar size; *Euconulus fulvus* is more glossy and has 1 to 2 fewer whorls in adult shells.

Habitat: A species found in mixed hardwood forests on hillsides and ravines living under layers of moist leaf litter.

Status: G5; Uncommon; but expected to be more common than records currently indicate.

Specimen: Kentucky, Pulaski County, Meece (FMNH 252087).

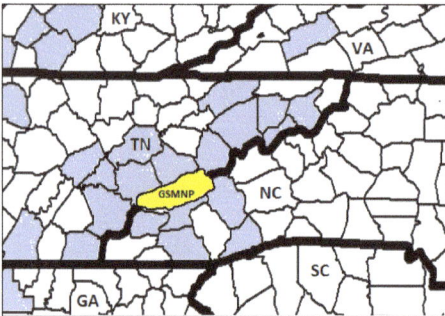

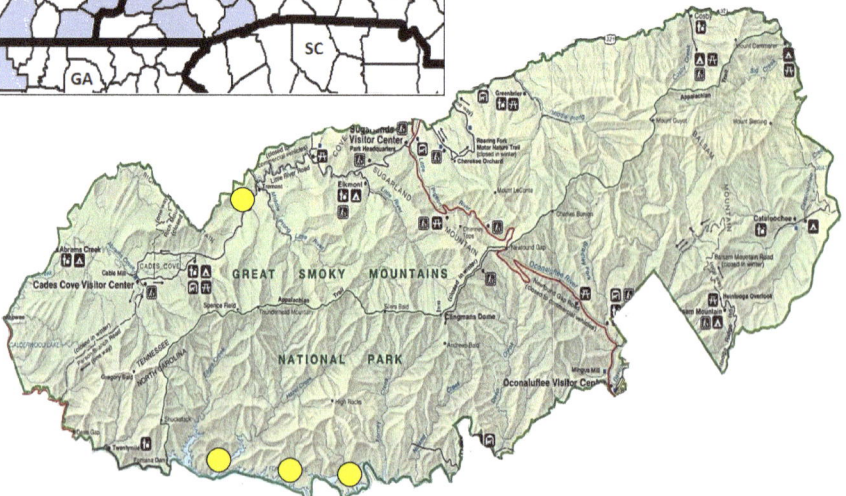

Brown hive   Euconulidae

Euconulus fulvus (Müller, 1774)

Diameter: 3.1-3.4 mm, height 2.4 mm

Description: Dome-shape; lip simple; shell with 4.5-6 tightly coiled whorls; umbilicus minutely perforate or nearly closed; pale-brown and glossy; shell thin, translucent in live snails and fresh dead; transverse striae are poorly developed (microscope required); spiral striae not well-defined but always present; without teeth; body whorl rounded.

Similar Species: *Euconulus chersinus* is more finely sculptured, has a more elevated shell, is less glossy and has two more whorls.

Habitat: A Holarctic species found in mixed hardwood forests on hillsides and ravines living under layers of moist leaf litter; also found in creek and river drift.

Status: G5; Locally Rare; in the GSMNP known only from around Fontana Lake but is expected to be more widespread than current records indicate.

Specimen: Michigan, Wayne County, Belle Isle, Detroit (FMNH 251565).

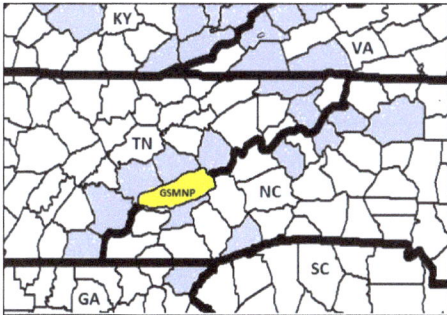

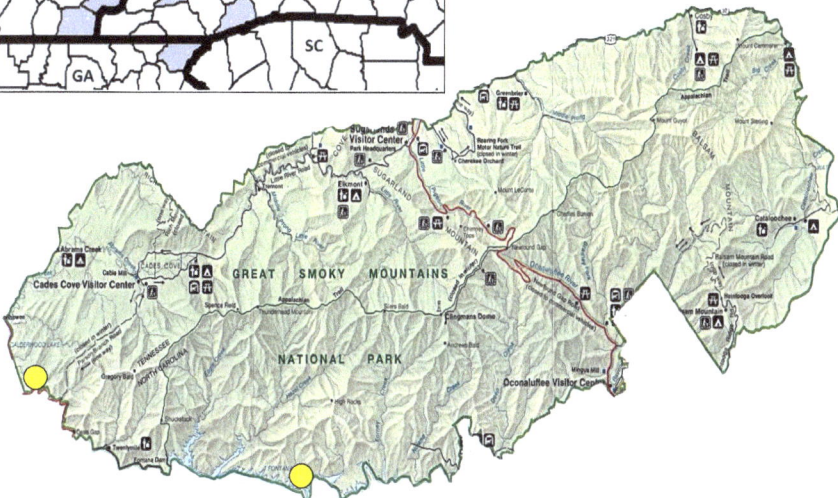

176

Cherrystone drop

Helicinidae

Hendersonia occulta (Say, 1831)

Diameter: 6-8 mm

Description: Heliciform; lip thickened (a); shell with 4.5-5 whorls; imperforate; with an operculum in live individuals; spiral striae are relatively well-developed and sometimes seen as fringes in young shells under 4 mm; periphery slightly angular occasionally weakly keeled, particularly in juvenile shells; eyes are located at the base of the tentacles; a primitive and largely tropical family of gastropods; this species is unique among park land snails by being dioecious (having the male and females separate).

Similar Species: *Stenotrema* species are around the same size but have a slit-like aperture and are without an operculum.

Habitat: A calciphile snail of river bluffs, mixed hardwood talus slopes, ravines and mountainsides.

Status: G4; Locally Rare; reported from scattered locations.

Specimen: Kentucky, Letcher County, Bad Branch Nature Preserve (author's collection).

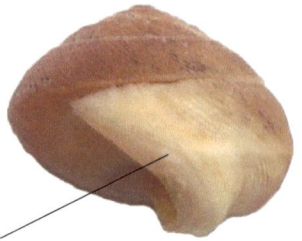

a

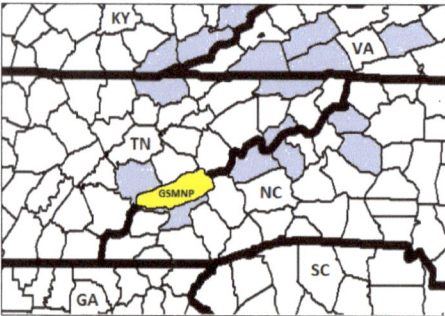

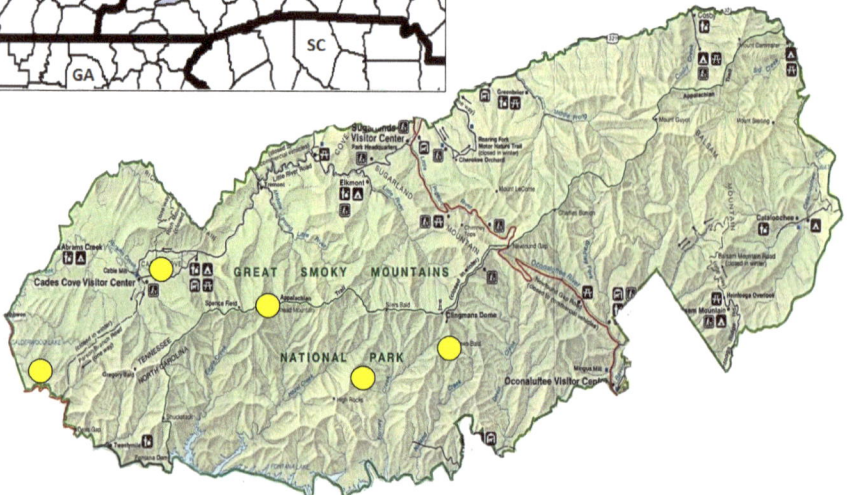

This section includes land snails that are more or less shaped like pills. All species of *Stenotrema* are under 15 mm in diameter (most species between 7 and 10 mm), the majority possessing hairs (see below picture) that cover the entire shell surface. These hairs collect forest debris, thought to help keep the snails well hidden from possible predators like birds and salamanders. The aperture of *Stenotrema* species are narrow (slit-like), denying easy access to predatory beetle larva. The long and wide parietal tooth of *Stenotrema* easily exceeds all other snails of similar size.

Genera Included:
(in order of appearance in text)

Stenotrema

Shells 4-12 mm, <u>pill-shape</u>, long parietal tooth (*Stenotrema*)

Stenotrema species are usually compact and pill-like in shape, have tightly coiled shells (the last whorl not expanded) and most are with fine or course hairs. All species have a long parietal (blade-like) tooth. *Stenotrema* species are best distinguished in frontal and bottom views, top views having very little diagnostic value. Examining the aperture is key (<u>figures d through l, page 180</u>). The aperture opens on the underside of the shell and is slit-like and narrow. The interdenticular sinus (page 180) is deep in some species and shallow in others. Another important trait is the basal notch (<u>figures d through l page 180)</u>, which can be nearly absent or a prominent feature of the lip. The fulcrum, an internal structure and part of the center supporting column of the shell, can be different lengths and can be seen through the bottom of live and fresh shells, but is difficult to see in old and badly weathered shells. *Stenotrema* differ from *Euchemotrema* in having a more closed aperture (a), a longer and wider parietal tooth (b), a completely sealed umbilicus (c) and in the park, average 2-3 mm smaller.

Stenotrema species of the GSMNP

*S. stenotrema
*Stenotrema species (undetermined)
*S. magnifumosum
*S. hirsutum
*S. altispira
*S. depilatum
*S. pilula

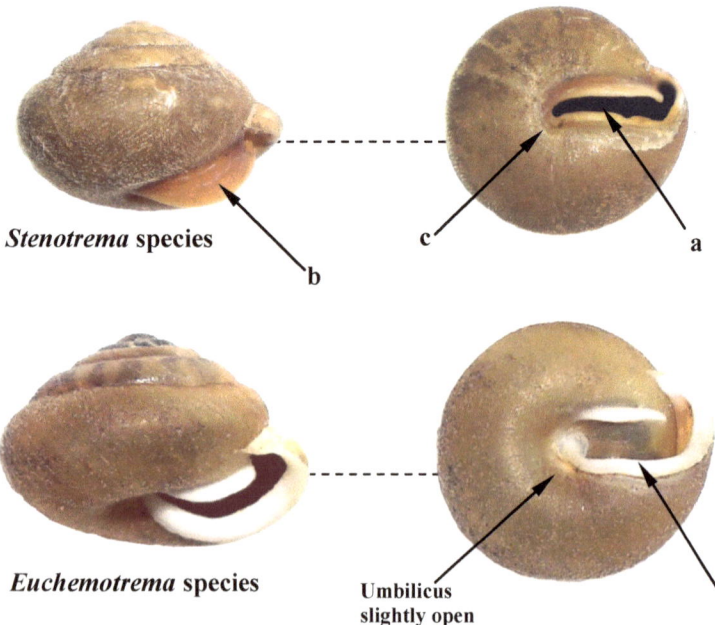

Stenotrema species   c   a
                           b

Euchemotrema species   Umbilicus
                            slightly open

Aperture Terminology in *Stenotrema*

Bottom View

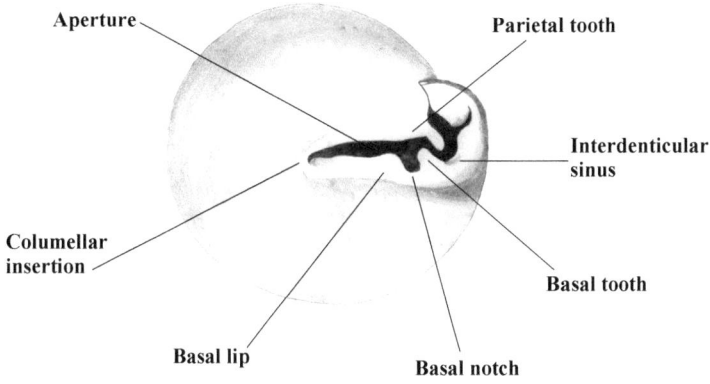

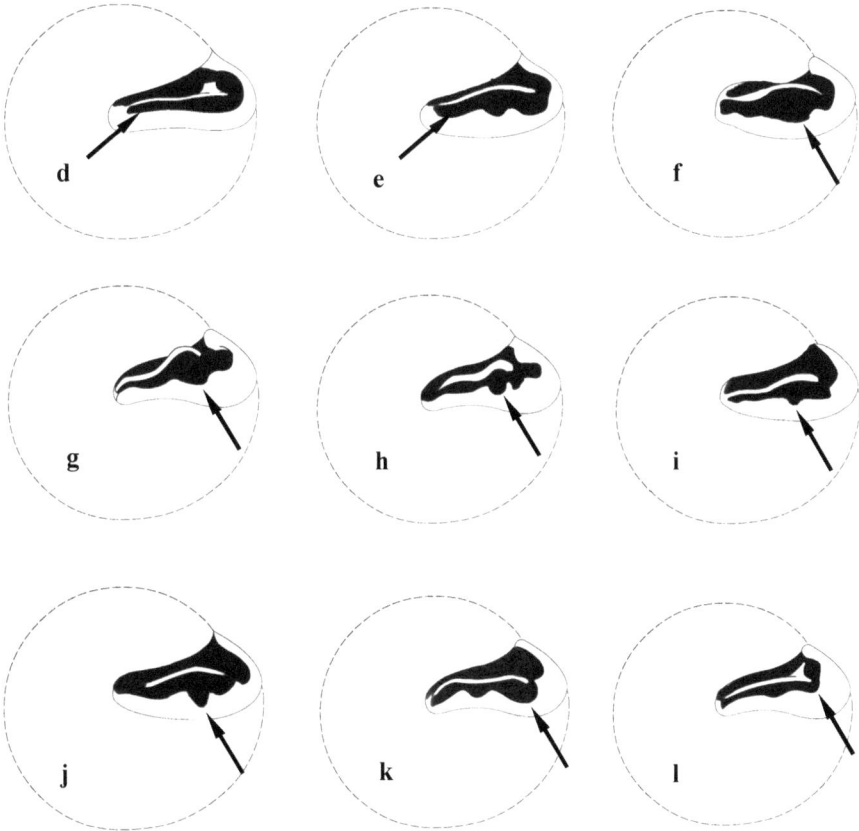

d) aperture narrow;) aperture wide; f) basal notch wide; g) basal notch medium; h) basal notch narrow; i) basal notch shallow; j) basal notch deep; k) interdenticular sinus deep; l) interdenticular sinus shallow

Inland slitmouth

Stenotrema stenotrema (L. Pfeiffer,1842)

Diameter: 7.8-15 mm

Description: Pill-shape; aperture narrow, especially near the columellar insertion (a); shell with 5-6 whorls; imperforate; brownish-tan or cinnamon-brown; minute hairs present on entire surface but are lost in aging shells; transverse striae poorly developed; basal notch is large and deep (b) the interdenticular sinus is deep (c); fulcrum is well-developed with a convex edge (seen through an opening made on the side of the shell); periphery well rounded.

Similar Species: The undetermined *Stenotrema* species (opposite page) averages 2 mm smaller, has a smaller basal notch and indistinct interdenticular sinus; other *Stenotrema* species of the GSMNP are typically 2-3 mm smaller.

Habitat: Usually found under leaf litter in lower elevation mixed hardwood forests.

Status: G5; Common; one of the most common low elevation *Stenotrema* species.

Specimen: Tennessee, Sevier County, Green Brier Picnic Area, GSMNP (GSMNP collection).

Polygyridae

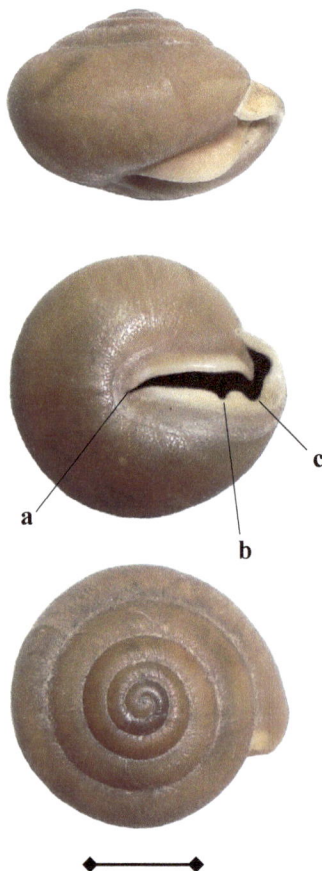

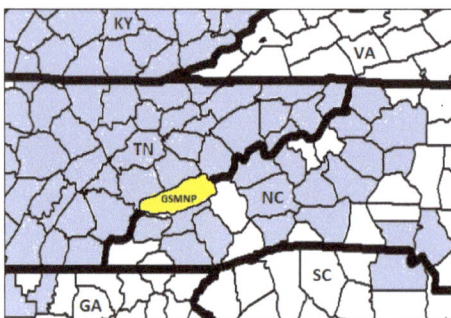

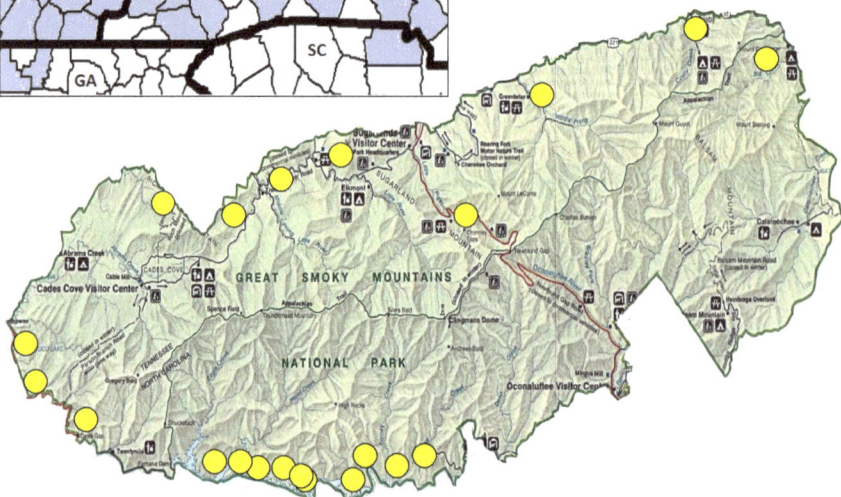

Cove slitmouth

Stenotrema species (undetermined)

Diameter: 6-9 mm

Description: Pill-shape and compact; parietal tooth long and wide; shell with 5-6 whorls; imperforate; cinnamon-buff; short stiff hairs present on entire surface but hairs are often lost in older shells; transverse striae poorly developed; basal notch shallow and the interdenticular sinus is indistinct in most specimens; fulcrum short; periphery rounded.

Similar Species: *S. stenotrema* is 2-3 mm larger, has a deeper basal notch, more-closed aperture, especially near the columellar insertion and the interdenticular sinus is notably deeper; the two species are occasionally found together.

Habitat: Found in lower elevation cove hardwood forests under forest litter, preferring mixed-mesophytic sites.

Status: G1/G2; Uncommon; Endemic to the GSMNP region; also found in the adjacent Nantahala and Cherokee National Forests; the species awaits further DNA investigation.

Specimen: Tennessee, Blount County, White Oak Sinks, GSMNP (GSMNP collection).

Polygyridae

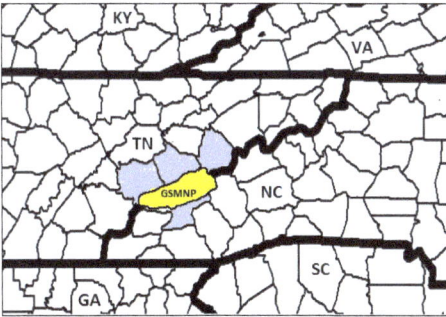

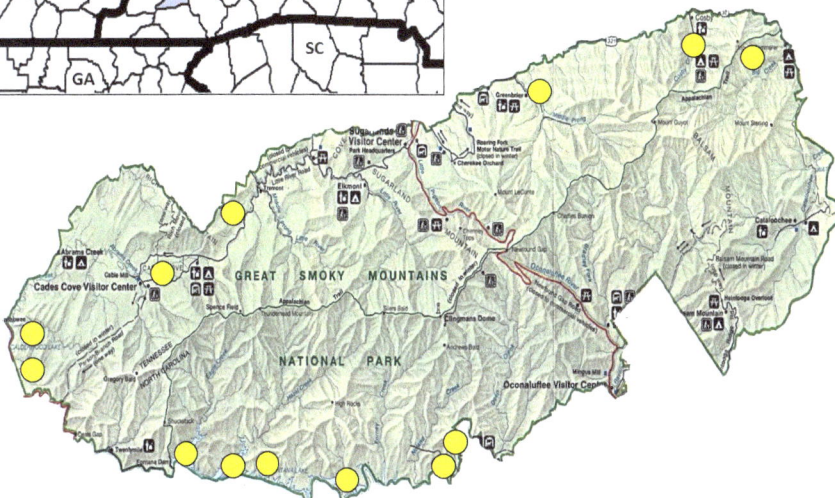

Hairy slitmouth

Polygyridae

Stenotrema hirsutum (Say, 1817)

Diameter: 6.2-9.6 mm

Description: Pill-shape; aperture generally wider than other *Stenotrema* species; shell with 5-5.5 whorls; imperforate; cinnamon-buff to tan; short, stiff hairs present on entire surface but hairs which are generally lost in older shells; transverse striae poorly developed; basal notch and the interdenticular sinus is moderately deep; the fulcrum (figure a) is well-developed, long and can be seen through the bottom view; shell periphery is boxy-shaped in frontal view.

Similar Species: *Stenotrema magnifumosum* is around the same size but has a bluntly angular form (not boxy).

Habitat: A species of dry upland mixed hardwood forests found under a variety of forest litter; also found hiding under logs and rocks.

Status: G5; Locally Rare; scattered locations across the GSMNP; Pilsbry reported this species in Cades Cove.

Specimen: Kentucky, Letcher County, Bad Branch (author's collection).

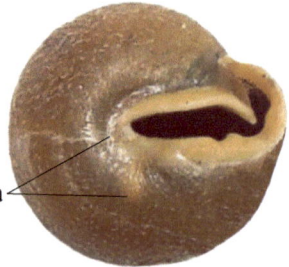

a

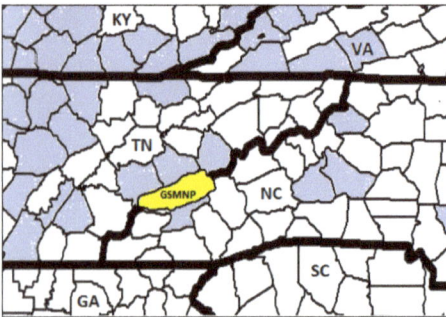

Appalachian slitmouth

Stenotrema magnifumosum (Pilsbry, 1920)

Diameter: 6-9 mm

Description: Pill-shape; shell with 5-5.5 whorls; brownish to cinnamon-brown; imperforate; without any discernible hairs; transverse striae poorly developed; basal notch and interdenticular sinus deep; fulcrum short; periphery is bluntly angular but this feature is somewhat variable.

Similar Species: *Stenotrema stenotrema* is 2-3 mm larger and has a more rounded periphery; *S. pilula* is smaller and like *S. hirsutum* is more boxy in shape; *S. altispira* is larger, has a higher shell profile and has a wider basal notch; *S. depilatum* is a species of higher elevation forests.

Habitat: Found in rich hardwood forests under forest litter, at elevations below 900 m.

Status: G4; Uncommon; records include southwestern portion of the park near Fontana Lake, Cades Cove, Snakeden Ridge and Rich Mountain.

Specimen: Tennessee, Blount County, White Oak Sinks, GSMNP (GSMNP collection).

Polygyridae

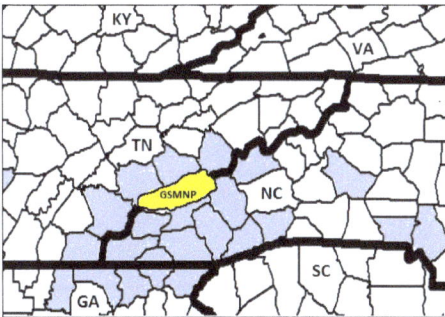

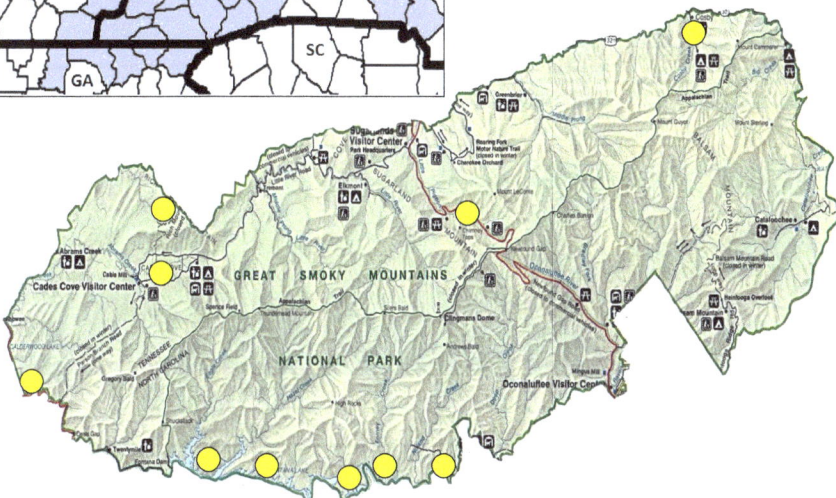

184

Highland slitmouth

Polygyridae

Stenotrema altispira (Pilsbry, 1894)

Diameter: 8-11.2 mm

Description: Pill-shape, globose; shell with 5-6 whorls; imperforate; tawny-olive to pale brown; short stiff hairs (a) present on entire surface but hairs are often lost in older shells; fulcrum short; transverse striae are poorly developed; basal notch wide and the interdenticular sinus is relatively deep; periphery rounded.

Similar Species: *Stenotrema stenotrema* is typically larger; *S. depilatum* is most like *S. altispira* in size and form but always without hairs.

Habitat: Found in mixed cove and northern hardwood forests under leaf-litter from around 1100 meters up to 2000 meters; this species is often found crawling on herbaceous vegetation in wet weather.

Status: G4; Relatively Common in upper elevations.

Specimen: North Carolina, Mitchell County, Roan Mountain (author's collection).

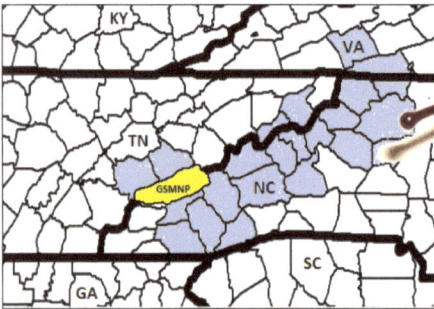

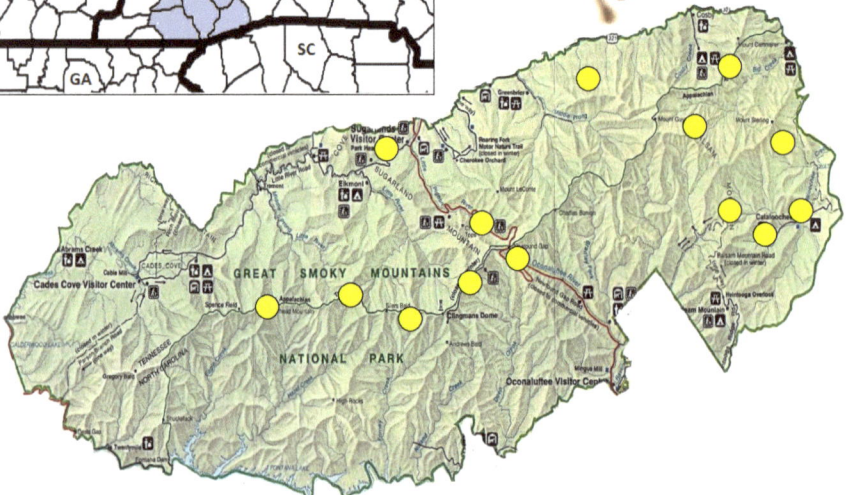

a

Great Smoky slitmouth

Polygyridae

Stenotrema depilatum (Pilsbry, 1895)

Diameter: 10-11.2 mm

Description: Pill-shape, globose; shell with 5-6 whorls; imperforate; tawny-olive to pale brown; without hairs seen in other species of *Stenotrema*; transverse striae poorly developed; fulcrum short; basal notch narrow and interdenticular sinus deep; periphery rounded.

Similar Species: *Stenotrema stenotrema* is 2-3 mm larger and has a more-closed aperture; *S. altispira* has short stiff hairs and the basal notch is notably wider; *S. pilula* is fully 5 mm smaller in shell diameter.

Habitat: Found in rich hardwood forests under wet leaf litter and moss at elevations generally above 1000 meters.

Status: G2; Globally Rare; Endemic to the GSMNP region; a species of limited distribution centering around the GSMNP; encountered much less frequency than *S. altispira*.

Specimen: North Carolina, Swain County, Clingmans Dome (GSMNP collection).

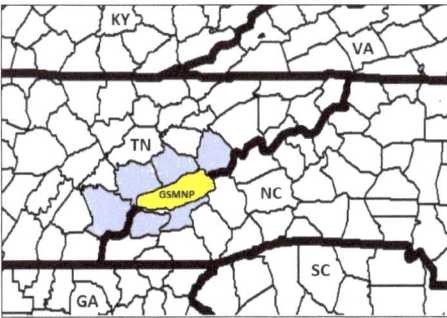

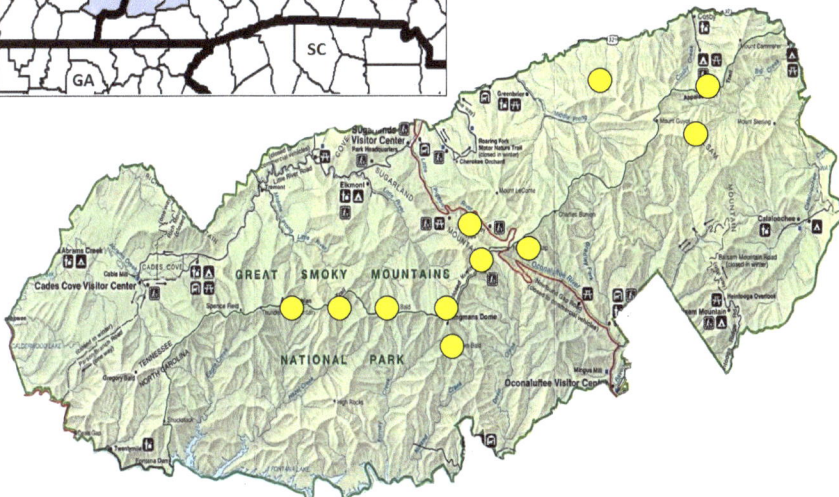

Pygmy slitmouth

Stenotrema pilula (Pilsbry, 1900)

Diameter: 5.7-6 mm

Description: Pill-shape and compact; shell with 5 whorls; imperforate; brownish-tan to cinnamon-brown; covered with short stiff hairs; transverse striae poorly developed; fulcrum long (a) and can be seen in the bottom view; basal notch and interdenticular sinus are deep; periphery more or less boxy.

Similar Species: This is the smallest *Stenotrema* in the GSMNP; *S. stenotrema* is larger and has a more-closed aperture; *S. altispira* usually has a more open aperture and the basal notch is wider; *S. hirsutum* has the same boxy-shape but is larger and the basal notch and interdenticular sinus are less well developed.

Habitat: Usually found in rich lower elevation hardwood forests under leaf litter but also a species of higher climes.

Status: G4; Relatively Common at lower elevations.

Specimen: Tennessee, Blount County, White Oak Sinks, GSMNP (GSMNP collection).

Polygyridae

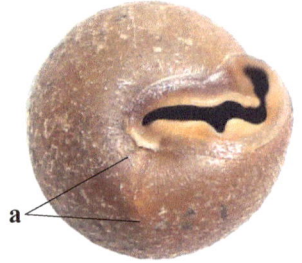

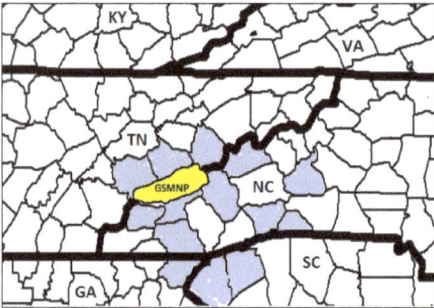

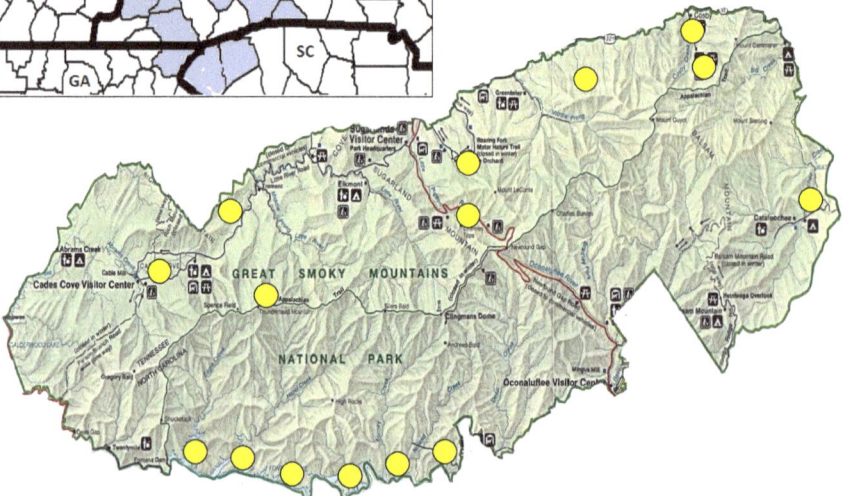

Stenotrema and *Euchemotrema* species compared (proportionate)

S. stenotrema

Stenotrema species (undetermined)

S. altispira

S. depilatum

S. magnifumosum

S. hirsutum

S. pilula

Euchemotrema fraternum

Euchemotrema fasciatum

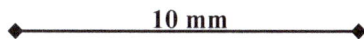

10 mm

The Protective Hairs of Land Snails

Above image of a highland slitmouth, *Stenotrema altispira,* covered in bits of soil and detritus. Below, a clean shell of same species showing the remarkable hairs that jacket the shell's exterior. The hairs collect material, thought to aide in crypsis, perhaps keeping the snail out of plain view of predators such as salamanders and birds. Both images from Roan Mountain, North Carolina.

This section includes land snails that possess shells that are wider than tall, heliciform or depressed heliciform, with or without teeth and most importantly, with reflected lips. Shells in this group can be small, 5 mm or less, such as *Vallonia*, but most are among the largest land snails found in the GSMNP, some reaching 45 mm in diameter. Most are cryptic in color, although several gastropods such as the *Allogona* are rather colorful.

Genera Included:
(in order of appearance in text)

Strobilops
Millerelix
Vallonia
Praticolella
Euchemotrema
Inflectarius
Patera
Fumonelix
Mesodon
Appalachina
Neohelix
Triodopsis
Xolotrema

I have to wake up at the crack of dusk to look this good!

Shells under 5.5 mm, <u>umbilicate</u>, reflected lip, with or without teeth (*Strobilops, Millerelix* and *Vallonia)*

Three land snails found in the GSMNP fall in this grouping, *Strobilops aeneus, Millerelix plicata* and *Vallonia excentrica.* These snails are generally uncommon gastropods and are easily separated from all snails of similar size and form by having a reflected lip; other snails of similar size will contain simple lips. The three species are dissimilar enough to make separation trouble-free and straightforward; they share no common attributes. Illustrated below are the three snails with a scale bar, showing actual size of each species.

<u>Grouping 1</u>: <u>shell 2.4-2.8 mm, top of shell notably ribbed, internal armature present</u>

*Strobilops aeneus, pg 192

<u>Grouping 2</u>: <u>shell 5.5 mm, depressed heliciform, large parietal tooth</u>

*Millerelix plicata, pg 193

<u>Grouping 3</u>: <u>shell 1.8-2.3 mm, lip widely reflected, widely umbilicate</u>

*Vallonia excentrica, pg 194

Bronze pinecone

Strobilops aeneus Pilsbry, 1926

Diameter: 2.4-2.8 mm

Description: Dome-shape; lip reflected; shell with 5.5 tightly coiled whorls; perforate to umbilicate; light to dark brown; elongated teeth or lamellae present in the aperture; teeth can be seen through the bottom of live and fresh dead shells or through the top; transverse striae modified into minute ribs; periphery angular.

Similar Species: *Strobilops labyrinthicus* (page 285) has a smaller umbilicus, higher shell profile and a more rounded periphery; *Euconulus* species are beehive-shaped, have smooth surfaces instead of ribs and have simple lips.

Habitat: Found in mixed hardwood forests, open glade-like areas especially around limestone outcrops but also commonly found in upland woods under the bark of rotting logs often in association with *Zonitoides arboreus*.

Status: G5; Uncommon, no doubt this species is more common than current records indicate.

Specimen: Kentucky, Letcher County, Bad Branch (author's collection).

Strobilopsidae

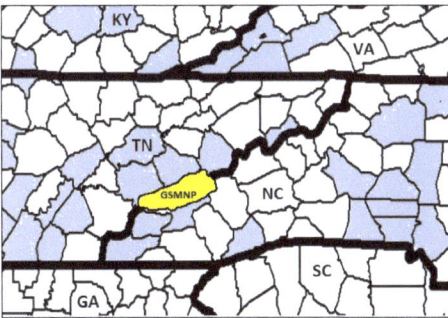

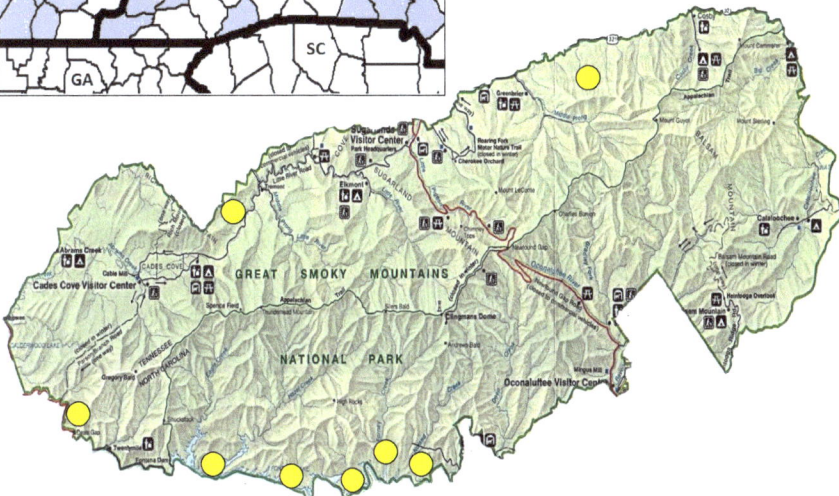

Cumberland liptooth

Polygyridae

Daedalochila plicata (Say, 1821)

Diameter: 5.5 mm

Description: Depressed heliciform and compact; lip warped and deformed-looking; shell with 5 whorls, first 4 whorls are tightly coiled but the last whorl is greatly expanded and appears twisted; widely umbilicate, tannish-brown; transverse striae well-developed; large parietal tooth and a deeply set palatal tooth; shell periphery squared or boxy, not rounded.

Similar Species: No other snail found in the GSMNP has the atypical umbilicus and warped aperture teeth of *Millerelix plicata.*

Habitat: A calciphile species found in dry hardwood forests and cedar glades around limestone outcrops; usually under leaf litter.

Status: G4; Locally Rare; recently documented in 2013 from the west end of TN side of the park and on a calcareous outcrop above Abrams Creek.

Specimen: Kentucky, Lyon County, Land Between the Lakes (author's collection).

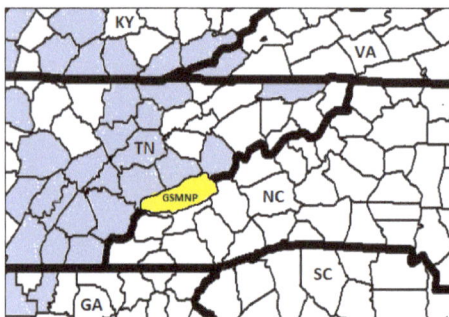

Iroquois vallonia

Valloniidae

Vallonia excentrica Sterki, 1893

Diameter: 1.8-2.3 mm

Description: Depressed heliciform; lip widely reflected; shell with 3-3.5 loosely coiled whorls, the last whorl greatly expanded (a); umbilicate; pale corneous or white; semi-transparent; shell surface smooth with only faint transverse striae (microscope required); no teeth present at any stage of growth.

Similar Species: The only small species in the GSMNP under 3 mm with a widely reflected lip.

Habitat: Found in open habitat such as glades, grassy edges and roadsides; it is frequently found with *V. pulchella* (Hubricht 1985) and is something of a "tramp" snail (Pilsbry 1948).

Status: G5; Locally Rare; only known from the Purchase Knob area of the park where it was likely introduced in soils surrounding planted apple trees.

Specimen: Kentucky, Powell County, Furnace Mountain (author's collection).

a

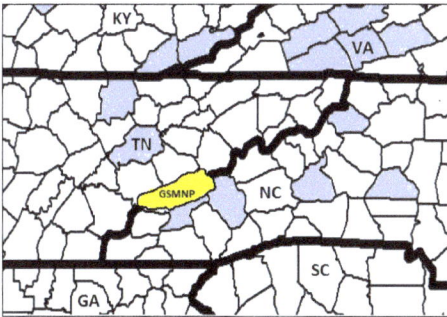

Shells 5-30 mm, <u>rimate</u>, reflected lip, with or without a parietal tooth (*Praticolella, Euchemotrema, Inflectarius* (in part) and *Mesodon* (in part)

Six land snails found in the GSMNP fall in this category; except for *Inflectarius downieanus, M. thyroidus* and *M. clausus,* these gastropods have a relatively long parietal tooth (a). The most important feature in this grouping is the rimate shell (b), the umbilicus not completely sealed. This character may be hard to see with the naked eye but can usually be distinguished with a hand lens of 10X. Two species, *P. lawae* and *I. downieanus,* are exceptionally rare and although reported from the park, their exact locations remain unknown.

<u>Grouping 1</u>: <u>shell 5-6 mm, rimate, long parietal tooth, hairs in diagonal rows, page 196</u>

*Praticolella lawae

<u>Grouping 2</u>: <u>shell 7-11 mm, rimate, long parietal tooth, hairs not in diagonal rows, page 197</u>

*Euchemotrema fraternum
*Euchemotrema fasciatum

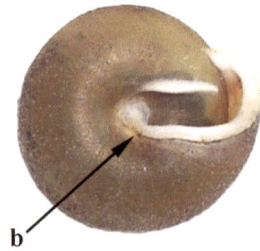

a

b

<u>Grouping 3</u>: <u>shell 10-30 mm, rimate, with or without teeth, shell globose, page 199</u>

*Inflectarius downieanus
*Mesodon clausus
*Mesodon thyroidus

Appalachian scrubsnail

Praticolella lawae (J. Lewis, 1874)

Diameter: 5.7-6.1 mm

Description: Heliciform; lip slightly reflected; shell with 4.5-5 whorls; rimate, the umbilicus never entirely closed; with or without an elongated parietal tooth; pale cinnamon-buff; apical whorls sculptured with papillae; short hairs are arranged in diagonal rows, an important diagnostic feature; indistinguishable growth wrinkles; shell periphery well-rounded.

Similar Species: *Mesodon clausus* is much larger and does not contain a parietal tooth; *Euchemotrema* species have a more closed umbilicus and lack the unique diagonally arranged hairs.

Habitat: A species of open pine woods, clearings and glades, typically on sandy soils (Hubricht 1984).

Status: G3; Locally Rare; although this species is listed in old park records, precise location information remains unknown; area of most likely occurrences (a).

Specimen: North Carolina, Clay County, Hayesville (FMNH 6135).

Polygyridae

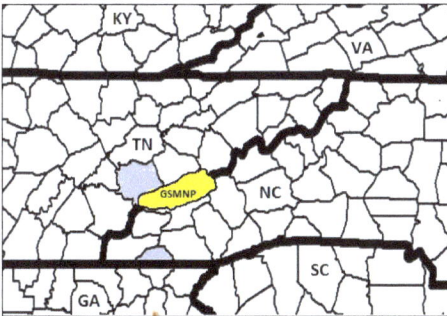

Upland pillsnail

Euchemotrema fraternum (Say, 1824)

Polygyridae

Diameter: 7.8-11.4 mm

Description: Heliciform; lip reflected; shell with 5-6 tightly coiled whorls; perforate to rimate, the umbilicus never completely closed; large parietal tooth that is blade-like; shell dull; pale-tan to cinnamon-buff; minute hairs present in last whorl but hairs are mostly lost in older shells; transverse striae poorly developed; shell periphery rounded.

Similar Species: Similar to *Stenotrema* species but without the basal notch and interdenticular sinus; *E. fasciatum* is more compressed and has a light color band (in most specimens) around its periphery.

Habitat: A species found under forest litter and around logs in upland mixed hardwood forests.

Status: G5; Locally Rare; in the GSMNP reported from the Big Creek area only, it is however expected to be found elsewhere in eastern portions of the park.

Specimen: Kentucky, Powell County, Furnace Mountain (author's collection).

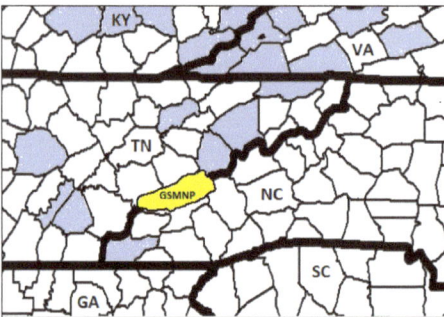

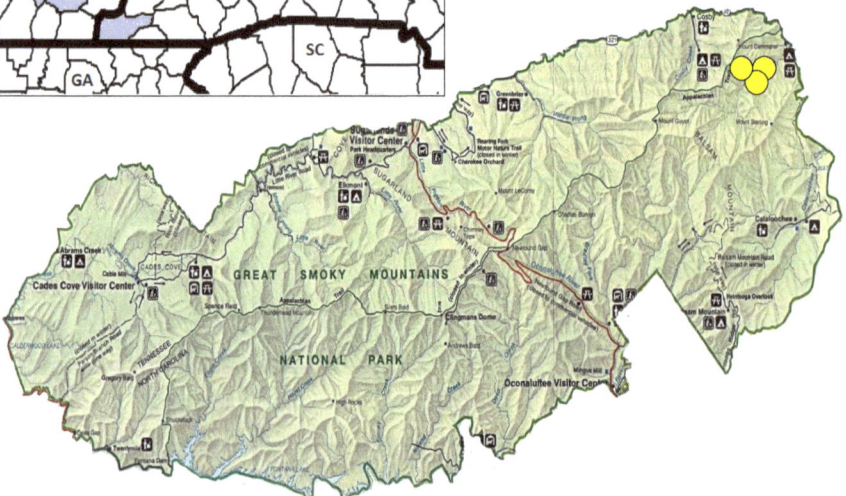

Mountain pillsnail

Polygyridae

Euchemotrema fasciatum (Pilsbry, 1940)

Diameter: 9.4-11 mm

Description: Depressed heliciform; lip reflected; shell with 5-6 tightly coiled whorl; perforate to rimate, the umbilicus never completely sealed; large parietal tooth that is blade-like; shell dull; pale-tan to cinnamon-buff; minute hairs present in last whorl but hairs are often lost in older shells; transverse striae poorly developed; shell periphery rounded with a light brownish band.

Similar Species: *Euchemotrema fraternum* is around the same size but has a higher shell profile and is without the faint periphery band.

Habitat: A species found under forest litter and around logs in upland mixed hardwood forests; often found resting above ground on trees between deep bark furrows.

Status: G3; Uncommon; scattered locations across southern portions of the park in lower elevations.

Specimen: North Carolina, Swain County, Forney Creek, GSMNP (GSMNP collection).

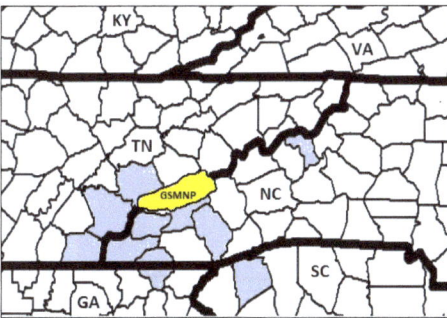

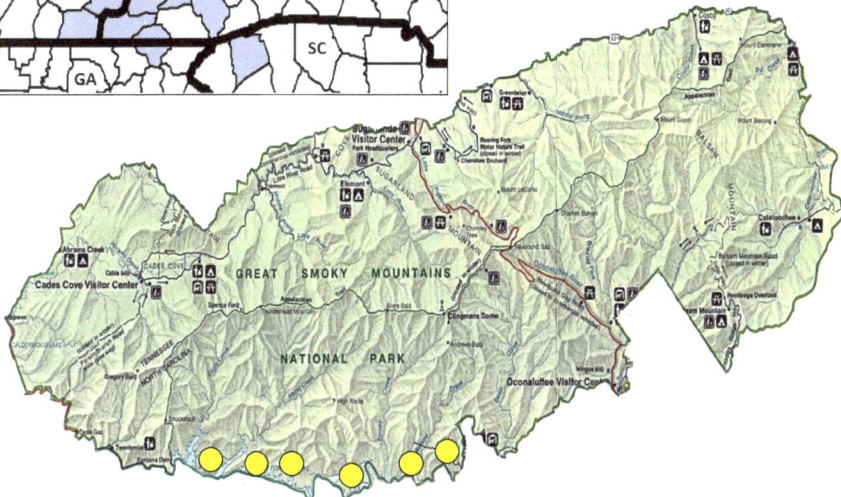

Dwarf globelet

Inflectarius downieanus (Bland, 1861)

Diameter: 10.5-14.8 mm

Description: Heliciform; lip only slightly reflected; shell with 5.5 whorls; rimate, umbilicus nearly but not completely closed; pale yellow to greenish horn colored; without a parietal, basal or palatal tooth; crowded spiral striae present on good shells but may be a barely detectible feature on badly weathered ones; periphery well-rounded.

Similar Species: Both *Inflectarius inflectus* and *I. rugeli* have well-defined palatal, basal and palatal teeth; *Mesodon clausus*, the species *I. downieanus* most resembles (page 200) is typically larger in size and has a slightly wider lip.

Habitat: Usually found under leaf litter on the summits of flat-top mountains in acidic soils.

Status: G3; Locally Rare; Hubricht (1956) documented this species below 800 m in the park but exact locations remain unknown, area of most likely occurrences (b).

Specimen: Alabama, Marshall County, near Gunterville (author's collection).

Polygyridae

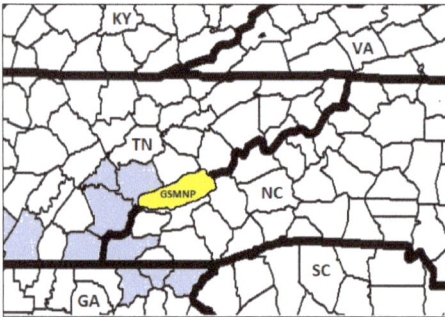

Yellow globelet

Polygyridae

Mesodon clausus (Say, 1821)

Diameter: 15-17 mm

Description: Heliciform; lip reflected; shell with 5-5.5 whorls; rimate, the umbilicus not entirely closed (a); without teeth; shell glossy and thin; pale-yellow to light tan; no hairs; transverse striae moderately developed; spiral striae present; shell periphery well-rounded.

Similar Species: *Mesodon thyroidus* is larger and has a small parietal tooth; by its color, the spiral striae and the nearly closed umbilicus *M. clausus* resembles *Inflectarius downieanus* in larger form.

Habitat: A calciphile species of low wet areas, marshes; swamps and floodplains; but also found along roadsides and waste places like old parking lots, damp roadside ditches, grassy slopes, and ditches along railroad tracks; in wet weather found climbing on herbaceous plants.

Status: G5; Uncommon except for White Oaks Sinks where it occurs in large numbers for reason that remain unknown

Specimen: Tennessee, Blount County, Cades Cove, GSMNP (GSMNP collection).

a

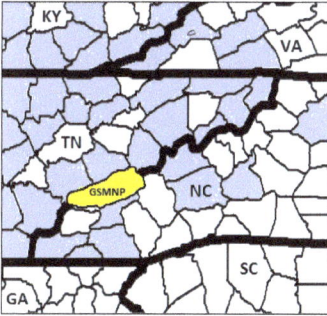

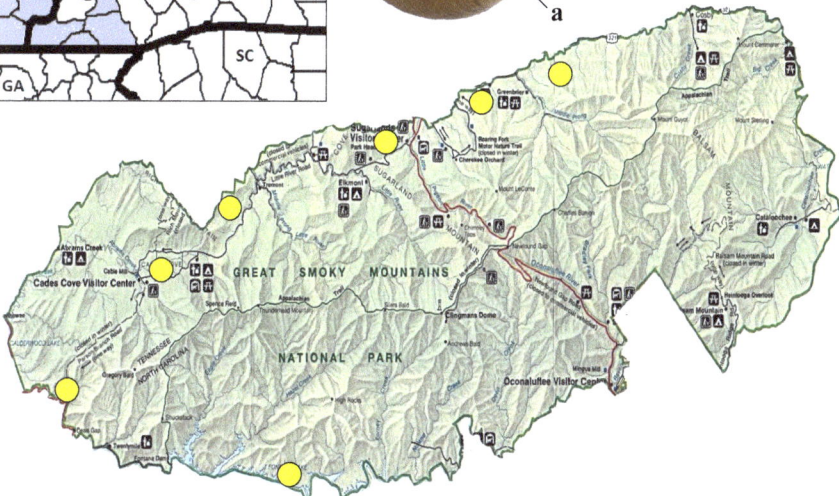

White-lip globe

Mesodon thyroidus (Say,1816)

Diameter: 15-30 mm

Description: Heliciform; lip reflected; shell with 5-5.5 whorls; umbilicus always rimate, never completely closed; small parietal tooth usually present; ivory yellow to pale yellowish-green; no hairs; transverse and minute spiral striae always present; shell periphery well-rounded.

Similar Species: *Mesodon clausus* is smaller and is without a parietal tooth; the umbilicus of *M. normalis* is always completely closed; *M. zaletus* has a thicker shell and a completely closed umbilicus.

Habitat: When found, this species can be exceptionally frequent especially along rights-of-ways; also in young floodplain forests, meadows and waste places; rarely found in mature intact woods.

Status: G5; Uncommon in the GSMNP; most likely sites will include areas around degraded habitats of the park.

Specimen: Kentucky, Powell County, Furnace Mountain (author's collection).

Polygyridae

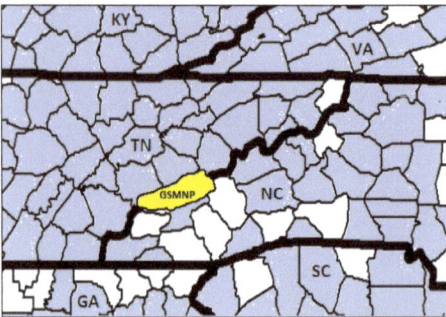

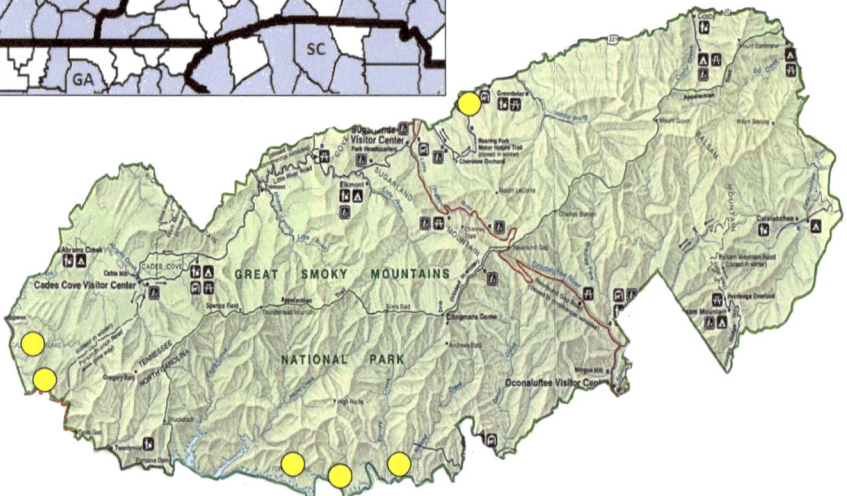

Shells under 15 mm (except for *I. ferrissi is* 19-26 mm), shells imperforate, reflected lip, teeth present (*Inflectarius*)

Most *Inflectarius* species are under 15 mm (with the exception of *I. ferrissi*, 19 -26 mm), are imperforate (with the exception of *Inflectarius downieanus* which is rimate, page 199) and contain lips not widely reflected (as seen in *Fumonelix* and *Mesodon* species). Some *Inflectarius* species have tightly coiled shells, while others are more loosely coiled. With such a wide range of shell shapes, no foolproof guideline exist for the group as a whole, short of DNA bar-coding and/or genital dissection. In the GSMNP, all species have a parietal tooth (a) with the exception of *I. downieanus*. *Inflectarius* species are best distinguished by frontal view and by examining the aperture for teeth. Several species have periostracal processes or fringes (see juvenile *Inflectarius rugeli* below) but these are often lost in older shells. The following *Inflectarius* species are ar-ranged as follows. **Group 1** are species that have three teeth, the parietal, basal and palatal and **Group 2** are species that have only one tooth, the parietal.

Grouping 1: shell 7-16 mm, having a parietal, basal and palatal tooth, page 203

*I. inflectus
*I. rugeli

a

Grouping 2: shell 7.8-20 mm, having a parietal tooth only, page 205

*I. verus
*I. ferrissi

A juvenile *Inflectarius rugeli* shell enlarged (around 3 mm in actual size), illustrating the periostracal processes or fringes seen on young shells under 5 mm, but often lost in older shells. Forest litter often sticks to these fringes and is believed to help camouflage the gastropod from potential predators such as birds and salamanders.

Juvenile *Inflectarius rugeli*

Shagreen

Polygyridae

Inflectarius inflectus (Say, 1821)

Diameter: 7.5-13.8 mm

Description: Depressed heliciform; lip reflected; shell with 4.5-5.5 whorls; imperforate; large parietal tooth present; basal tooth small; palatal tooth small and only slightly recessed into the aperture (a); cream buff-colored to light yellowish-horn; shell surface with short periostracal processes and fine transverse striae or wrinkles; periphery rounded.

Similar Species: *Inflectarius rugeli* is slightly larger and the palatal tooth is larger and more deeply recessed into the aperture (see figure b, next page), making the aperture more crowded in frontal view; intermediate forms between the two species are not uncommon.

Habitat: This is a common land snail of roadside ditches, railroad rights-of ways, grassy areas and shale banks of road cuts.

Status: G5; Locally Rare; Thompson (1981) documented this species below 800 m in two locations shown below.

Specimen: Kentucky, Powell County, Furnace Mountain (author's collection).

a

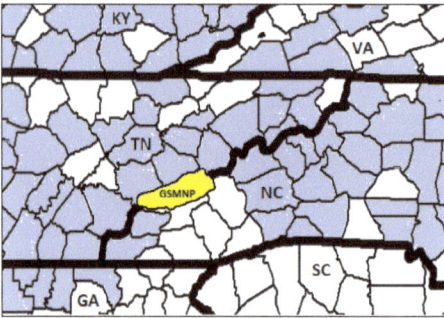

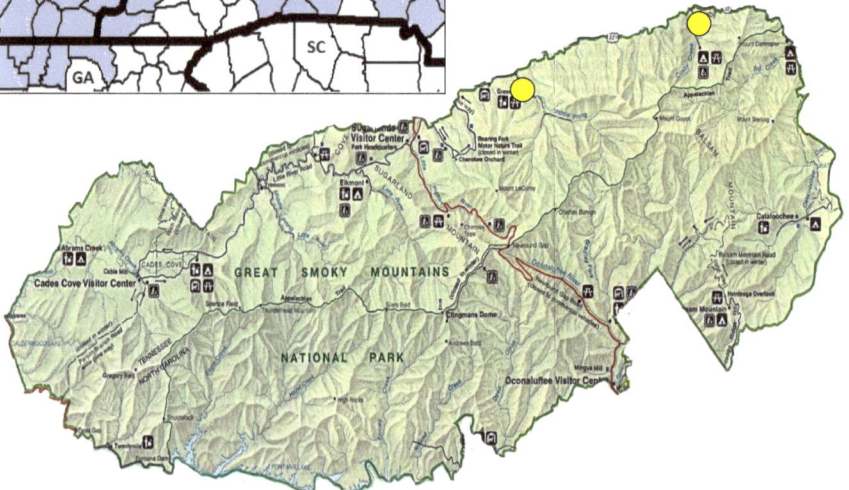

Deep-tooth shagreen

Polygyridae

Inflectarius rugeli (Shuttleworth, 1852)

Diameter: 7.8-16.4 mm

Description: Depressed heliciform; lip reflected; shell with 5.5 whorls; imperforate; large parietal tooth present; basal tooth small; palatal tooth large and deeply recessed into the aperture (b); waxy horn-colored; shell surface with short periostracal processes that may at times appear as scales especially in older shells.

Similar Species: *Inflectarius inflectus* is slightly smaller, has a smaller palatal tooth which is less deeply recessed into the aperture. There are several distinct size forms, two of which are found together at Roaring Fork (GSMNP). They display no external separating features except for a constant and reliable shell size difference of around 3 mm. This has been observed elsewhere in the Southern Appalachians (pers. comm. John Slapcinsky, 2012).

Habitat: A habitat generalist of mixed hardwood forests.

Status: G5; Common.

Specimen: Kentucky, Letcher County, Bad Branch (author's collection).

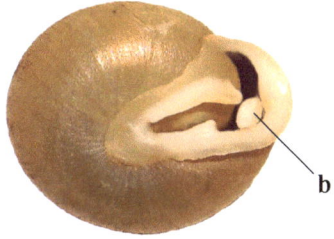

b

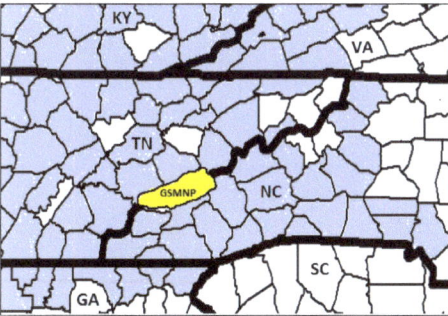

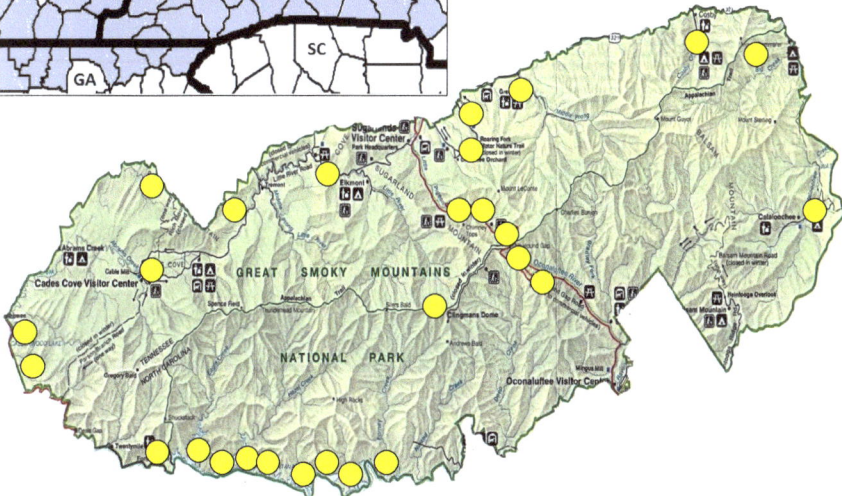

Fuzzy covert

Polygyridae

Inflectarius verus Hubricht 1954

Diameter: 12-14.6 mm

Description: Depressed heliciform; lip reflected; shell with 5.5 whorls; imperforate; large parietal tooth with a wide base; with or without (usually without) a small basal tooth; palatal tooth absent; shell surface dull with fine erect periostracal processes or fringes (refer to young *Inflectarius rugeli* image on page 202); periphery well-rounded.

Similar Species: Both *Inflectarius inflectus* and *I. rugeli* have well-defined basal and palatal teeth; *I. subpalliatus* (illustrated in Part II of the book) has a different shape and has a greater space between the parietal tooth and lower lip.

Habitat: Found at elevations between 450 m and 1100 m in well-developed rock talus; also under leaf litter and around rotting logs.

Status: G1; Globally Rare; a species Endemic to GSMNP; currently reported from eastern portions of the park and one location at Martins Gap trail.

Specimen: North Carolina, Haywood County, Big Creek, GSMNP (GSMNP collection).

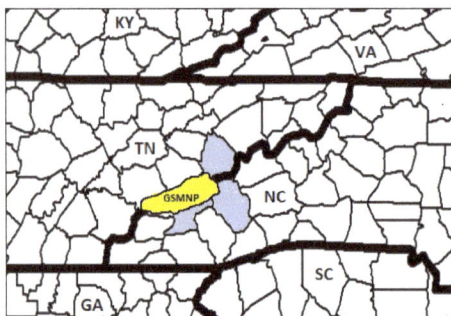

205

Smoky Mountain covert

Polygyridae

Inflectarius ferrissi (Pilsbry, 1897)

Diameter: 19-26 mm

Description: Depressed heliciform; lip reflected; shell with 5.5 whorls; imperforate; parietal tooth small; without any discernible basal tooth; palatal tooth absent; shell surface glossy with weakly developed transverse striae; spiral striae well-developed both on top and base of shell; periphery well- rounded.

Similar Species: *Inflectarius ferrissi sericeus* (Ferriss, 1905) has a silky, not glossy, luster, is densely papillose, without spiral striae and is known only from Plott Balsam, Jackson County, North Carolina.

Habitat: Usually found at elevations above 600 to 2000 m in well-developed rock talus, occasionally around rotting logs.

Status: G1; Locally Rare; a species Endemic to the GSMNP; Sevier and Cocke County, Tennessee and Swain County, North Carolina, distribution centered in the GSMNP; the species is even rarer outside park boundaries.

Specimen: North Carolina, Swain County, Newfound Gap, GSMNP (GSMNP collection).

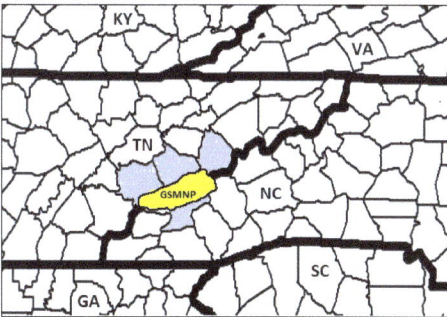

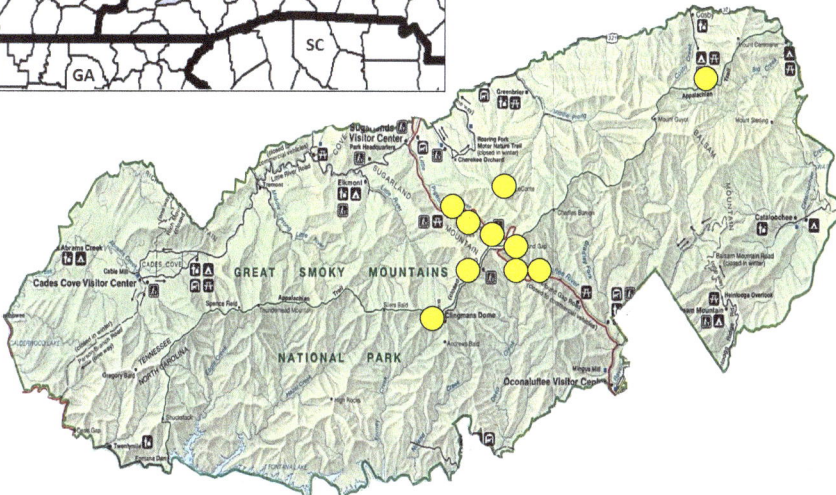

Shells 15-24 mm, imperforate, <u>with long, low basal tooth</u>, reflected lip (*Patera*)

Patera species are usually depressed heliciform with the oddball exception of *P. clarki* (page 208). This species defies the standard by being heliciform, looking very much like a small *Mesodon*. All adult *Patera* shells exhibit reflected lips and in the GSMNP, display a large parietal tooth. While the long, low basal tooth (a) is a wanting feature on some shells, it is an important diagnostic mark for the group as a whole and should not be discarded. There are either spirally arranged papillae (b) or spiral striae (c). These diagnostic features may be an obscure feature on aged shells that have worn surfaces. An area that generally remains protected from abrasions is located just behind the reflected lip of mature shells (d). All *Patera* species are imperforate. Immature *Patera* shells will not have reflected lips, and so, are sometimes confused with young *Mesomphix* shells. The following *Patera* species are arranged according to these features and are as follows: **Group 1** shells that are heliciform, **Group 2** shells that are depressed heliciform and have spiral papillae and **Group 3** shells that have spiral striae in the form of incised lines.

Grouping 1: shells that are heliciform, page 208

P. clarki clarki

Grouping 2: shells with spiral papillae, page 209

P. appressa

Grouping 3: shells with spiral striae, page 210

P. perigrapta

d

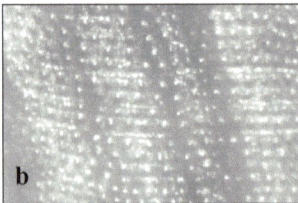

b

Spirally arranged papillae

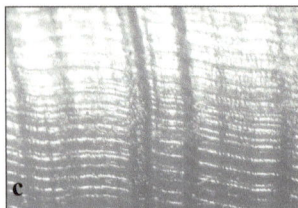

c

Spiral striae

a

Dwarf proud globe

Polygyridae

Patera clarki clarki (I. Lea, 1859)

Diameter: 13-18.2 mm

Description: Heliciform; lip reflected; shell with 5.5-6 whorls; imperforate; large parietal tooth present; basal tooth small but is a constant and reliable diagnostic feature; pale buff to tan; no hairs in adult shells; transverse striae well-developed; spiral striae strongest on top becoming more obsolete on sides and base of shell.

Similar Species: *Mesodon elevatus* is larger with a thicker shell; *P. clarki nantahala* has a lower shell profile (but neither species is currently reported from the GSMNP).

Habitat: Rich woods in upper elevation forests under leaf litter, rock talus and around logs; sometimes found crawling on the undersides of rock features such as rock outcrops.

Status: G3; Relatively common; Endemic to the GSMNP region; although this species has a somewhat restricted distribution in the southern mountains, it is generally common where found.

Specimen: North Carolina, Swain County, Nantahala National Forest (author's collection).

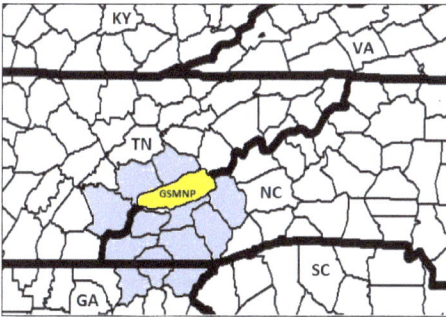

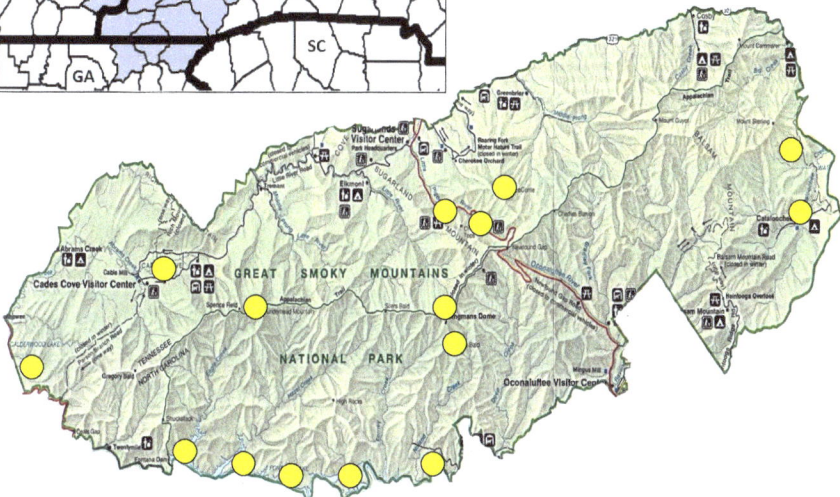

Flat bladetooth

Patera appressa (Say, 1821)

Diameter: 13-19.5 mm

Description: Depressed heliciform; lip reflected; shell with 4.5-5 whorls; imperforate; large parietal tooth present with a wide base (a); basal tooth small and sometimes poorly defined; cinnamon-buff to brownish-horn; shell surface somewhat glossy; adult shells without hairs; transverse striae well-developed on top and sides but weakly defined on the base; spirally arranged papillae typically present; shell periphery rounded but some populations may have a slightly angular periphery.

Similar Species: *Patera perigrapta* has strongly developed spiral striae and the parietal tooth has a narrow base.

Habitat: Limestone and sandstone rock outcrops and cave entrances.

Status: G5; Locally Rare; known only along limestone cliffs adjacent to the Little River near the Foothills Parkway.

Specimen: Kentucky, Powell County, Red River Gorge (author's collection).

Polygyridae

a

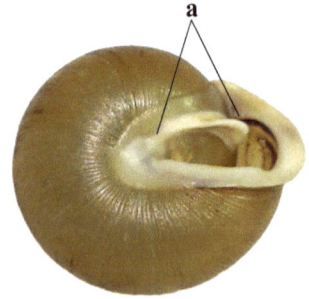

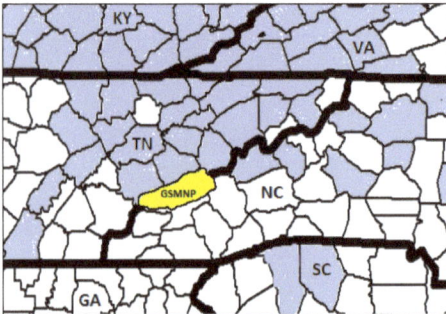

Engraved bladetooth

Polygyridae

Patera perigrapta (Pilsbry, 1894)
Diameter: 15.8-23.4 mm
Description: Depressed heliciform; lip reflected; shell with 5.5-6 whorls; imperforate; parietal tooth medium with a narrow base (b); basal tooth small and sometimes poorly defined; shell tannish in color and dull-glossy; no hairs in adults; transverse striae well-developed; spiral striae are deeply engraved and remain a strong and constant feature.
Similar Species: *Patera appressa* is generally smaller, has spirally arranged papillae and the base of the parietal tooth is clearly wider; *Inflectarius ferrissi* has a smaller parietal tooth and is glossier.
Habitat: Commonly found on the underside of downed logs, sometimes in hollow trees in mixed hardwood forests, also found around rock outcroppings.
Status: G5; Common; the most common *Patera* species in the park.
Specimen: North Carolina, Swain County, Nantahala National Forest (author's collection).

b

210

Shells 8-20 mm, imperforate, <u>without long, low basal tooth,</u> reflected lip (*Fumonelix*)

Fumonelix species in the GSMNP range from heliciform to depressed heliciform, most species having widely reflected lips at maturity and all species being imperforate. Shells are with or without a parietal tooth and without basal and palatal teeth. The transverse striae are usually a strong character in this group and one species is with spiral striae. *Mesodon species are similar in shape but are generally 8-10 mm larger in diameter*. *Fumonelix* centers around the GSMNP and becomes less common farther from this region. The group as a whole are indicative of quality habitat and where forests are highly degraded the species may be negatively affected. The following *Fumonelix* include **Group 1** shells with spiral striae and **Group 2** shells without spiral striae.

Grouping 1: shells with spiral striae, parietal tooth present, page 212

F. langdoni

Grouping 2: shells without spiral striae, parietal tooth typically present, page 213

F. jonesiana
F. wetherbyi
F. wheatleyi
F. clingmanica
F. christyi

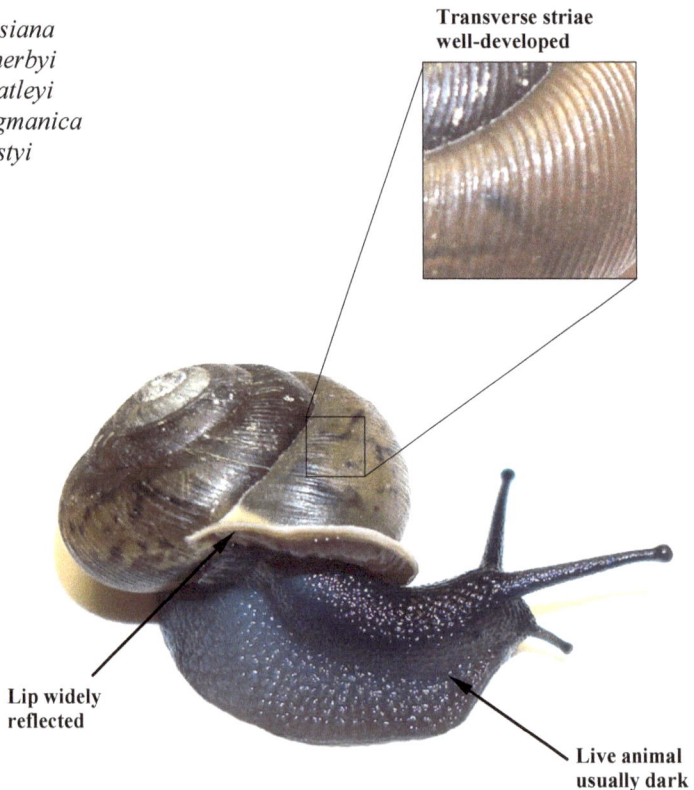

Transverse striae well-developed

Lip widely reflected

Live animal usually dark

Talus covert

Polygyridae

Fumonelix langdoni Dourson 2012

Diameter: 17-19 mm

Description: Depressed heliciform to heliciform; lip reflected; shell with 5-6 whorls; imperforate; small parietal tooth with a narrow base; shell yellowish-brown; well-developed transverse striae; spiral striae present that appear as closely-spaced incised lines, strongest on the base but only seen as traces on top; without hairs in adults; periphery rounded.

Similar Species: *Fumonelix wheatleyi* has a higher shell profile, a smaller parietal tooth and, most importantly, lacks any spiral striae; *I. ferrissi* is more glossy, has less developed transverse striae but better developed spiral striae.

Habitat: Found in well-developed talus and under leaf litter in beech-gap and northern hardwood forests (>1500 m).

Status: G1; Globally Rare; a species Endemic to the GSMNP, its entire global range thought to centered around Starkey Gap.

Specimen: North Carolina, Swain County, Starkey Gap, GSMNP (GSMNP collection).

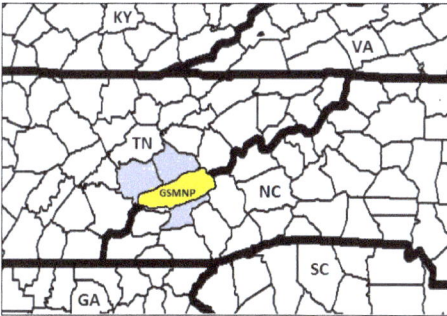

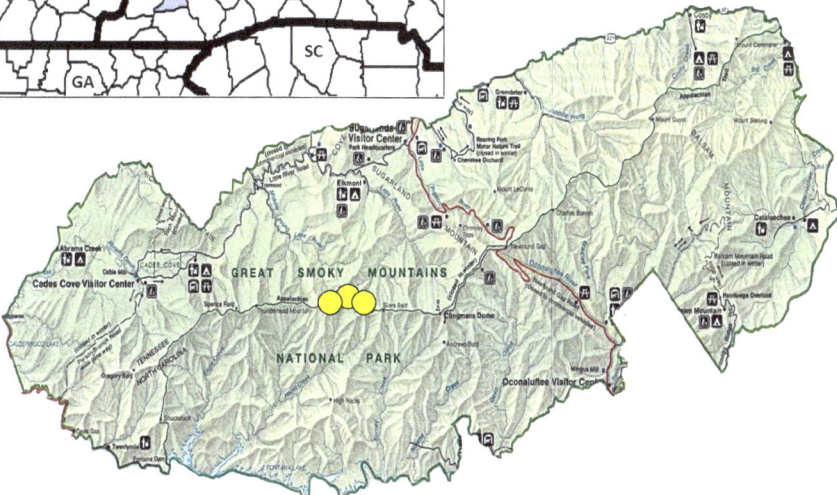

Big-tooth covert

Polygyridae

Fumonelix jonesiana (Archer, 1938)

Diameter: 12.8-13.5 mm

Description: Heliciform; lip reflected; shell with 5-5.5 whorls; imperforate; large parietal tooth; shell dull-glossy chestnut-color; sculptured with oblique, prostrate hairs, often lost in old shells trap a thin layer of soil, making it appear dirty; shell has fine riblets and longitudinal pits; periphery well-rounded; darker shells illustrated with live animal inside shell; lighter shell (fig. a) without.

Similar Species: *Mesodon* species are larger and are without prostrate hairs; *Patera* species are more compressed and are usually with a poorly developed basal tooth.

a

Habitat: Found under leaf litter in the beech, yellow birch and hemlock woods between 1400 and 1650 m.

Status: G1; Globally Rare; a species <u>Endemic to the GSMNP</u>, its entire global range contained in only a few square miles around Newfound Gap; this is one of the rarest land snails in the GSMNP, indeed, North America.

Specimen: North Carolina, Swain County, Newfound Gap, GSMNP (GSMNP collection).

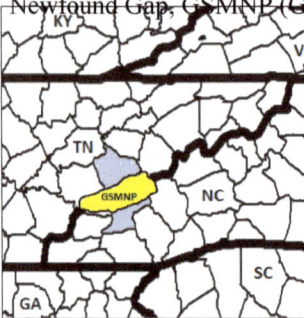

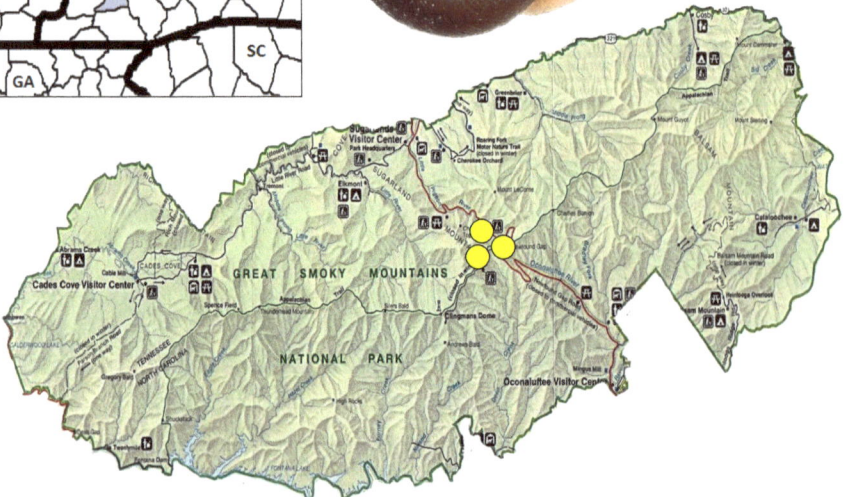

Clifty covert

Polygyridae

Fumonelix wetherbyi (Bland, 1873)

Diameter: 17-20 mm

Description: Heliciform; lip widely reflected; shell with 5-6 whorls; imperforate; large parietal tooth; dull cinnamon-buff; sculptured with oblique, prostrate hairs but these are often lost in old shells; the rough shell texture can easily be felt in the fingers and these hairs often trap a thin layer of soil, making it appear dirty.

Similar Species: *Mesodon* species are larger and have distinct transverse and spiral striae on their shell surface and are without prostrate hairs; *F. jonesiana* is smaller, has a glossier shell surface and found in higher elevation forests around Newfound Gap.

Habitat: Most common on steep slopes below sandstone outcrops or in woods strewn with sandstone talus or blocks of rock; in the park a species of lower elevation old growth forests.

Status: G4; Locally Rare; in the park known only from the Albright Grove.

Specimen: Tennessee, Sevier County, Albright Grove, GSMNP (GSMNP collection).

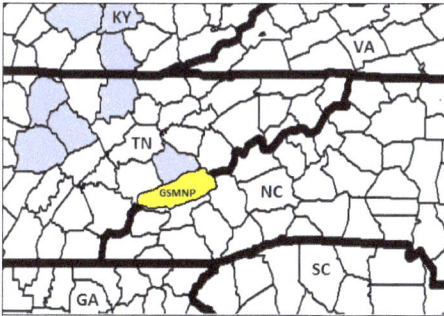

214

Cinnamon covert

Fumonelix wheatleyi (Bland, 1860)

Polygyridae

Diameter: 14-18 mm

Description: Heliciform; lip widely reflected (a); shell with 5-5.5 whorls; imperforate; typically with a small parietal tooth but sometimes without; reddish horn-colored; well-developed transverse striae; without any traces of spiral striae, periphery well-rounded.

Similar Species: *Fumonelix clingmanica* is smaller and has a more delicate shell;. *F. jonesiana* is smaller with a larger parietal tooth; *Mesodon* species average 10 mm larger.

Habitat: Found under leaf litter in the mixed hardwood forests at all elevations.

Status: G4; Common; In 1985 Hubricht discussed two *Fumonelix wheatleyi* forms from the park having very different genitalia, but identical shell morphology. He failed to described the new *Fumonelix* formally and so the species has remained unnamed. The two forms are apparently both found at all elevations.

Specimen: North Carolina, Macon County, Sugar Cove Creek, Nantahala National Forest (author's collection).

a

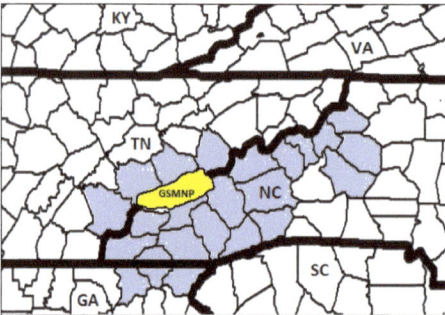

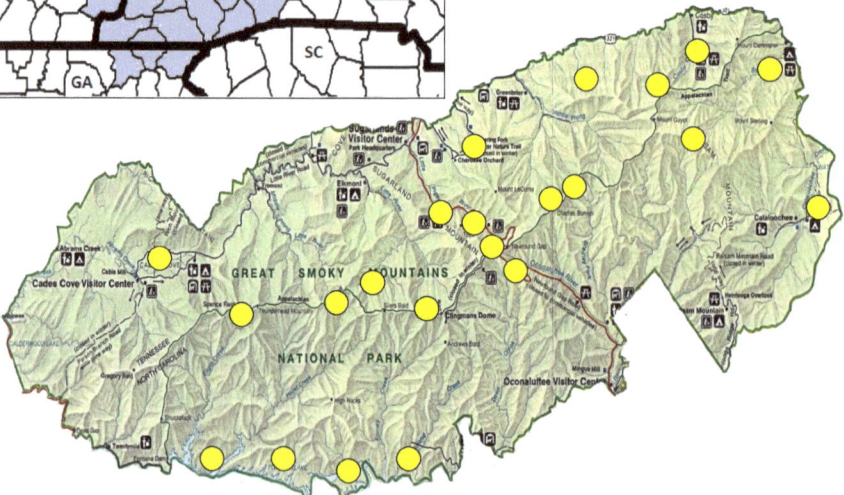

Summit covert

Polygyridae

Fumonelix clingmanica (Bland, 1860)

Diameter: 12-13.2 mm

Description: Heliciform; lip reflected; shell with 5-5.5 whorls; imperforate; typically without a parietal tooth but in some specimens a minute tooth is present; light horn-colored; transverse striae are much weakened or nearly effaced; without any spiral striae, with or without short fine hairs, these often lost in older shells; periphery well-rounded.

Similar Species: *Fumonelix wheatleyi* is larger with better developed transverse striae and has a parietal tooth.

Habitat: Found on the highest summits (above 1500 m) under leaf litter in spruce/fir and northern hardwood forests.

Status: G1; Globally Rare; a species <u>Endemic to the GSMNP</u>; this species remains a questionable taxa, not recognized in Turgeon et al. (1998).

Specimen: North Carolina, Swain County, around Clingmans Dome, GSMNP (GSMNP collection).

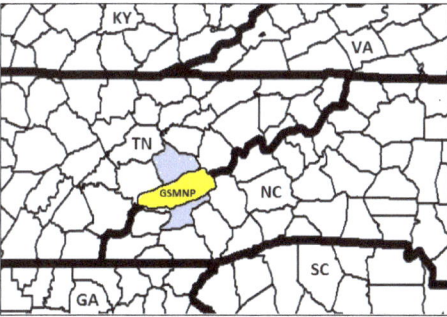

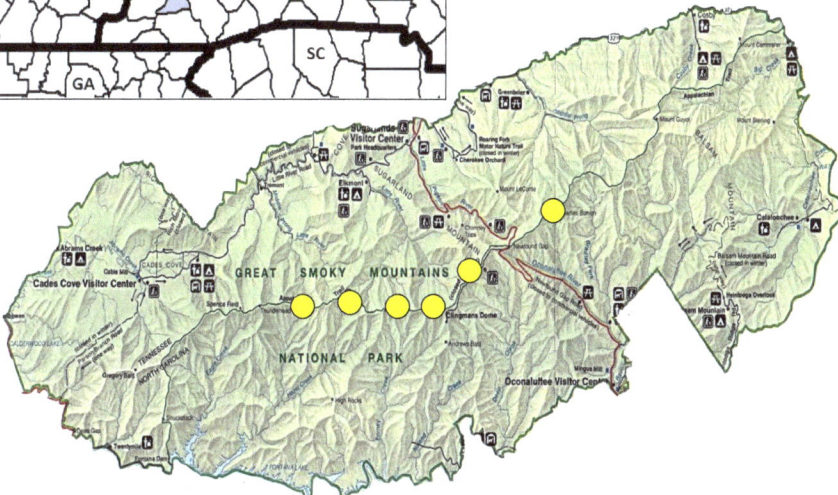

216

Glossy covert

Polygyridae

Fumonelix christyi (Bland, 1860)

Diameter: 8.3-10 mm

Description: Depressed heliciform; lip only slightly reflected (a); shell with 4.5 whorls; imperforate; large parietal tooth with a wide base; dark horn-colored; well-developed transverse striae without any notable spiral striae, with or without short fine hairs but these are often lost in older shells; periphery is well-rounded.

Similar Species: *Mesodon* and *Patera* species are much larger and have lips that are more widely reflected; other *Fumonelix* species have notably reflected lips; *Inflectarius rugeli* and *I. inflectus* have well-developed basal and palatal teeth and poorly developed transverse striae.

Habitat: Found under leaf litter in lower elevation hardwood forests.

Status: G3; Uncommon; found in the southwestern portion of the park near the north side of Fontana Lake and White Oak Sinks area.

Specimen: North Carolina, Graham County, Nantahala National Forest (author's collection).

a

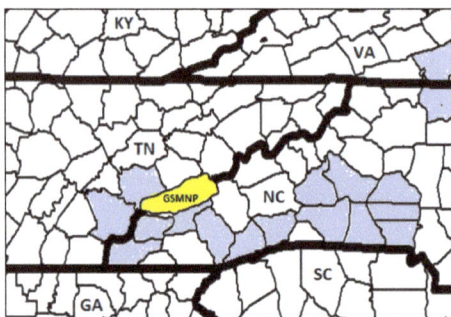

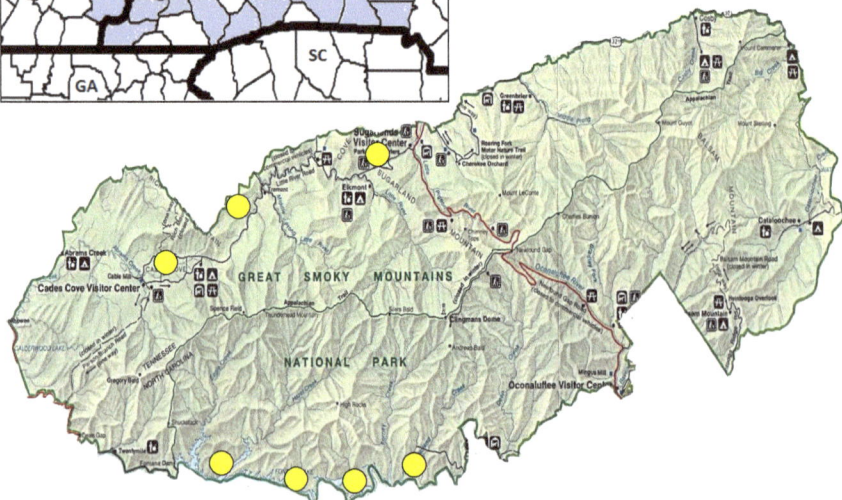

Shells large, 20-45 mm, imperforate (except for *Appalachina*), with or without a parietal tooth, reflected lip (*Mesodon, Appalachina and Neohelix*)

Mesodon, Appalachina and *Neohelix* species are among the largest and most conspicuous land snails in the park. Mostly heliciform and always having reflected lips in mature shells, several species have a small parietal tooth but none contain basal or palatal teeth. Transverse and spiral striation are a constant and well-developed feature with the umbilicus generally imperforate. Top views of these genera are not typically diagnostic for separation. <u>*Fumonelix* species are similar in shape but in general, half the size.</u> The following species are arranged according to some of these features and are as follows: **Group 1** large shells that typically contain a parietal tooth, **Group 2** large shells that are usually without a parietal tooth and **Group 3** other large shells that are similar looking to *Mesodon* species.

Grouping 1: shells typically with a parietal tooth, page 219

*Mesodon. zaletus

Grouping 2: shells usually without a parietal tooth, page 221

*Mesodon normalis
*Mesodon altivagus

Grouping 3: large shells similar to *Mesodon* species, page 223

*Appalachina chilhoweensis
*Neohelix albolabris

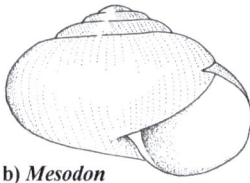

b) *Mesodon*

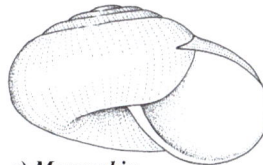

c) *Mesomphix*

Important Reminder: Immature *Mesodon* shells are often confused with immature and mature *Mesomphix* species (see above pictures b and c). Note that the immature **Mesodon** has a less oval aperture (b), being more rounded while the mature **Mesomphix** aperture is oval in shape (c) signifying that it has reached maturity. *Mesodon* species will have a reflected lip only when mature.

Toothed globe
Polygyridae

Mesodon zaletus (A Binney, 1837)

Diameter: 19-31 mm

Description: Heliciform; lip reflected; shell with 5.5-6 whorls; imperforate; small parietal tooth typically present but may be lacking in some individuals; shell solid (thick–walled); cream color to cinnamon-buff; no hairs in adults; transverse striae well-developed; spiral striae always present; shell periphery rounded.

Similar Species: *Mesodon normalis* is larger with a slightly higher profile, has a thinner shell and never contains a parietal tooth; *M. thyroidus* has a thinner shell and is rimate to perforate; *Neohelix albolabris* is generally a larger, species, having a thicker shell and is without a parietal tooth.

Habitat: Most common on rich, mesic, wooded slopes with mature to old growth forest cover.

Status: G5; Relatively common; likely more widespread than records currently indicate, it is expected to occur throughout the park at most elevations under 1500 m.

Specimen: Tennessee, Sevier County, Roaring Fork, GSMNP (GSMNP collection).

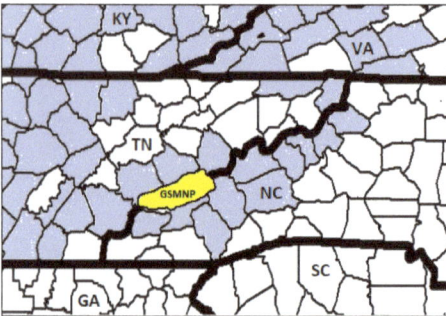

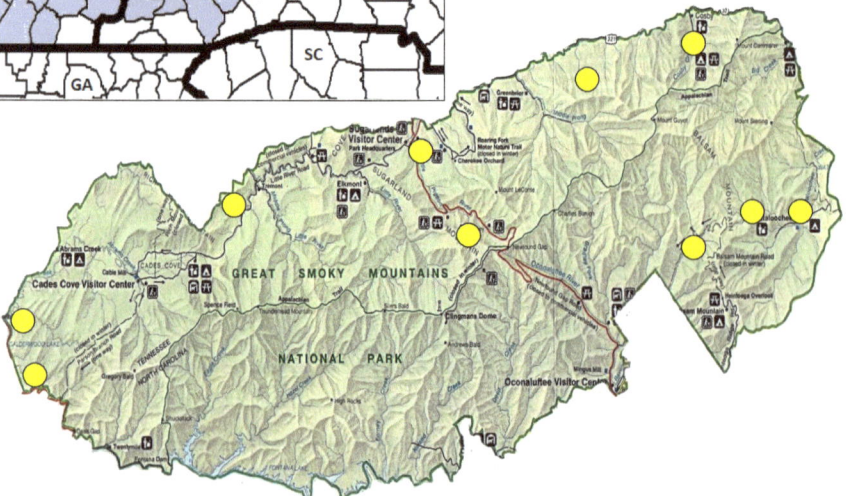

Different Genera Compared (figures are life-size and proportionate)

1 *Praticolella lawae*
2 *Triodopsis hopetonensis*
3 *Euchemotrema fraternum*
4 *Stenotrema stenotrema*
5 *Inflectarius verus*
6 *Fumonelix jonesiana*
7 *Inflectarius downieanus*
8 *Mesodon clausus*
9 *Fumonelix wheatleyi*
10 *Patera perigrapta*
11 *Mesodon thyroidus*
12 *Mesodon altivagus*
13 *Mesodon normalis*
14 *Appalachina chilhoweensis*
15 *Neohelix albolabris*
16 *Neohelix major*

Grand globe

Polygyridae

Mesodon normalis (Pilsbry, 1900)

Diameter: 21-38 mm

Description: Heliciform; lip reflected; shell with 5.5-6 whorls; imperforate; no teeth present; horn-colored to tannish-olive; no hairs in adult shells; transverse and spiral striae always present; shell periphery well rounded; shell height will vary from different populations; Despite its large size, *M. normalis* is reported to be an annual species (Hubricht, 1985).

Similar Species: *Mesodon thyroidus* has a rimate umbilicus that is slightly open; *M. zaletus* has a thicker shell and has a parietal tooth. *N. albolabris* has a thicker shell and lip and a lower shell profile.

Habitat: A species of acidic soils; found in upland habitats (up to 1400 m), usually in mixed hardwoods but also found in pine woods around logs.

Status: G5; Common, the most commonly observed large land snail in the GSMNP.

Specimen: Figure from Tennessee, Cocke County, Cosby Campground, GSMNP (GSMNP collection).

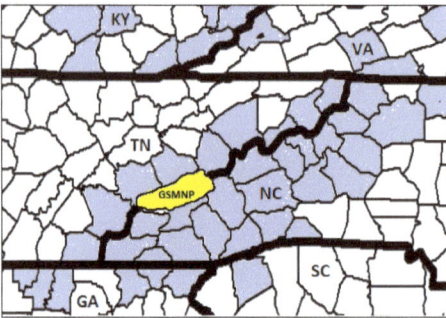

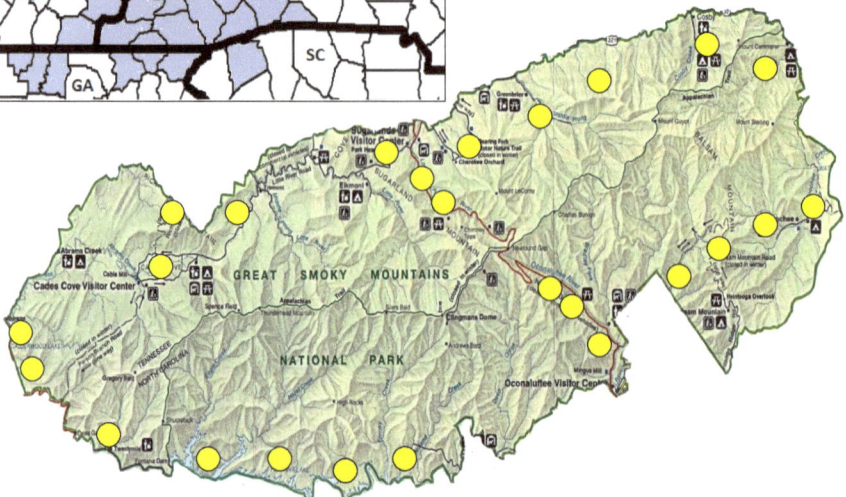

221

Wandering globe

Mesodon altivagus (Pilsbry, 1900)

Diameter: 23.4-27.5 mm

Description: Heliciform; lip reflected; shell with 5.5-6 whorls; imperforate; with or without a small parietal tooth located high on the parietal wall; color of this species varies considerably; with or without a light color band; transverse and spiral striae always present; periphery well rounded.

Similar Species: *Mesodon thyroidus* has a rimate umbilicus that is slightly open; *M. normalis* is larger, does not have a parietal tooth and is found below 1400 m; *M. zaletus* has a thicker shell and a larger parietal tooth.

Habitat: A species of high elevation (from about 1400 to 2000 m) acidic soils, northern hardwood and spruce/fir.

Status: G1; Globally Rare; a species Endemic to the GSMNP, its entire global range within the boundaries of the park; this species may be in decline (pers. comm. John Slapcinsky, 2012).

Specimen: Figure from North Carolina, Swain County, GSMNP (GSMNP collection).

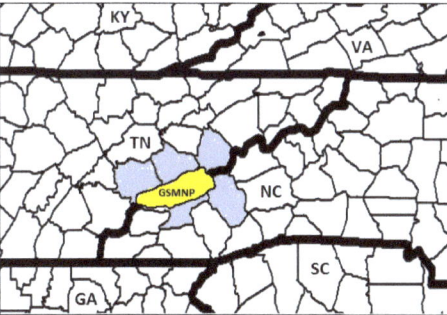

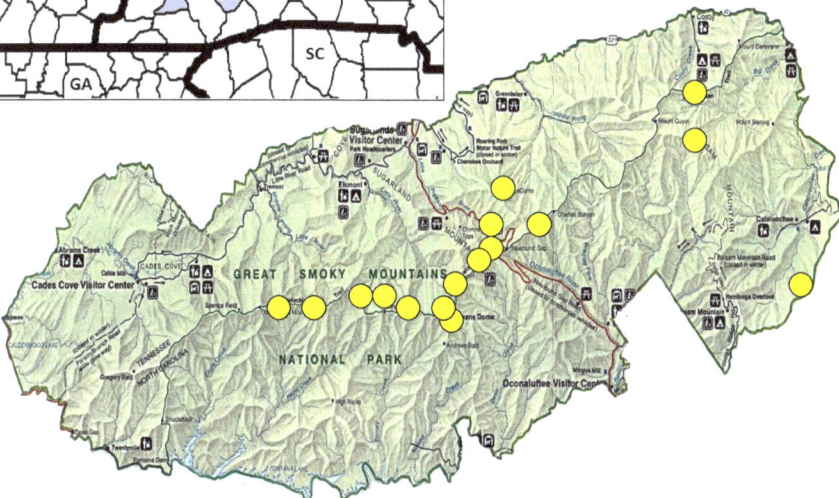

Queen crater

Polygyridae

Appalachina chilhoweensis (Lewis, 1870)

Diameter: 26-42 mm

Description: Heliciform; lip reflected but narrow; shell with 6-6.5 whorls; shell thin for its large size; umbilicate or rimate; usually without any teeth but some specimens will contain a barely detectable parietal tooth; cream-color; no hairs; transverse and minute spiral striae always present; shell periphery well-rounded; this is one of the largest terrestrial land snails found in the GSMNP.

Similar Species: *Appalachina sayana* (page 303) has a thinner wire-like lip and has both a parietal and basal tooth. *Mesodon* species are generally more globose.

Habitat: A species of upland mixed hardwood forests, found under leaf litter; also found on wooded slopes with sandstone boulders or talus and along wooded roadsides.

Status: G4; Uncommon; typically found in low numbers at mid elevation to 1500 m throughout the GSMNP.

Specimen: Tennessee, Blount County, Cades Cove (GSMNP collection).

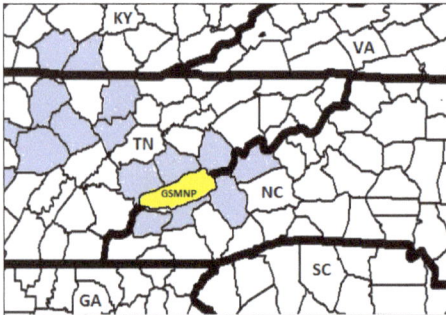

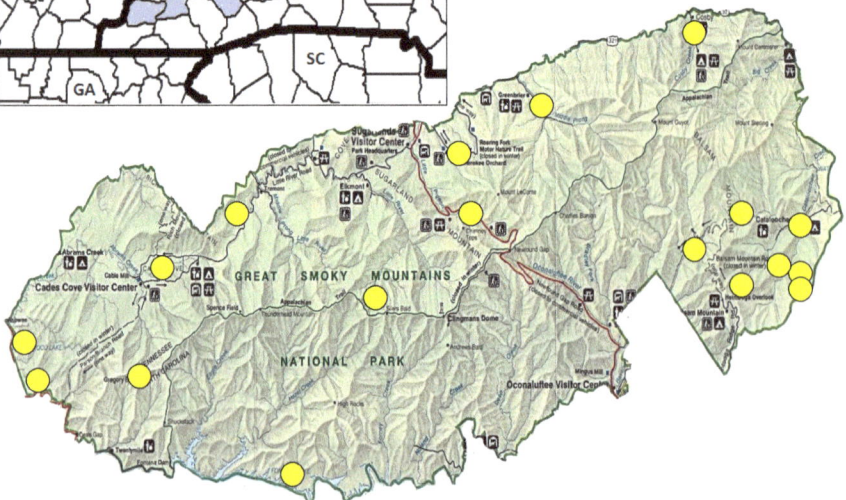

Whitelip

Polygyridae

Neohelix albolabris (Say, 1817)

Diameter: 17.6-45.3 mm

Description: Heliciform; lip reflected, thickened and wide; shell with 5-6 whorls; shell solid with a dull surface; imperforate; without teeth; cream-buff to pale-tan; transverse striae well-developed, impressed wavy spiral striae well-developed; no hairs; periphery well-rounded; one of the largest land snails in eastern north America.

Similar Species: *Mesodon normalis* has a higher shell profile and thinner shell; *M. zaletus* has a higher shell profile and usually contains a parietal tooth.

Habitat: A species found in a wide range of upland mixed hardwood sites; at the base of limestone cliffs but also found on dry acidic ridge tops, in waste places and urban areas.

Status: G5; Locally Rare; this species does not appear to be common in the park for reasons that remain unknown.

Specimen: North Carolina, Swain County, around Fontana Lake (GSMNP collection).

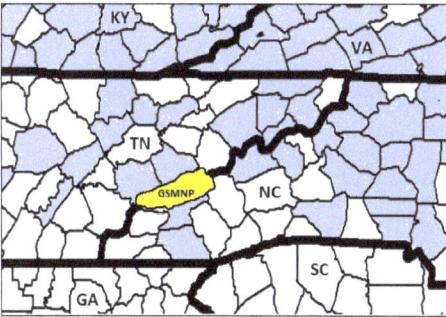

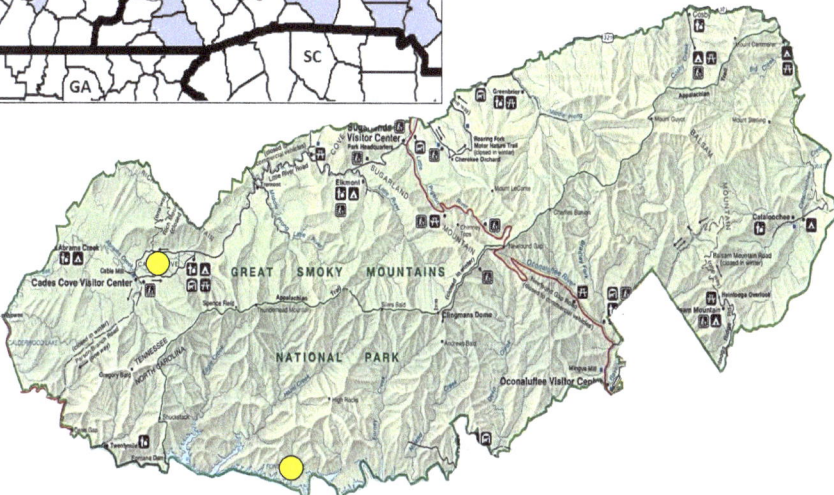

1) Shell 9-22 mm, 3 noticeable teeth, <u>umbilicate</u> (*Triodopsis*)
2) Shell 19-25 mm, 3 noticeable teeth, <u>imperforate</u> (*Xolotrema*)

Triodopsis species in the GSMNP are depressed heliciform, most species having widely reflected lips at maturity and all are umbilicate. Shells will typically contain three teeth, the parietal, basal and palatal; the parietal tooth being the largest. In some species, the basal and palatal teeth may be wanting features and very rarely, specimens are found where these teeth are missing altogether. It is important to determine if the parietal tooth points above or below the palatal tooth (figures a and b). Always use the bottom view to determine this feature. The transverse striations are usually a strong character in this group. The lip may turn downward abruptly (c opposite page) or more gently. Be careful not to confuse *Triodopsis* with *Xolotrema* species (which are imperforate, having a completely sealed umbilicus). The following *Triodopsis* and *Xolotrema* species found in the park include: **Group 1** three teeth, parietal tooth points above the palatal tooth, <u>umbilicate</u>, **Group 2** three teeth, parietal tooth points at or below the palatal tooth, <u>umbilicate</u> and **Group 3** three teeth and <u>imperforate</u>.

<u>Grouping 1: umbilicate, parietal tooth points above palatal tooth (a), page 226</u>

T. vulgata
T. hopetonensis
T. fallax

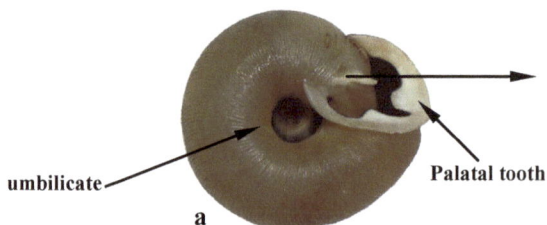

umbilicate

Palatal tooth

a

<u>Grouping 2: umbilicate, parietal tooth points at or below palatal tooth (b), page 229</u>

T. tridentata

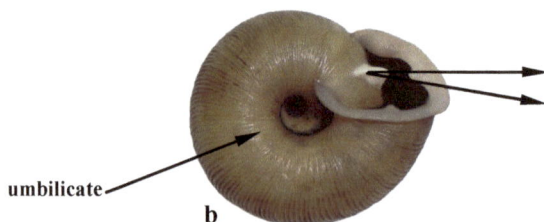

umbilicate

b

<u>Grouping 3: imperforate, three teeth as in *Triodopsis* species, page 230</u>

Xolotrema denotatum

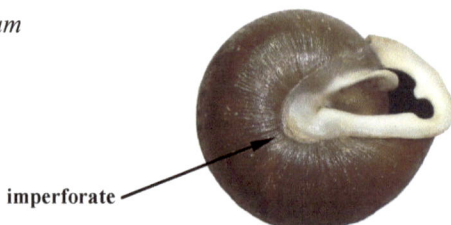

imperforate

Dished threetooth

Polygyridae

Triodopsis vulgata Pilsbry, 1940

Diameter: 13.5-19.5 mm

Description: Depressed heliciform; lip reflected and abruptly turning downward (c); shell with 5-6 whorls; umbilicate; parietal tooth large, points well above the palatal tooth; basal tooth small; palatal tooth larger and with a wider base, set deeper into the aperture than that of other *Triodopsis* species; the three teeth crowd the aperture; cream-buff to cinnamon buff; transverse striae are well-developed on the top, sides and bottom of the shell; scattered papillae are present but generally a weak feature, strongest around the umbilicus.

Similar Species: The lip of other *Triodopsis* species lack the steep downward curve as seen in side view (c).

Habitat: Usually found on soil and rubble among limestone in mixed hardwood forests.

Status: G5; Locally Rare; found in southern and western portions of the park.

Specimen: Tennessee, Blount County, sink hole at Bull Cave, GSMNP (GSMNP collection).

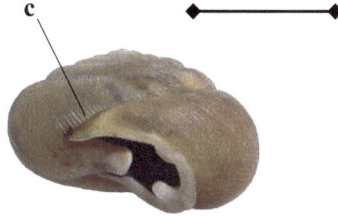

c

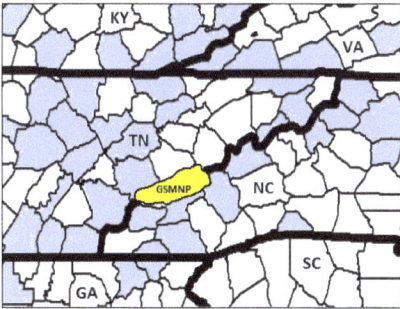

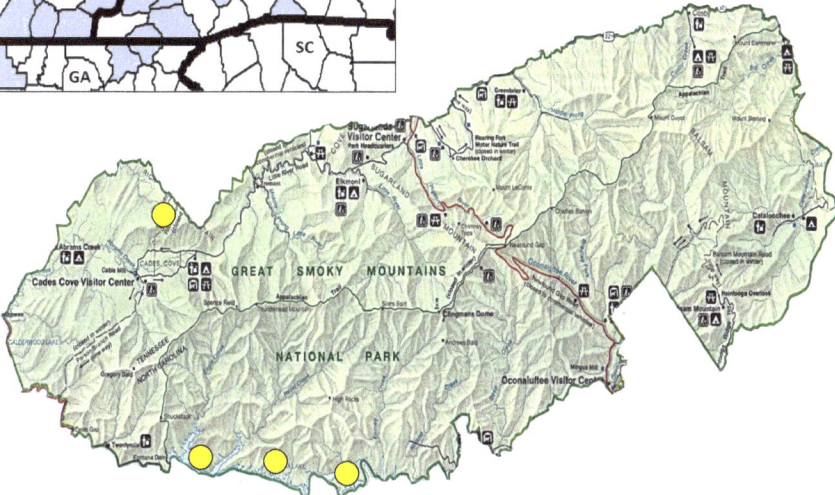

Magnolia threetooth

Polygyridae

Triodopsis hopetonensis (Shuttleworth, 1852)

Diameter: 9.2-13 mm

Description: Depressed heliciform, shell more compact than other *Triodopsis* species; lip reflected; shell with 4.5-5.5 whorls, the last whorl not expanded much; umbilicate; parietal tooth small, points well above the palatal tooth; palatal tooth not situated on a buttress and is recessed in the aperture; olive-horn; transverse striae are well-developed on the top and sides of the shell continuing well into the umbilicus region; adult shells without hairs; periphery rounded.

Similar Species: *Triodopsis tridentata* is larger, its parietal tooth points at or below the palatal tooth.

Habitat: A species of waste places, roadsides, and scrap piles of treated lumber (the main transportation of this species to new locations); natural habitats include low wet woodlands.

Status: G5; Locally Rare; known only from Purchase Knob.

Specimen: Kentucky, Martin County, a parking lot in Inez (author's collection).

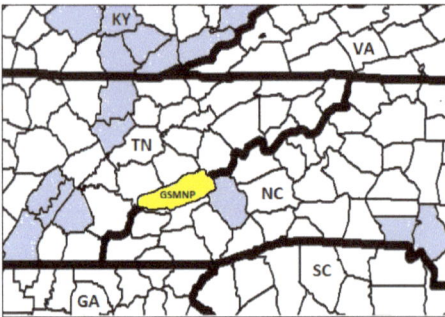

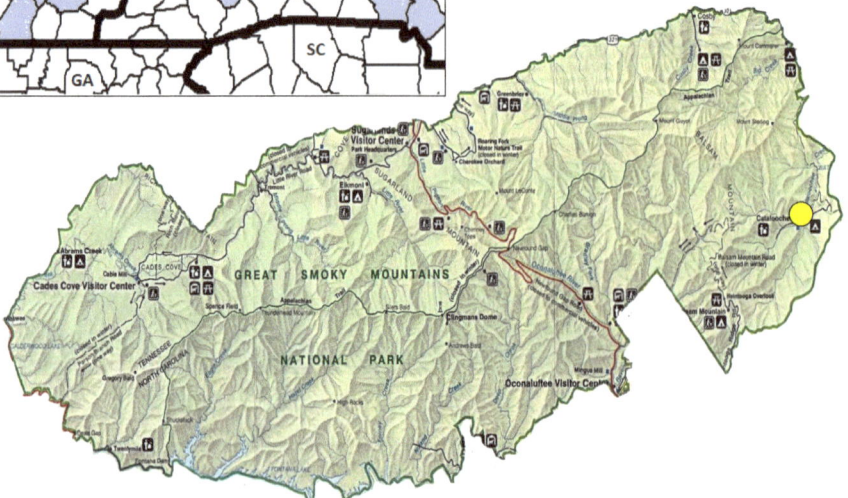

Mimic threetooth

Polygyridae

Triodopsis fallax (Say, 1825)

Diameter: 9.9-13.5 mm

Description: Depressed heliciform to heliciform, shell more compact than other *Triodopsis* species; lip reflected; shell with 5.5 whorls, the last whorl not expanded greatly; umbilicate, in frontal view the large, curving parietal tooth crowds the aperture and, in bottom view, points well above the palatal tooth; basal tooth sits on a buttress; palatal tooth deeply recessed in the aperture; olive-buff; transverse striae are not well developed; periphery rounded.

Similar Species: *Triodopsis tridentata* is larger, its parietal tooth points at or below the palatal tooth.

Habitat: A species of mixed hardwood forests under leaf litter but also found along roads and other disturbed places.

Status: G5; Locally Rare; there are old records of this species in the park, but exact locations remain unknown; area of most likely occurrence (a).

Specimen: North Carolina, Cumberland County, Fort Bragg (author's collection).

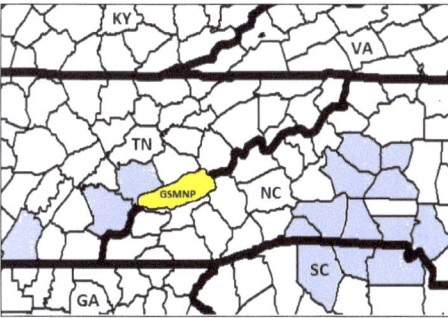

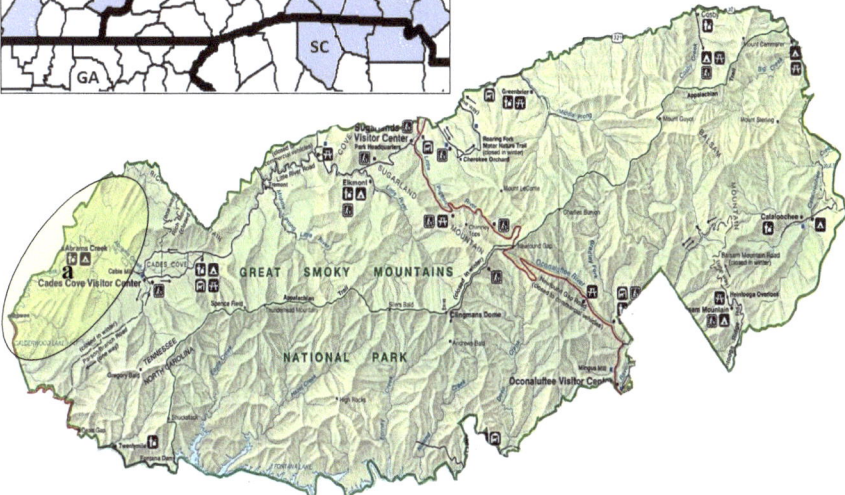

Northern threetooth

Triodopsis tridentata (Say, 1816)

Diameter: 12-22 mm

Description: Depressed heliciform; lip reflected, not abruptly turning downward; shell with 5-6 whorls; umbilicate; parietal tooth points at or below the palatal tooth; basal and palatal teeth small and not crowding the aperture in frontal view; light-cream to pale cinnamon-buff; transverse striae are well-developed on the top, sides and bottom of the shell; like most *Triodopsis* species there are scattered papilla on top and on the base; without hairs in adult shells; periphery rounded.

Similar Species: *Triodopsis vulgatus* has larger more crowded teeth and a aperture that turns abruptly downward.

Habitat: Found in rich upland woods under forest litter, rocks and logs, but also a snail of roadsides and urban areas.

Status: G5; Common; the most common *Triodopsis* in the GSMNP; specimens from Cades Cove can reach 22 mm in size.

Specimen: Tennessee, Cocke County, Cosby Campground, GSMNP (GSMNP collection).

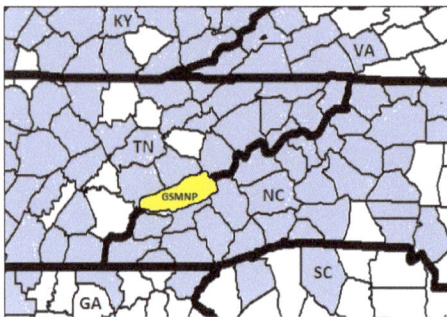

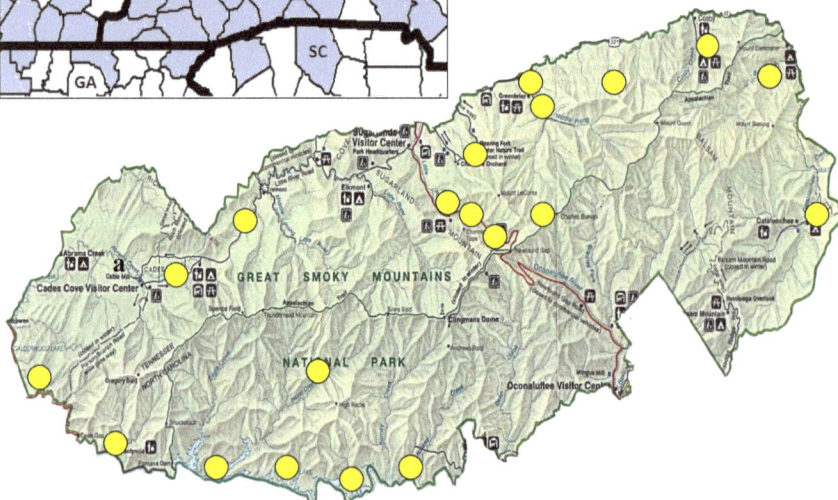

Velvet wedge   Polygyridae

Xolotrema denotatum (Fèrussac, 1821)

Diameter: 19-25.6 mm

Description: Depressed heliciform; lip reflected; shell with 5.5 whorls; imperforate; parietal tooth large, points well above the palatal tooth; basal tooth small and may be absent or so poorly defined that it is not detectable in some specimens; palatal tooth of medium build; tawny olive to snuff brown; transverse striae weakly developed, shell surface with close set periostracal processes or thickened-hairs (a) on live and fresh shells; periphery rounded in most specimens, but in some populations shell peripheries may be weakly angular.

Similar Species: Although *Triodopsis* species also contain three teeth, they differ by having a open not closed umbilicus.

Habitat: Often associated with rotting hardwood trees in advanced stages of decay in a variety of upland mixed hardwood sites.

Status: G5; Uncommon in scattered location across the GSMNP.

Specimen: Tennessee, Sevier County, Roaring Fork, GSMNP (GSMNP collection).

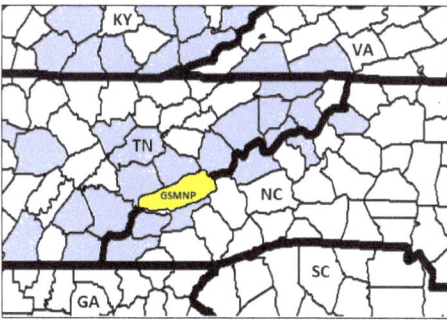

a

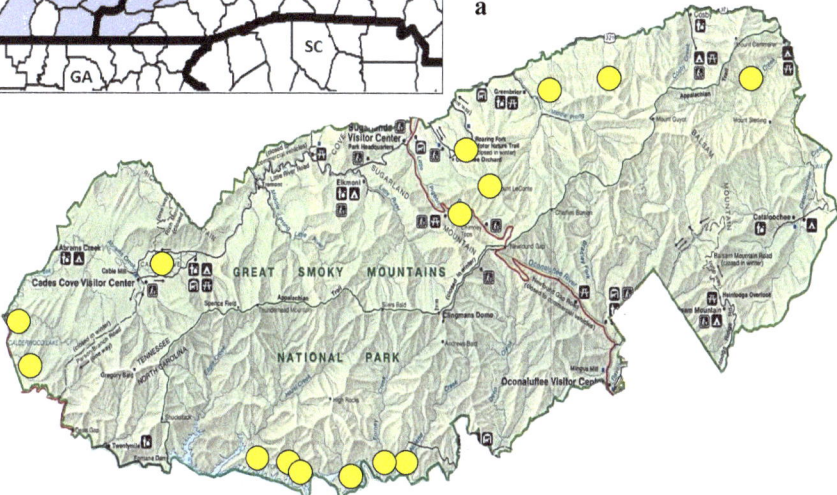

230

The Junkyard Bug

The notorious "Junkyard Bug" was first discovered in the Great Smoky Mountains National Park in 2006. It is a larva of a lacewing, in the family Chrysopidae. What was amazing about this particular larva was its luggage. Six species of land snails: *Punctum vitreum* (new record for North Carolina), *Punctum minutissimum, Punctum blandianum, Carychium clappi, Gastrocopta contracta,* and *Gastrocopta pentodon* were attached to the larva's back (figure a). Remarkably one species, *G. pentodon,* was still alive but unable to detach itself from the back of the insect larva. The bug is reported to use forest litter, insect parts and now land snails to camouflage itself from potential predators. What the lacewing doesn't known is that many salamanders like to eat snails. Figure a showing four of the six land snails (two remain hidden from view). Figure b is the naked lacewing larva showing the hairs that hold the junk in place. Other snail researchers have also reported finding lacewing larva with attached land snails.

a

b

I normally charge admission for this sort of thing, Bub!

In general, native slugs are the least studied land snails in many regions of the world and the GSMNP is no exception. These under appreciated gastropods are often all categorized as pests, yet nothing could be farther from the truth. Native slugs are rarely a nuisance and typically become scarce or disappear altogether when the native vegetation has been eliminated. Even in ecosystems that are more or less pristine, native slugs can be hard to find. In contrast, introduced slugs are quick to overpopulate an area, particularly in degraded habitat, and without natural controls in place. These inadvertent guests find themselves at home in places like Cades Cove around old settlement structures which are also occupied by non-native vegetation such as Kentucky fescue.

Considered excellent indicators of a healthy forest, native slugs by and large are only common where there is significant decomposing structure (logs), a habitat also required by other groups of organisms such as small mammals, salamanders and micro-invertebrates. All slugs, native or otherwise, can vary considerably in color and mantle patterns even within the same population. Many gastropods in the family *Philomycidae* can only be reliably separated by dissection or DNA sequencing (Fairbanks pers. comm. 2010).

Genera Included:
(in order of appearance in text)

Philomycus
Megapallifera
Pallifera
Deroceras

Shells? I'm much too cool for all that baggage.

Key to the *Philomycidae* Family

Philomycidae are native terrestrial slugs found primarily in the eastern half of the United States and there remain a number of undescribed species (Fairbanks 1998). Many of the more conspicuous *Philomycidae* are large, reaching crawling lengths of more than 100 mm. There are three North American genera in this family (see below): *Philomycus, Megapallifera,* and *Pallifera. Philomycus* species are large (75-100 mm), obese slugs. In terms of their reproductive anatomy, the species of *Philomycus* are the most interesting, characterized by the presence of a dart sac and dart (referred to as love-darts). Some *Philomycus* species, in isolation, become self-fertilizing (Nicklas and Hoffman 1981). In *Philomycus* species the mantle covers the entire body, including the head (a). *Megapallifera* species are smaller in size (approximately 80 mm long), are also obese but lack the dart sac and dart (Fairbanks 1998) and the head is not entirely covered (b). *Pallifera* species are the smallest among the group (approximately 30 mm long), are slender and, like *Megapallifera,* lack the dart sac and dart. The head remains uncovered by the mantle. Great color variation exists in Philomycidae slugs. Young slugs of all species will generally have the

same mantle patterns as the adults. One last detail for separating slugs includes their defensive slime. To view this feature, first gently irritate the animal (in one spot) with a small stick until a small mass of slime accumulates (d). In natural lighting, determine the color of the slime. This is an important diagnostic mark and varies by species. The crawling slime of slugs is typically clear.

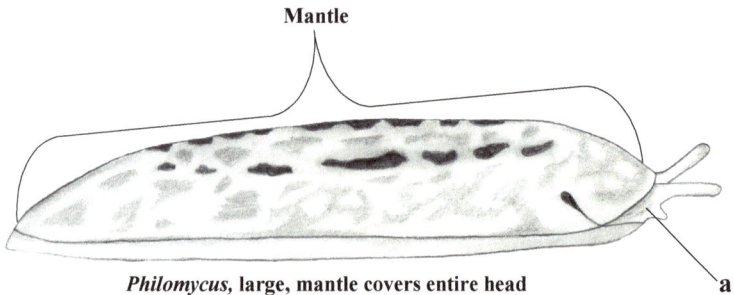

Philomycus, large, mantle covers entire head

Megapallifera, medium, mantle does not cover entire head

Pallifera, small, mantle does not cover head area

Native Slugs of the GSMNP Compared (Proportionate)

80 mm

Philomycus carolinianus, KY

Philomycus togatus, NC

Philomycus venustus, NC

Megapallifera mutabilis, KY

Pallifera dorsalis, KY

Pallifera hemphilli, NC

Pallifera fosteri, NC

Pallifera secreta, NC

Philomycus flexuolaris, KY

Deroceras laeve

Carolina mantleslug   Philomycidae
Philomycus carolinianus (Bosc, 1802)

Length: Adults 50-100 mm while crawling

Description: Mantle uniformly mottled with brown; usually two dorsal longitudinal rows of black spots (the best diagnostic feature for the species); contains a dart sac and dart; defense mucus whitish (d); when touched with a dry finger, the mucus immediately seeps out of the entire mantle having a bitter taste much like the skin secretions of a spring peeper (MacGregor pers. comm. 2010); sole of foot white (b); SEM of love dart (c) by Jodi White–McLean.

Similar Species: *P. flexuolaris* is smaller, without the dark rows of black spots, the two species are often found together.

Habitat: Found in floodplain forests over much of its range, it becomes an upland species in the mountains, but does not occur much over 600 m; found hiding under exfoliating bark of large rotting hardwood logs in advanced stages of decay, especially logs forming log bridges over small ravines, also found inside hollow trees like beech.

Status: G5; Uncommon but likely occurs throughout the park.

Specimen: Figures (a & b) from Powell County, Red River Gorge, Kentucky; figures (d) stick showing defense mucus color and both slugs in retracted, defensives posture, Nolichucky River Gorge, Tennessee.

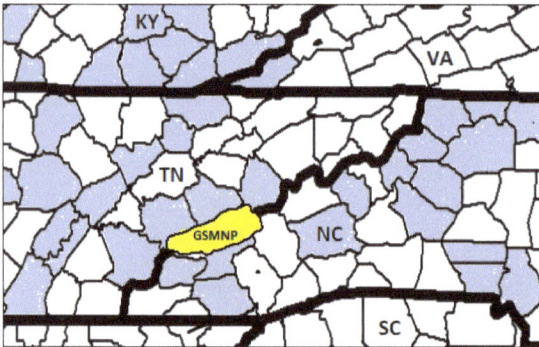

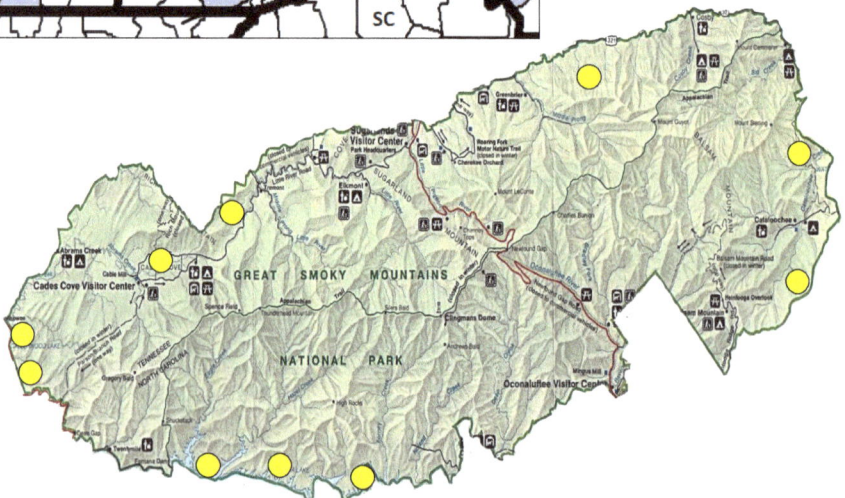

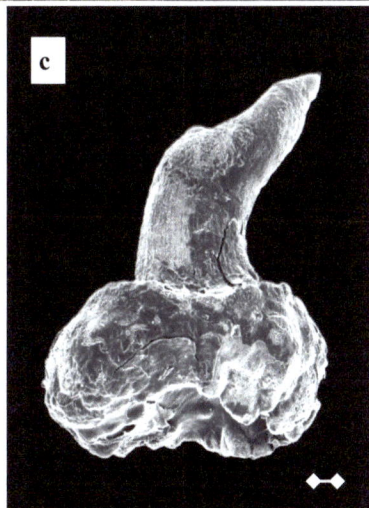

Winding mantleslug   Philomycidae

Philomycus flexuolaris Rafinesque, 1820

Length: Adults 50-80 mm while crawling

Description: Three longitudinal mottled bands (dorsal and lateral), are usually distinct, but sometimes markings are so pale as to be scarcely discernible and in some specimens they may be nearly lost in a general dusky shade (figure a, specimen on the right); although the base color of the mantle varies considerably (as can be seen in the images on opposite page) the darker color patterns remain more or less the same; this species possesses a dart sac and dart; defense mucus pale-yellow (b); foot margin pale, sometimes with a hint of yellow; sole of foot cream to flesh color (c).

Similar Species: *Philomycus carolinianus* is generally larger, having two longitudinal rows of black or brown spots running down the center of mantle or back; *P. carolinianus* and *P. flexuolaris* can often be found living on the same large beech log.

Habitat: A species of upland mixed hardwood forests up to 1500 m; as with most native slugs during the day it goes into hiding under the exfoliating bark of large rotting logs in advanced stages of decay, especially large beech trees.

Status: G5; Relatively common in southern portions of the park only.

Specimen: Figures (a, b and c)from Powell County, Red River Gorge, Kentucky and figure (d) from Pike County, Kentucky.

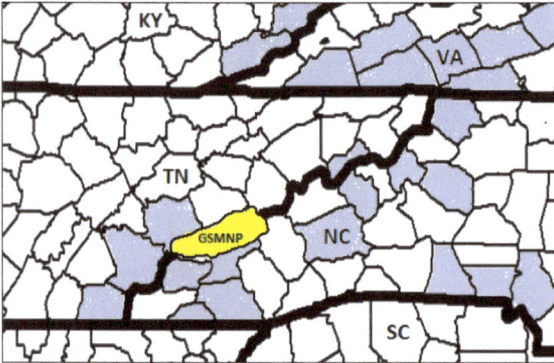

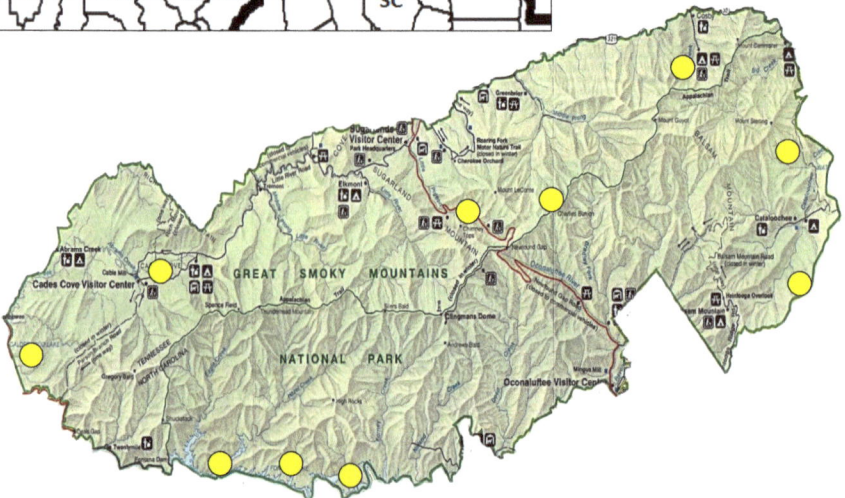

Two *P. flexuolaris* interested in proliferation.

Variable mantleslug   Philomycidae

Philomycus togatus (Gould, 1841)

Length: Adults 60-100 mm while crawling

Description: Top of animal heavily mottled , consisting of a broad dorsal band (sometimes broken into two bands) and narrower lateral bands (bands in some specimens may be hard to detect at times) on each side and scattered, irregular, small spots between the bands (Hubricht 1951); as can be seen on these two pages there is considerable variation in the species; defense mucus orange (b); foot margin orange-red to orange (c), probably one of the best external features of the species.

Similar Species: Differs from typical *P. carolinianus* in reaching a somewhat larger size, in being darker brown without the two row of black spots (Hubricht 1951) and having orange not whitish defense mucus.

Habitat: An upland (to 1200 m) species found on wooded hillsides and in ravines, under loose bark of logs; in wet weather found on trunks of smooth-barked trees at night.

Status: G5; Locally Rare; only reported from southern portions of the park.

Specimen: Figures (a & b) are from Dickenson County, Breaks Interstate Park, Virginia (author) and figure (d) from Surry County, North Carolina (Wayne Van Devender).

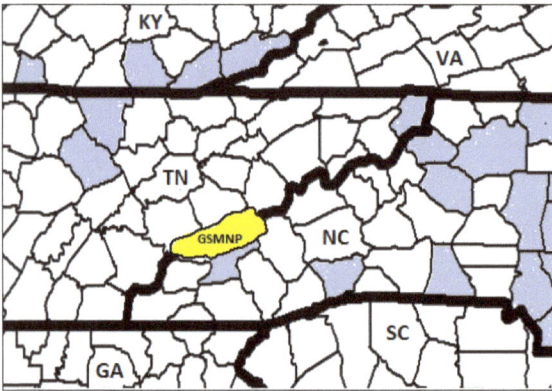

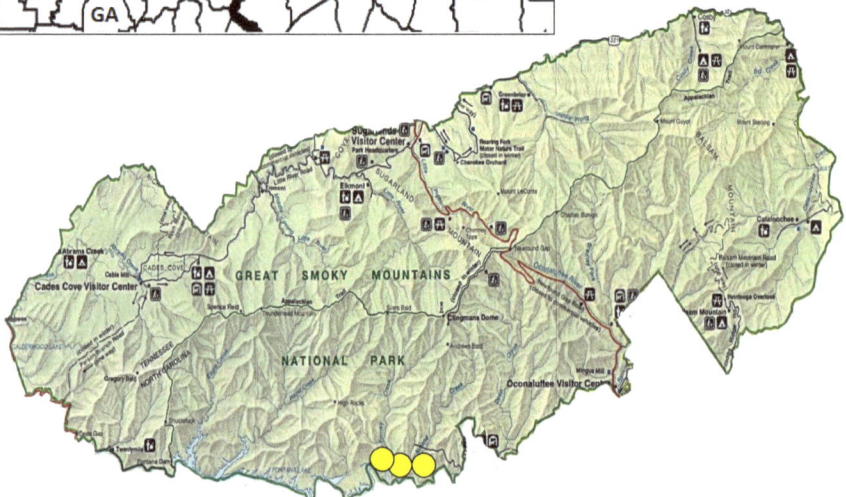

Size of slug crawling

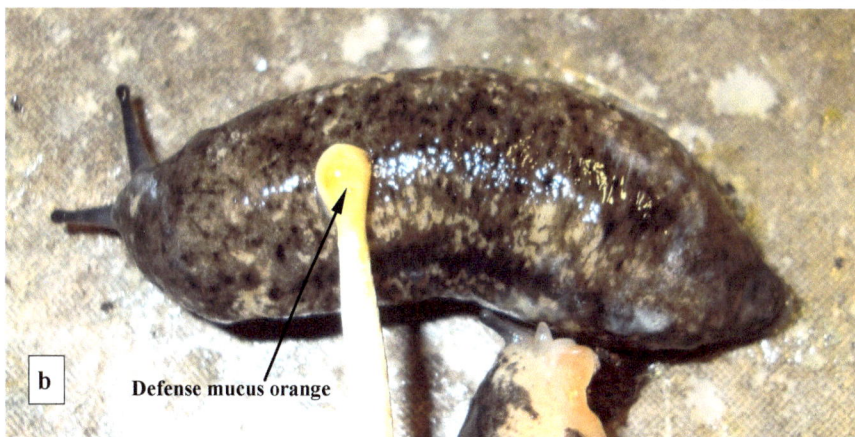

Defense mucus orange

Juvenile

Brown-spotted mantleslug   Philomycidae

Philomycus venustus Hubricht, 1953

Length: Adults 75-120 mm while crawling

Description: The color pattern varies from individuals having two contrasting dark chestnut brown to black dorsal bands, a narrow lateral band on each side, connected to the dorsal bands by a series of oblique bands or spots (a and c), forming chevrons; to individuals in which this pattern is broken up into to a series of spots (d); defense mucus white; sole of foot creamy white (b).

Similar Species: *Philomycus virginicus* (illustrated in Part 2)is darker brown in color and has lighter longitudinal and oblique bands; *P. carolinianus* does not have the oblique bands or spots (forming chevrons) seen in *P. venustus*.

Habitat: An upland species (up to 1800 m on Roan Mountain in North Carolina) found on wooded hillsides and mountains; during the day it can be found hiding under the exfoliating bark of large rotting logs in advanced stages of decay; also found in crevices and cavities of beech trees.

Status: G4; Uncommon, the true distribution of most slugs remains ambiguous, a result of to few land snail surveys that collect these interesting slugs.

Specimen: Figure (a) from Forney Creek, GSMNP, North Carolina; figures (b and c) from Pike County, Kentucky and figure (d) from Mitchell County, Roan Mountain, North Carolina.

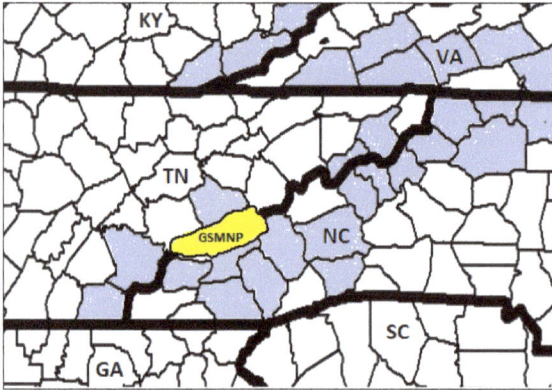

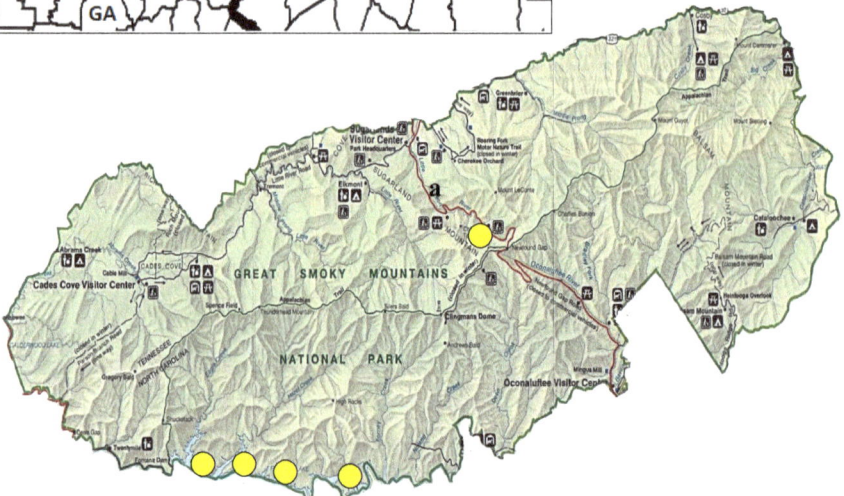

241

Changeable mantleslug   Philomycidae
Megapallifera mutabilis (Hubricht, 1951)

Length: Adults 60 mm while crawling

Description: Mantle color is fawn or tan, not covering entire head; but this feature may be difficult to discern, requiring an observer to rely on the use of a hand lens of at least 10X power; mantle usually has two (on occasions three) interrupted dorsal dark gray-brown longitudinal pigment bands (more often spots) and a continuous wavy lateral band on each side (figures a and c), small scattered pigment spots between the bands; spots sometimes producing vague chevrons on the mantle; defense mucus tannish-white or white figure (b); margins of the foot olive, gray or sometimes whitish; sole of foot cream (a).

Similar Species: *Philomycus* species are slightly larger in length and notably more robust when crawling and the mantle covers the entire body.

Habitat: Found in upland woods; during warm wet weather at night, it can be found crawling on the trunks of smooth bark trees like the American beech;

Status: G5; Locally Rare; Hubricht (1956) documented this species at the Sugarlands Visitor Center.

Specimen: Figure (a) from North Carolina, Halifax County, Medoc Mountain SP (Wayne Van Devender), figure (b) from Rowan County, Kentucky and figure (c) from West Virginia, Monongalia County, Coopers Rock State Forest.

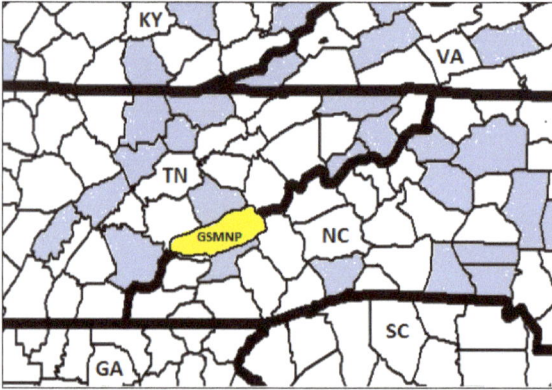

Size of slug crawling

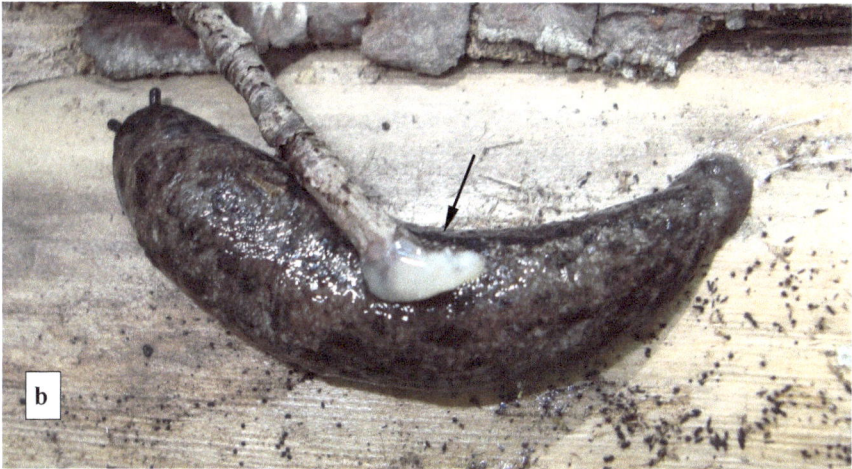

Severed mantleslug   Philomycidae

Pallifera secreta (Cockerell, 1900)

Length: Adults 20-30 mm while crawling

Description: Mantle very dark gray or gray-blue colored with numerous and well scattered small, round, black spots, often more profuse at the mantle edges near the foot margins and on the anterior end of the mantle; sole whitish with an ochreous (yellowish) tint (Cockerell, 1900); defense mucus whitish.

Similar Species: *Pallifera dorsalis* is smaller; with or without an interrupted black line down the center of the mantle; *P. fosteri* is smaller, is flesh colored, has interrupted clusters of spots down the center (dorsal) of the mantle and the anterior margins of its foot is brownish red.

Habitat: This slug lives deep down in drifts of damp leaves that are located next to logs and found in small depressions in upland mixed hardwood forests; up to 1500 m in the southern Appalachians.

Status: G4; Uncommon, but likely more common than records currently indicate, but like other species of *Pallifera,* these slugs are secretive and remain well hidden during daytimes hours.

Specimen: Figures (a and b) from North Carolina, Transylvania County, Rosman (John Slapcinsky) and figure (c) from North Carolina, Mitchell County around Carvers Gap, Roan Mountain (author).

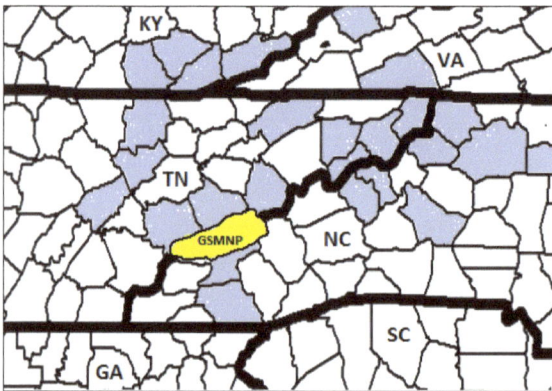

Size of slug crawling

Pale mantleslug   Philomycidae

Pallifera dorsalis (A. Binney, 1842)

Length: Adults 20-30 mm while crawling

Description: The live animal is ashy-blue, gray or brownish with or without (more often without) an interrupted black line down the center of the mantle; defense mucus translucent amber to whitish; margins and sole of the foot are white (c).

Similar Species: *Pallifera fosteri* larger with distinct black spots, which are more prominent at the mantle edges, its mantle slightly humped in front making the neck and head longer than seen in *P. dorsalis*; *P. ohioensis* (fully illustrated in Part 2) has distinct red edges running along the sole of its foot (d).

Habitat: A snail of humid forests, generally found in deep, moist leaf litter in and around rock talus or large piles of rotting wood.

Status: G5; Uncommon, but the species is likely more widespread in the GSMNP and southern mountains than current records indicate.

Specimen: Figure (a, b & c) from Buncombe County, North Carolina, slug shown in a typical defensive posture (b); figure (d) comparing *P. dorsalis,* from Powell Co, Kentucky and *P. ohioensis* from Blackwater Falls State Park, West Virginia.

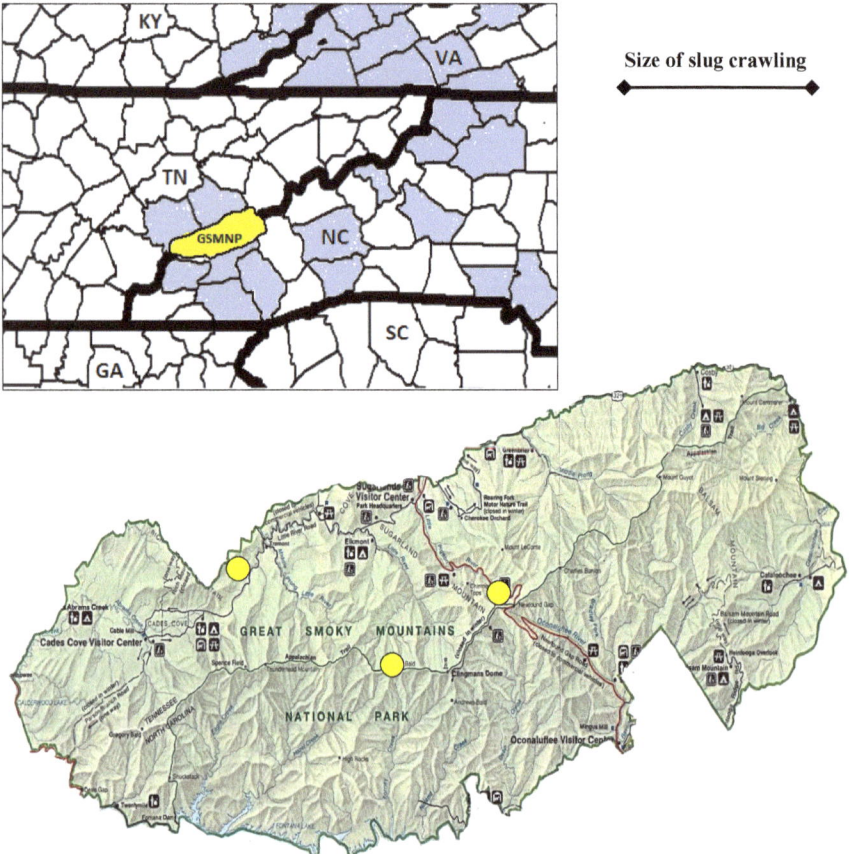

Size of slug crawling

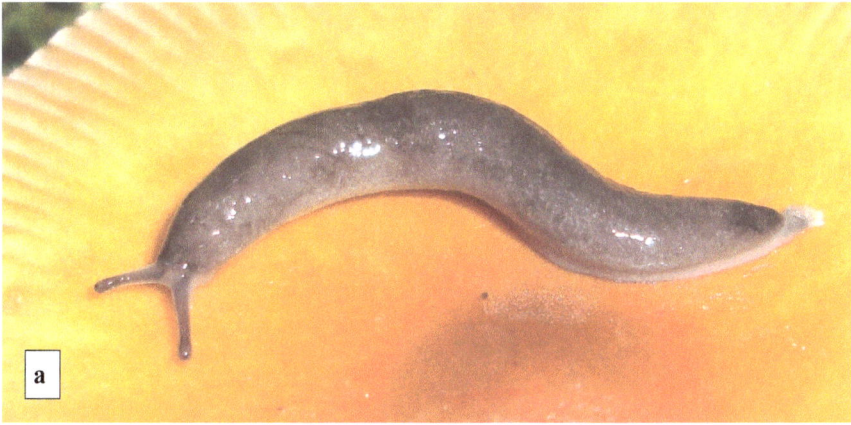

a

Defensive posture

b

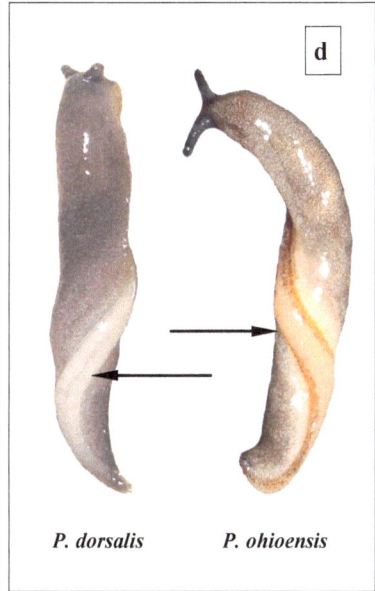

d

P. dorsalis   *P. ohioensis*

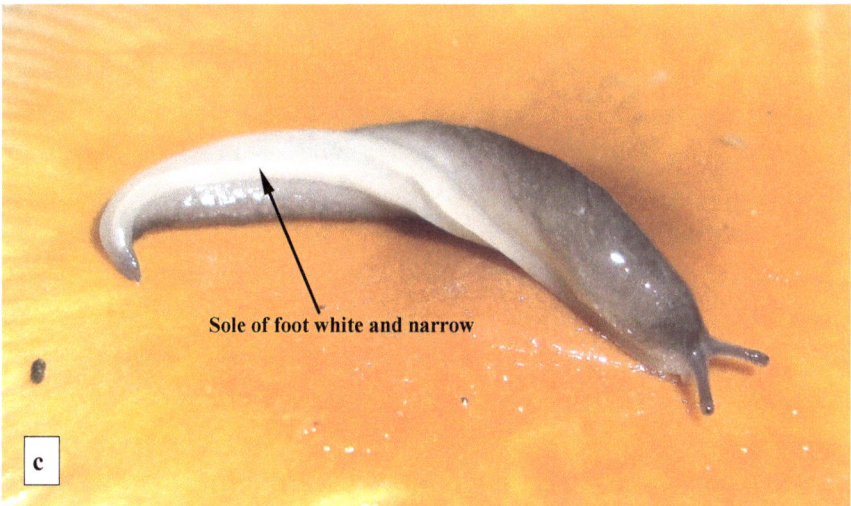

Sole of foot white and narrow

c

Black mantleslug   Philomycidae
Pallifera hemphilli (W.G. Binney, 1885)
Length: Adults 30-50 mm while crawling
Description: The live animal is uniformly black on top, figure (a) with a lighter bottom figure (b); defense mucus grayish, figure (c); margins and sole of the foot slightly lighter than the top of the slug, figure (b) showing both the slugs bottom and top sides.
Similar Species: *Pallifera fosteri* is smaller with distinct black spots, *P. dorsalis* is smaller and lighter in color; *P. ohioensis* (illustrated in Part 2) has red margins along the foots edge.
Habitat: A snail of high elevation, wet spruce/fir forests, during the day the species is found hiding under exfoliating bark of dead trees and rotting logs in advanced stages of decay; also a species found climbing trees; this species has been observed resting in sunny spots on chilly morning, perhaps its dark color being used to thermal–regulate its body temperature for food digestion or some other metabolic function (pers. obs.); most active at night especially during wet weather when temperatures are above 40 degrees Fahrenheit.
Status: G3; Uncommon to Rare.
Specimen: All figures from Roan Mountain at High Knob, Mitchell County, North Carolina.

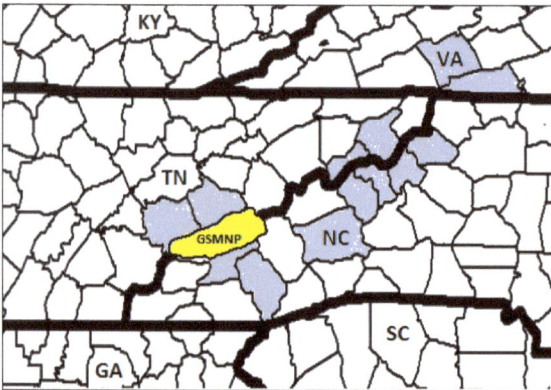

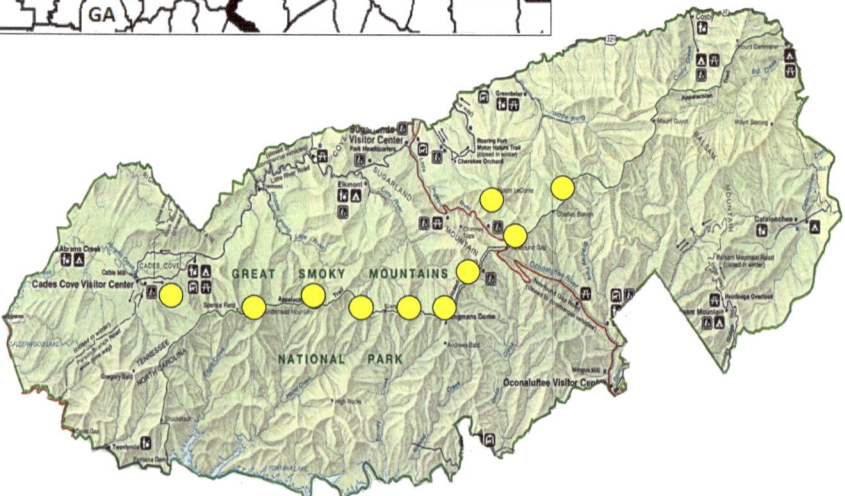

Foster mantleslug   Philomycidae

Pallifera fosteri F. C. Baker, 1939

Length: Adults 15-25 mm while crawling

Description: The species may be recognized by the blackish spots on a whitish or flesh-colored mantle; these spots or blotches may form interrupted, irregular longitudinal lines, especially near the base of the mantle, or irregularly spaced clusters of small dot-like spots scattered over the dorsal surface; in some specimens the black spots form coalescing blotches elongated in form (Pilsbry 1948); Grimm (1961) reported that the mantle is very light tan, spotted and reticulate with dark brownish gray, the reticulations heaviest in the middle of the back (figure d) but no dorsal line is formed; at the sides, the reticulations omay form two broken lines; tentacles slate gray and anterior margins of the foot is brownish red (figure b, white arrows); other characters include a slightly humped mantle at the anterior with a neck and head slightly longer than other *Pallifera* species; defense mucus usually transparent-amber.

Similar Species: *Pallifera dorsalis* is smaller, has a different color and is without the spots on the mantle edges.

Habitat: A species found in a variety of habitats from floodplains to mountain tops under forest litter and around rotting logs in advanced stages of decay.

Status: G5; Uncommon; although only one site is reported for this species in the GSMNP, it is likely more common than records currently indicate.

Specimen: Figures (a, b and c) from North Carolina, Swain County, Nantahala Gorge (all Wayne Van Devender).

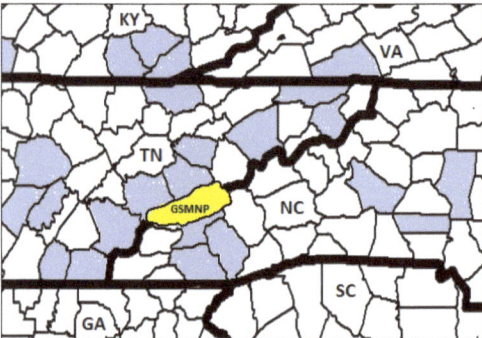

Juvenile *P. fosteri*, Mitchell County, North Carolina

Meadow slug   Agriolimacidae

Deroceras laeve (Müller, 1774)

Length: Adults 25 mm while crawling

Description: Color of animal varies from yellowish to pale gray, dark gray to nearly black, sometimes flecked with gray; the mantle (b) is situated on the anterior (front) end of the dorsal (back); pneumostome (breathing hole) behind middle of mantle (as opposed to *Arion* species); defense mucus colorless and watery; margins of the foot tan; this slug is a favorite food of the brown snake, *Storeria dekayi* (MacGregor pers. comm. 2010).

Similar Species: Exotic *Arion* species are similar (refer to 310); native slugs of the eastern US have mantles that cover or mostly cover the entire animal back.

Habitat: A Holarctic species often found under cardboard and rubbish along roadsides and in vacant lots, also found in meadows and natural glades; even though *D. laeve* is considered native to eastern North America, it has become a garden pest.

Status: G5; Locally Rare; reported from the west prong of Chimney Top trailhead.

Specimen: Figures (a, c and d) from Madison County, Kentucky (all Wayne Van Devender).

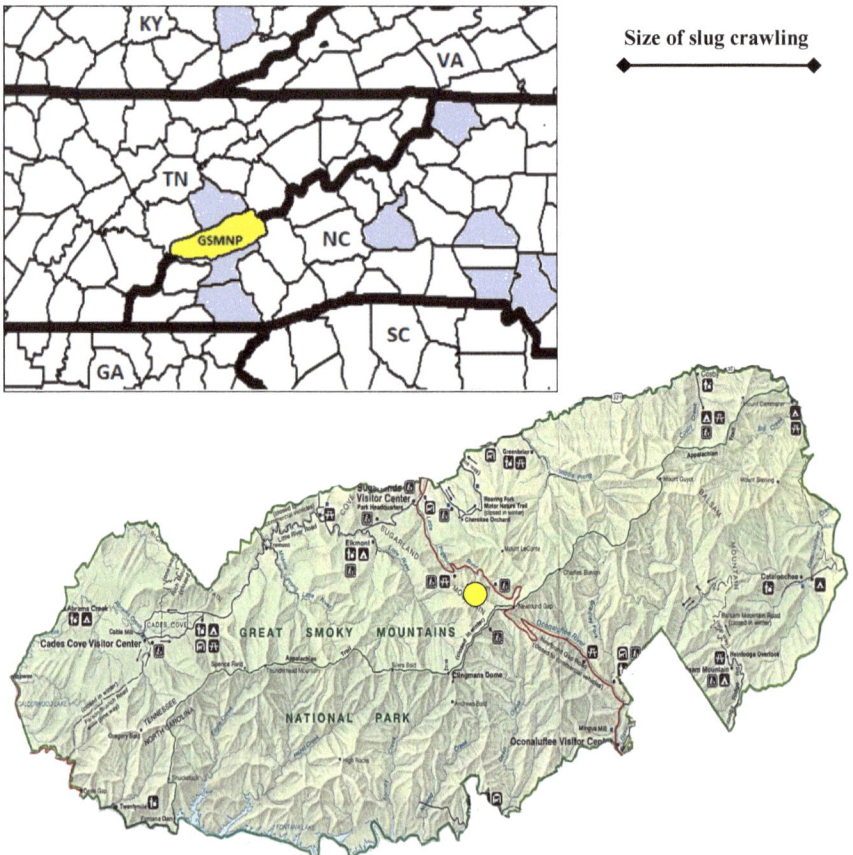

Size of slug crawling

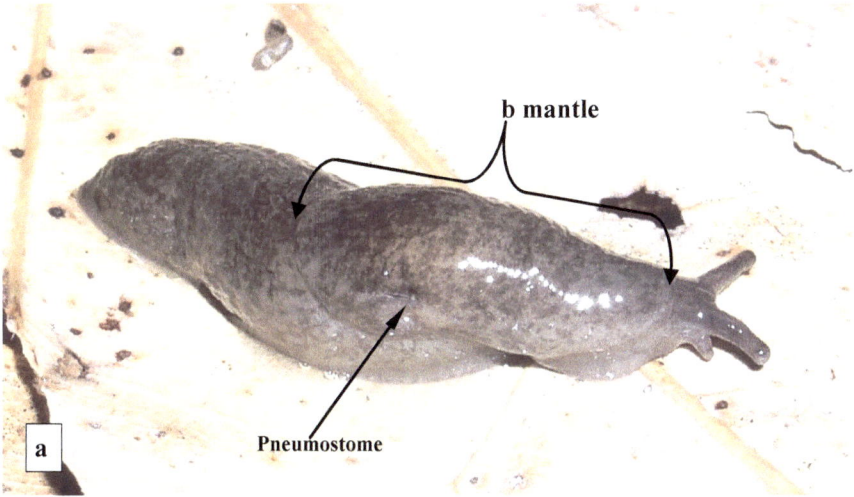

b mantle

Pneumostome

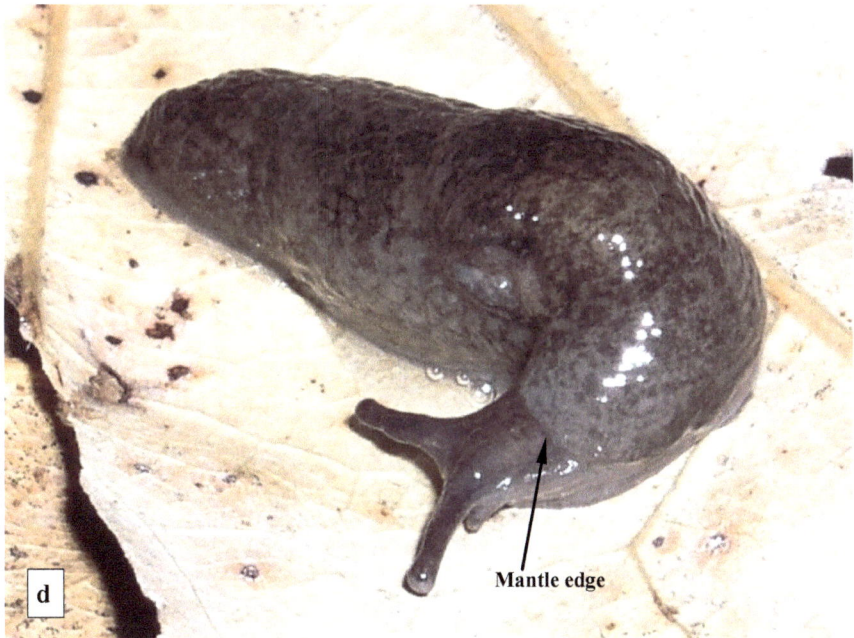

Mantle edge

Beyond the Identification of Gastropods
Help to Push the Frontiers of Science!

The Southern Appalachian Mountains are a national center of diversity for land snails. With this book, a reader can now identify almost any gastropod found in the park down to the species level. *But that's just the beginning.*

In 1916, a momentous act by Congress mandated that the National Park Service (NPS) protect all parks and the vast number of native species that call them home "...unimpaired for future generations...". But the Great Smoky Mountains National Park cannot protect these beautiful, interesting and ecologically important animals without knowing more about *each species'* distribution, rarity, habitat requirements, interactions with other species, growth rates and many other facets of "natural history". Just knowing what species are in the park is an important first step, but that alone does not help much when decisions need to be made that affect these species. We have many questions about these creatures and YOU can help discover their secrets!

The Park is a changing landscape, and not all the changes are for the good. With some of the highest rates of acid precipitation measured in North America, large wild-land fires, invasive non-native plants and animals that alter natural forests, pollutants like methyl mercury which are documented as accumulating in parts of the park, changing moisture and temperature regimes, and other threats to species' survival, there is much that needs to be discovered— and little time to do so. Unfortunately, the NPS has very limited funding for such scientific inquiries. Some types of research are more appropriately undertaken by large research institutions, such as genetic studies and "cause and effect" research about pollutants; however, there are many other critical questions about the mollusk fauna that can be answered by "citizen scientists" working under a park issued permit in groups or even as individual efforts.

Below a list of some of questions/issues for which more data are needed:

*There are number of rare and/or endemic land snails in the park. What are the specific distributions of each in the park, and how can we estimate their population number or density? What are their specific habitat requirements and threat? Adopt a snail species and become the world's authority on it!

*What is the growth rate of different snail species? How long does each species take to mature and how long do they live? Some very large species are reported to grow and live for only about a year. Can snails be individually marked with no effect to the animal in order to facilitate tracking and data collection?

*Some snail species populations are believed to be declining in areas of known high acid deposition. Can repeated transects in favorable weather (or other sampling techniques) estimate populations for a site?

*What is the short term and long term impact of wild-land fire affect land snail populations?

*What is the status of non-native snails in the park? Are they expanding their range or are they relegated to specific habitats? What habitat characteristics favor them? Are native species excluded by exotic species?

*What snails (and in what densities) are typical of each vegetation type, or geological or moisture type? How are all snail densities related to the relative availability of calcium and other basic cations in the substrate?

Some activities will be better suited to areas outside the park. For additional questions that need to be answered, guidance on approaches to be used, current research projects, other questions and permit applications, contact the park's research coordinator at the website: http://www.nps.gov/grsm/naturescience/ahslc_research_home.htm or phone them at 828.926.6251

GSMNP Snails That Warrant Further Investigation

The snails listed below are in need of further taxonomic study using additional characters like reproductive morphology or molecular data. Within the GSMNP, there are currently four forms (Hubricht, 1985; Pilsbry, 1945) and three that may represent species new to science. They are as follows:

Distinct forms recognized by past collectors

1) *Paravitrea umbilicaris dentata* (Pilsbry, 1945)
2) *Ventridens suppressus magnidens* (Pilsbry, 1945)
3) *Mesomphix latior monticola* (Pilsbry, 1945)
4) *Mesomphix cupreus politus* (Pilsbry, 1945)

Undetermined forms (new species?)

1) Form resembling *Fumonelix wheatleyi* (Hubricht, 1985)
2) Form resembling *Inflectarius rugeli* type (Dourson, 2006)

Four species: *Paravitrea umbilicaris dentata, Ventridens suppressus magnidens, Mesomphix latior monticola* and *Mesomphix cupreus politus* are illustrated in this text. *Paravitrea u. dentata* can be found with *P. umbilicaris* and the two are distinct. This occurs around the west end of the TN side of the park. *Ventridens suppressus magnidens* occurs in Cades Cove and is common around low places and springs. In the GSMNP, *Mesomphix latior monticola* is currently known from the south side of the park above Fontana Lake. *Mesomphix cupreus politus* has been collected at Bull Cave Sink Hole and around the Calderwood area at the west end of the TN side of the park near the Tennessee River.

The remaining two forms: one resembling *Fumonelix wheatleyi* and one resembling *Inflectarius rugeli* are not illustrated in the book. Hubricht (1985) found and discussed two *Fumonelix wheatleyi* forms found in the park that had very different genitalia, but could not separate the shells based on external appearances. Hubricht did not describe the new *Fumonelix* formally and so the second form has remained unnamed. The two forms are apparently found at all elevations.

Two *Inflectarius rugeli*-forms have been found by the author and other investigators (Slapcinsky 2012, pers. comm.) at multiple locations including around the parking lot of the trail to Grotto Falls. Here, two *I. rugeli* forms live under leaf litter, side by side, with no discernible external shell differences except for a constant and reliable variation in diameter of approximately 2-3 mm.

Anguispira jessica and its orange defense slime, Roan Mountain, NC

257

Part II
Land Snails of the Southern Appalachians
Tennessee and North Carolina
(Not Yet Recorded in the GSMNP)

This section provides detailed information (including range maps) on land snails that have a reasonable chance of occurring in the Great Smoky Mountains National Park, due to their close proximity, or their similar ecological requirements. Also included are those land snails found in the Southern Appalachian Mountains (from around Mount Rogers, Virginia south to Chilhowee Mountain, Tennessee (see map on page 11), making the text more comprehensive. If you find species in the park that do not match gastropods reported from the GSMNP, this section should cover all possible candidates, unless the species is new to science or a peculiar and isolated ecological form. Range maps are included to illustrate proximity to the park based on Hubricht (1985), other researcher's distribution records and my own knowledge of these species elsewhere in the southern mountains. Land snails in this section are arranged on the next two pages and illustrated in two simple categories: A) land snails with simple lips, not reflected and B) land snails that have reflected lips. The basic keys begin with the smaller species progressing to the larger taxa

Sorry, the park is all filled up!

Tourists.

Pictured Key to Land Snails Found Outside the GSMNP

A. Land Snails with simple Lips

Mediappendix oklahomarum, Ambersnail (8mm), pg. 261
Mediappendix vermeta Ambersnail (12mm), pg. 262
Succinea concordialis Ambersnail (15mm), pg. 263

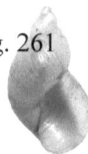

Lucilla singleyana, Coil (3mm), pg. 265

Striatura exigua, Striate (3mm), pg. 266

Pilsbryna vanattai, Bud (4mm), pg. 267
Pilsbryna aurea, Bud (3mm), pg. 268
Pilsbryna nodopalma, Bud (3mm), pg. 269
Pilsbryna quadrilamellata, Bud (3mm), pg. 270

Paravitrea blarina, Supercoil (4mm), pg. 271
Paravitrea varidens, Supercoil (7mm), pg. 272
Paravitrea tridens, Supercoil (5mm), pg. 273
Paravitrea ternaria, Supercoil (7mm), pg. 274
Paravitrea lacteodens, Supercoil (5mm), pg. 275
Paravitrea reesei, Supercoil (4mm), pg. 276

Glyphyalinia solida, Glyph (7mm), pg. 277
Glyphyalinia ocoae, Glyph (4mm), pg. 278
Glyphyalinia clingmani , Glyph (4mm), pg. 279

Anguispira strongylodes, Tigersnail (18mm), pg. 281

Helicodiscus bonamicus, Coil (5mm), pg. 282

Mesomphix inornatus, Button (18 mm), pg. 283

Ventridens coelaxis, Dome (7mm), pg. 284

B. Land Snails with Reflected Lips

Strobilops labyrinthicus, Pinecone (2mm), pg. 285

Stenotrema cohuttense, Slitmouth (7mm), pg. 286
Stenotrema edvardsi, Slitmouth (7mm), pg. 287
Stenotrema spinosum, Slitmouth (12mm), pg. 289
Stenotrema barbigerum, Slitmouth (10mm), pg. 290

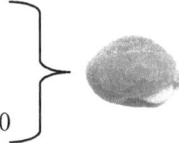

Patera clarki nantahala, Globe (18mm), pg. 291

Mesodon elevatus, Globe (20mm), pg. 292
Mesodon andrewsae, Globe (20mm), pg. 294

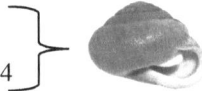

Inflectarius subpalliatus, Covert (13mm), pg. 295

Fumonelix archeri, Covert (15mm), pg. 296
Fumonelix cherohalaensis, Covert (20mm), pg. 297
Fumonelix orestes, Covert (15mm), pg. 298
Fumonelix roanensis, Covert (15mm), pg. 299

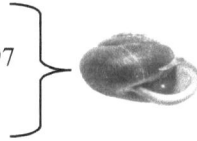

Neohelix dentifera, Whitelip (30mm), pg. 300
Neohelix major, Whitelip (40mm), pg. 301

Allogona profunda, Forestsnail (30mm), pg. 302

Appalachina sayana, Crater (25mm), pg. 303

Triodopsis anteridon, Threetooth (14mm), pg. 305
Triodopsis pendula, Threetooth (14mm), pg. 306
Triodopsis tennesseensis, Threetooth (20mm), pg. 307

Philomycus virginicus, Mantleslug (100mm), pg. 309
Pallifera ohioensis, Mantleslug (30mm), pg. 311

Detritus ambersnail   Succineidae

Mediappendix oklahomarum (Webb, 1953)

Height: 7-8.8 mm tall

Description: Succiniform; lip simple; shell compact and obese, with 2-2.5 whorls; no teeth or lamellae present in the aperture in any stage of growth; shell surface weakly striate and glossy, often coated with soil particles; shell color is a greenish-yellow, but this varies; sides of the foot smoky.

Similar Species: *Novisuccinea ovalis* is taller with a notably larger aperture; *M. vermeta* is larger and less compact in form; *Succinea concordialis* is larger with a capacious aperture.

Habitat: Usually found in the leaf litter of wooded hillsides or in pine woods, on acidic soils; the species is rarely very abundant (Hubricht 1985).

Status: G5; Uncommon; not yet recorded from the GSMNP, but known from counties in Tennessee and North Carolina that border the park.

Specimen: Kentucky, Wayne County, Monticello (FMNH 236387).

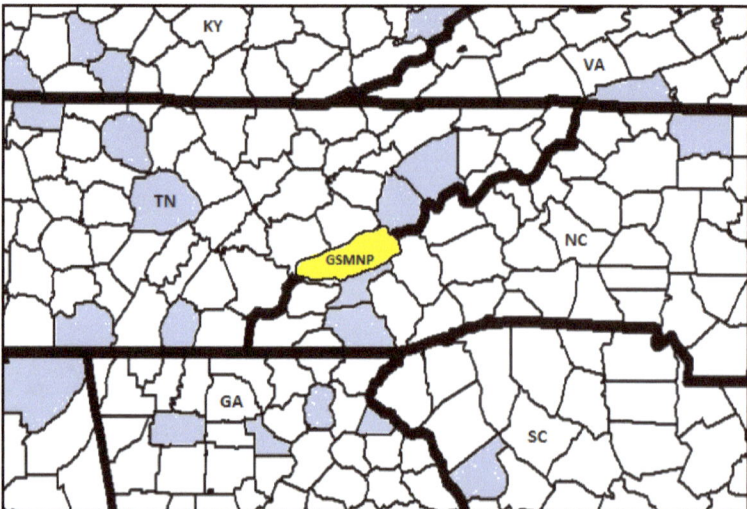

Suboval ambersnail   Succineidae

Mediappendix vermeta (Say, 1829)

Height: 7-13 mm tall

Description: Succiniform; lip simple; shell with 2.5-3.5 whorls; perforate; shell surface weakly striate, pale yellow-olive, often coated with soil particles.

Similar Species: *M. oklahomarum* is more compact and generally found in drier situations; *Novisuccinea ovalis* is larger with a notably larger aperture

Habitat: An amphibious land snail that appears restricted to wet ground found around ponds, marshes, muddy banks of open ditches and swamps in both open and shaded conditions; in upland sites under stones.

Status: G5; Locally Rare; Many authors (e.g. Burch 1962; Patterson and Burch 1966) recognized *M. vermeta* as the name applicable to specimens from the USA, which were formerly identified as *Catinella avara* (Turgeon et al. 1998).

Specimen: Pennsylvania, Allegheny County, Leetsdale near Little Sewickley Creek, Carnegie Museum (CM 75744).

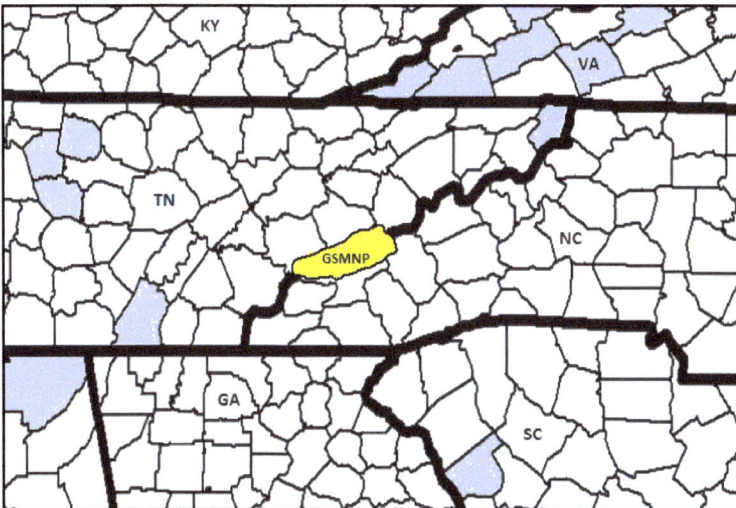

Spotted ambersnail   Succineidae

Succinea concordialis I. Lea, 1864

Height: 12-16.8 mm tall

Description: Succiniform; lip simple; aperture copious; shell with 3 whorls; perforate; no teeth or lamellae present in any stage of growth; shell surface weakly striate and near paper thin but firm; color of shell a pale honey-yellow; the foot, including head and eye-stalks is a grayish-white, speckled with irregular grayish-black spots; sole of foot pale yellow (Pilsbry 1948). Formerly known as *S. concordialis.*

Similar Species: *Novisuccinea ovalis* is larger and usually more inflated in its form, but this feature varies among populations.

Habitat: An amphibious gastropod found on moist earth near water (not in it), on banks of small creeks and gravel bars.

Status: G5; Locally Rare.

Specimen: Kentucky, Crittenden County, near Preacher Creek, 3 M northwest of Crayne (FMNH 235656).

Shells are propionate to each other

m. vermeta, **PA**   *M. oklahomarum*, **KY**

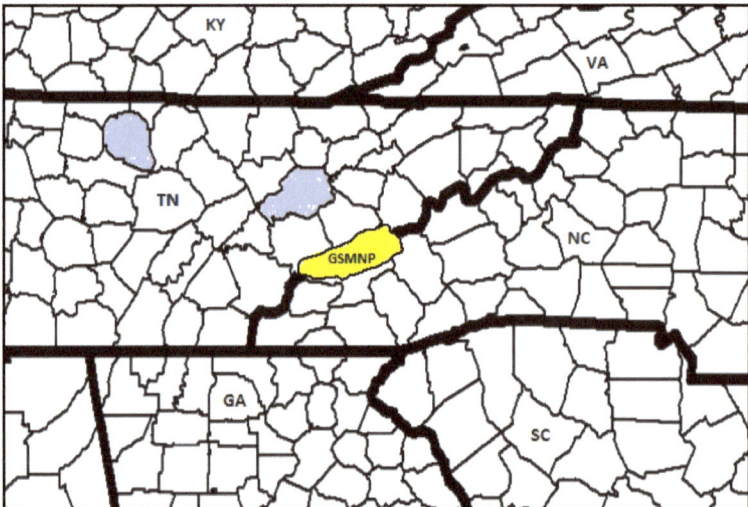

S. concordialis,

263

An Oval ambersnail, *Novisuccinea ovalis* climbing a ghost pipe, *Monotropa uniflora*, Roan Mountain, North Carolina.

Smooth coil

Helicodiscidae

Lucilla singleyana (Pilsbry, 1889)

Diameter: 2.4-3 mm

Description: Depressed heliciform; lip simple, aperture oval; shell with 3-4 whorls; umbilicate; fresh shells are glossy and translucent (as these three images to the right illustrate); no teeth present; scant trace of spiral ornamentation (strong microscope required), but this faint feature is usually present in fresh shells only; deceased and dried animal seen in bottom view, causing some parts of the shell to appear darker.

Similar Species: Bottom and top views of *Lucilla* species are indistinguishable except that *L. scintilla* is smaller and displays (under a strong lens) no microscopic spiral lines; *Hawaiia* species are very similar but have a notably wider umbilicus and better developed shell surface transverse and spiral striae.

Habitat: Usually a species of open grassy areas, roadsides, railroads and occasionally in caves.

Status: G5; Locally Rare; not yet recorded in the GSMNP; this species has been wat under collected and likely occurs in more locations than current records would suggest.

Specimen: Tennessee, Franklin County, Dry Cave (authors collection)

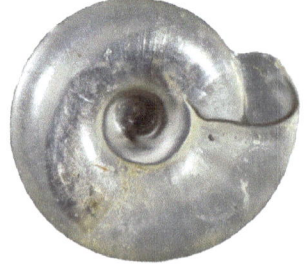

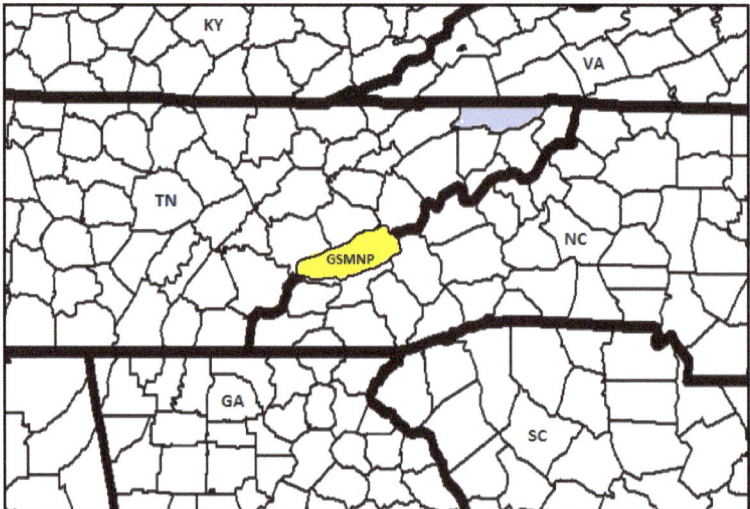

Ribbed striate

Gastrodontidae

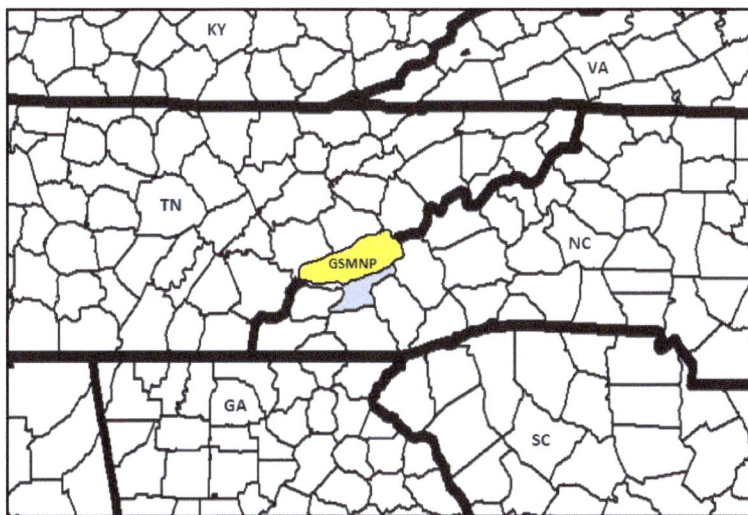

Striatura exigua (Stimpson, 1847)

Diameter: 2.2-2.4 mm

Description: Depressed heliciform; lip simple; aperture roundish; shell with 3.5 whorls; widely umbilicate; a coppery color; shell translucent in live snails and fresh dead; without teeth; transverse striae are well-developed forming minute paper-thin riblets; the most widely-spaced of any *Striatura*; spiral striae also well defined but are an extremely small micro-feature of the shell surface.

Similar Species: *Striatura ferrea* is larger in diameter, has a much smaller umbilicus and has a finer shell surface.

Habitat: Living on wet ground around head water seeps in upper elevation sphagnum bogs (over 1000 meters) surrounded by mixed hardwood forests.

Status: G5; Locally Rare; not yet recorded in the GSMNP, but known from a sphagnum bog (1100 meters) in the Nantahala National Forest in Swain County, North Carolina. This population is approximately 250 miles farther south of reported populations in southeastern West Virginia (Hubricht 1985).

Specimen: North Carolina, Swain County, Nantahala National Forest (author's collection).

266

Honey bud

Oxychilidae

Pilsbryna vanattai (Walker and Pilsbry, 1902)

Diameter: 3.8-4.4 mm

Description: Depressed heliciform; lip simple; shell with 4.5-5 loosely coiled whorls, the last whorl expanded; perforate to umbilicate; shell glossy, honey colored; internal teeth (figures a and b) present in shells under 3 whorls, in mature shells all traces of teeth are re-absorbed; the delicate indented transverse striae are irregular; with spiral striae; as in all *Pilsbryna* species juvenile shells containing less than three whorls are necessary for an proper ID.

Similar Species: *Glyphyalinia* species are without internal armature; *Paravitrea* have tightly coiled whorls, their last whorls generally not expanding much as in *Pilsbryna*.

Habitat: Rocky, wooded mountainsides, most common in wet leaf litter along streams, seeps but also found in deeper leaf litter at the base of limestone and other sedimentary rocks next to stream banks (Slapcinsky and Coles 2004).

Status: G3; Rare; Locally Endemic.

Specimen: Tennessee, Carter County, Cherokee National Forest (author's collection).

a

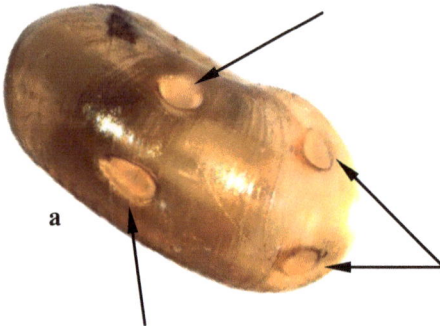

Internal lamellae as seen through bottom of juvenile shells under three whorls

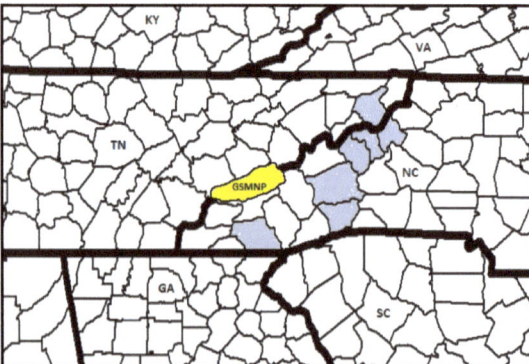

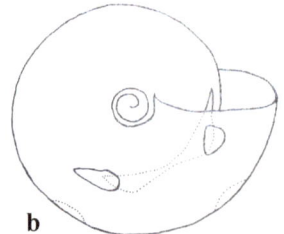

b

(Slapcinsky and Coles 2004)

Ornate bud

Pilsbryna aurea (H.B. Baker, 1929)

Diameter: 2.9-3.6 mm

Description: Depressed heliciform; lip simple; shell with 3 loosely coiled whorls, the last whorl expanded; umbilicate; shell glossy; internal lamellae (c & d) are present in shells under 3 whorls but are reduced or more commonly completely absent in adults; transverse striae are weakly developed and irregular and shell is with fine spiral striae.

Similar Species: *Glyphyalinia* species are without internal armature in all stages of growth; Juvenile shells *P. aurea* have distally expanded parietal lamella and do not possess the mid-basal and sutural lamellae of *P. quadrilamellata*.

Habitat: Rocky wooded hillsides along small streams in wet leaf litter along streams & seeps but also found in deep leaf litter at the base of limestone and other sedimentary rocks along stream banks; (Slapcinsky and Coles 2004).

Status: G1; Globally Rare; <u>Locally Endemic</u>.

Specimen: Tennessee, Washington County, Dry Creek (FLMNH 408000).

c

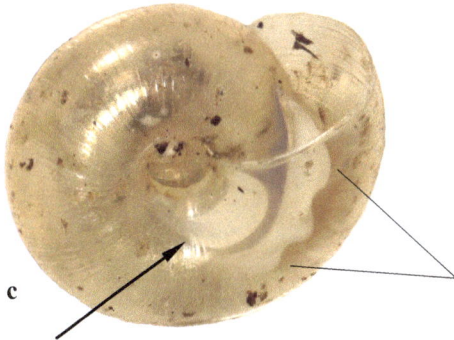

Internal lamellae as seen through bottom of juvenile shells under three whorls

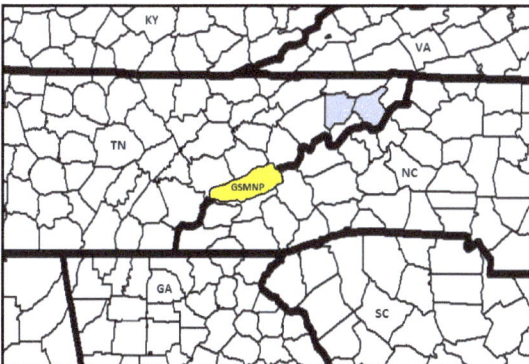

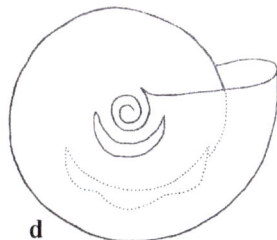

d

(Slapcinsky and Coles 2004)

Oar tooth bud

Oxychilidae

Pilsbryna nodopalma (Slapcinsky and Coles, 2004)

Diameter: 2.7-3.2 mm

Description: Depressed heliciform; lip simple; shell with 4.5 loosely coiled whorls, the last whorl expanded; umbilicate; shell clear and glossy, there are undulate parietal lamella and 2 to 4, paired, subcolumellar and lower palatal teeth, all grouped near the aperture (figures a and b) present in shells under 4 whorls, in mature shells all traces of these apertural teeth are re-absorbed; delicate indented transverse striae are irregular; with fine spiral striae.

Similar Species: *P. nodopalma* differs from all other species of *Pilsbryna* in having the whorls widest above the midpoint rather than below; *Glyphyalinia* species are without internal armature.

Habitat: Usually wooded rocky hillsides in leaf litter and on relatively dry rock outcrops; Slapcinsky and Coles 2004).

Status: G1; Globally Rare; <u>Locally Endemic</u>.

Specimen: Tennessee, Greene County, Hurricane Gap Road, CNF (FLMNH 406758).

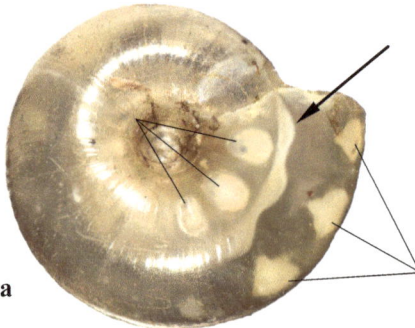

Internal lamellae and teeth as seen through bottom of juvenile shells under four whorls

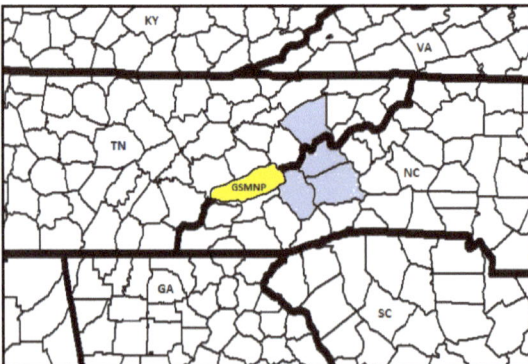

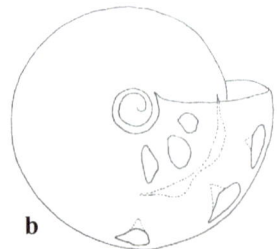

a

b

(Slapcinsky and Coles 2004)

Four blade bud

Oxychilidae

Pilsbryna quadrilamellata (Slapcinsky and Coles, 2004)

Diameter: 2.8-3.2 mm

Description: Depressed heliciform; lip simple; shell with 4.5-5 loosely coiled whorls, the last whorl expanded; umbilicate; shell glossy with four lateral lamellae (figure c & d showing 3 of these); all four lamellae are reduced as individuals reach maturity, however some trace of lamellae, especially the basal lamella, remains in many adults; irregular and dense indented transverse striae and fine spiral striae.

Similar Species: Adult *P. quadrilamellata* are unusual in having the sutural margin of the final third of the body whorl flattened; other *Pilsbryna* species do not have this feature.

Habitat: Leaf litter within approximately 20 meters of the base of a cold talus slope on a northeast facing slope of Nolichucky River; (Slapcinsky and Coles 2004).

Status: G1; Globally Rare; Endemic to Unicoi County Tennessee.

Specimen: Tennessee, Unicoi County, Unaka Springs, CNF (FLMNH 408014).

c

Internal lamellae as seen through bottom of juvenile shells under four whorls

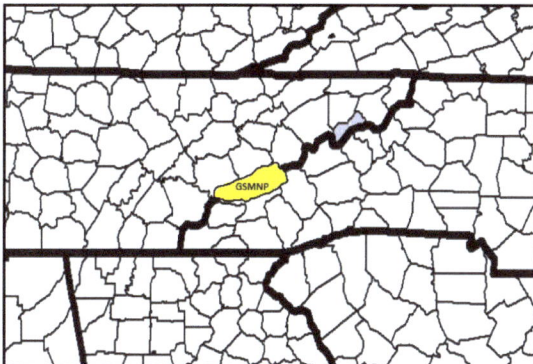

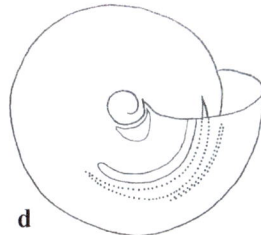

d

(Slapcinsky and Coles 2004)

Shrew supercoil

Pristilomatidae

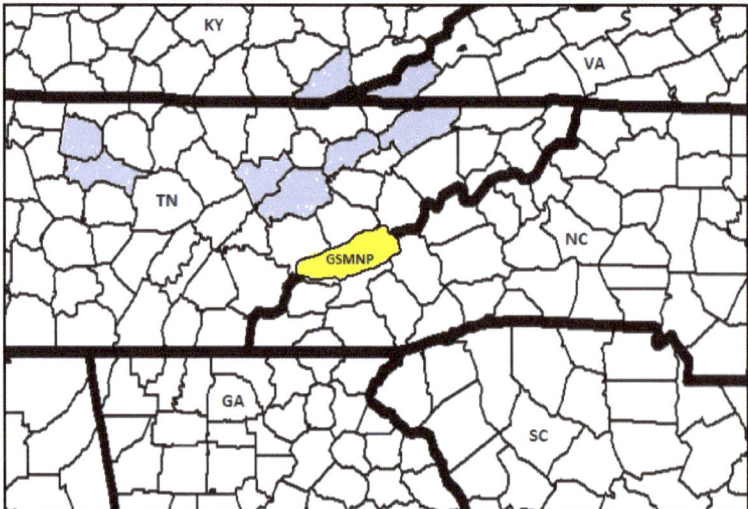

Paravitrea blarina Hubricht, 1963

Diameter: 3.9 mm

Description: Depressed heliciform; lip simple; shell with 5.5 tightly coiled whorls; umbilicate; shell glossy, thin and translucent in fresh and live shells; embryonic whorl smooth, later whorls sculptured with numerous irregularly spaced transverse striae, impressed growth lines, distinct above but becoming obsolete below; there are no teeth at any stage of growth.

Similar Species: *Paravitrea blarina* differs from all other small (under 4 mm) *Paravitrea* species in the absence of teeth in the shell at all stages of growth (Hubricht 1985); differs from *P. capsella* in the notably wider umbilicus.

Habitat: This is a secretive species, found in the lower layers of leaf mold; at one location, it was found crawling about in shrew burrows under the leaves on lower elevation wooded hillsides and in ravines (Hubricht 1985).

Status: G3; Uncommon; not yet recorded in the GSMNP, but known in nearby Tennessee counties; if in the park it would likely occur along the park's northern borders.

Specimen: Kentucky, Bell County, Pine Mountain Gap, Pineville (FMNH 248873).

271

Roan supercoil

Paravitrea varidens (Hubricht, 1978)

Diameter: 7.6 mm

Description: Depressed heliciform and compact; lip simple; aperture roundish; shell with 8.5 tightly coiled whorls; umbilicate; shell thin, glossy and translucent in fresh and live shells; indented transverse striae are faint and irregularly spaced; two or three rows of 3 to 4 rather large teeth seen through the bottom of fresh and live shells of juveniles (figure a), these teeth sometimes fuse to form a radial lamella; adults are typically without these teeth.

Similar Species: *Paravitrea capsella* is smaller and is without teeth in both juvenile and adult shells; *P clappi* has a perforate umbilicus; *P. andrewsae* of the same region has five teeth in each row which are present in adults and has a smaller umbilicus.

Habitat: Higher elevation mixed hardwood forests, rock talus, small seeps and in tree crotches that have ample leaf litter fill.

Status: G2; Locally Rare; Endemic to the Roan Mountain region of North Carolina & Tennessee.

Specimen: North Carolina, Mitchell County, Roan Mountain (FMNH 248752).

Pristilomatidae

a

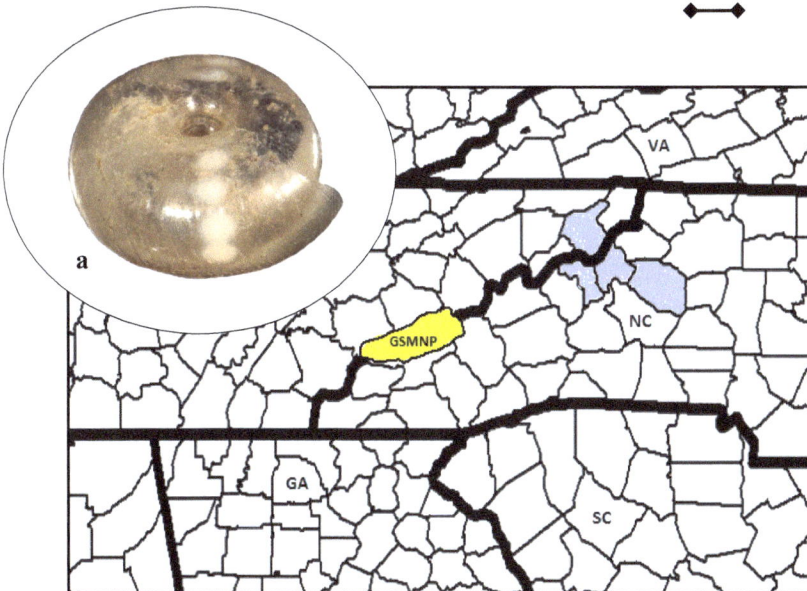

White foot supercoil

Pristilomatidae

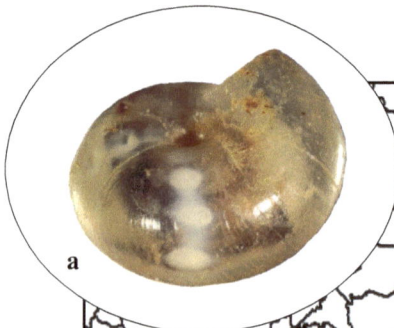

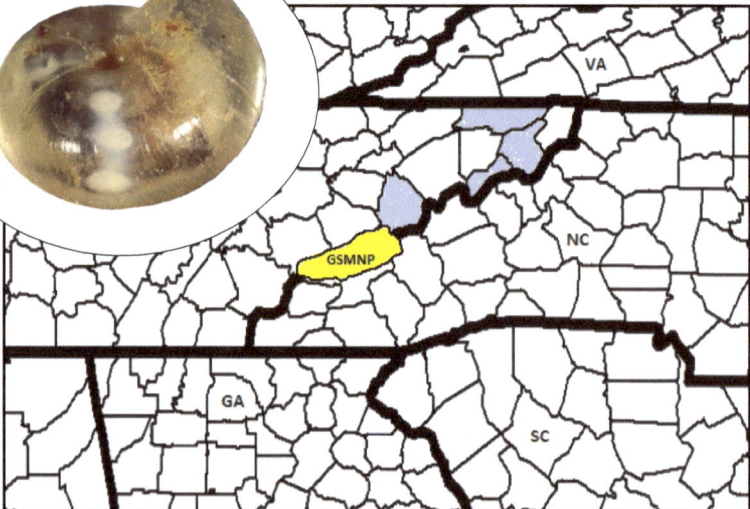

Paravitrea tridens (Pilsbry, 1946)

Diameter: 5.6-6.2 mm

Description: Depressed heliciform; lip simple; shell with 5.6-6.2 tightly coiled whorls; perforate to umbilicate; shell glossy, thin and translucent in fresh and live shells, sculptured with numerous irregularly spaced transverse striae; adult shells are toothless but young shells up to 3.3 mm in diameter have 1-3 (usually 2) radial sets of three teeth (a), but some juvenile shells of similar size are without teeth.

Similar Species: This species stands close to *P. reesei*, but the toothed young are smaller and it attains a toothless adult larger than any *reesei* known (Pilsbry 1945) and is a larger species; *Paravitrea capsella* has a more rounded aperture and is usually without teeth in all stages of growth; *P. ternaria* has a notably larger umbilicus.

Habitat: Found among moist leaf litter in mixed hardwood forests.

Status: G2; Rare to uncommon; Locally Endemic; although not yet recorded in the GSMNP, it occurs close to park boundaries.

Specimen: Tennessee, Sullivan County, Holston River 1 mile NE of Spurgeon (FMNH 24728).

273

Sculpted supercoil

Pristilomatidae

Paravitrea ternaria (Hubricht, 1978)

Diameter: 6.8 mm

Description: Depressed heliciform or heliciform; lip simple; shell with 8.5 tightly coiled whorls; umbilicate, umbilicus deep and well-like, 1.4 mm wide; shell pale brownish, thin, shining and translucent in fresh and live shells; shell sculptured with numerous irregularly spaced impressed transverse striae; within the last whorl there are usually 2-3 radial rows of 3 rather large teeth which are present at all stages of growth.

Similar Species: *Paravitrea ternaria* is most closely related to *P. tridens,* differing by having a larger umbilicus and by having radial rows of 3 teeth present at all stages of growth; the animal is dark gray while the animal of *P. tridens* is nearly white (Hubricht 1978); *P. capsella* is without teeth at all stages of shell growth; *P varidens* has 3 or 4 teeth in juvenile shells which are typically absent in adults.

Habitat: Found under moist leaf litter in ravines of higher elevation (1100 meters) northern hardwood forests.

Status: G1; Globally Rare; Locally Endemic; not yet recorded in the GSMNP.

Specimen: Tennessee, Unicoi County, 5.2 miles S of Flag Pond (FMNH 24718).

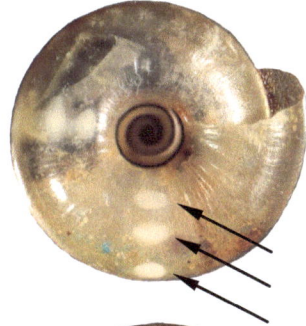

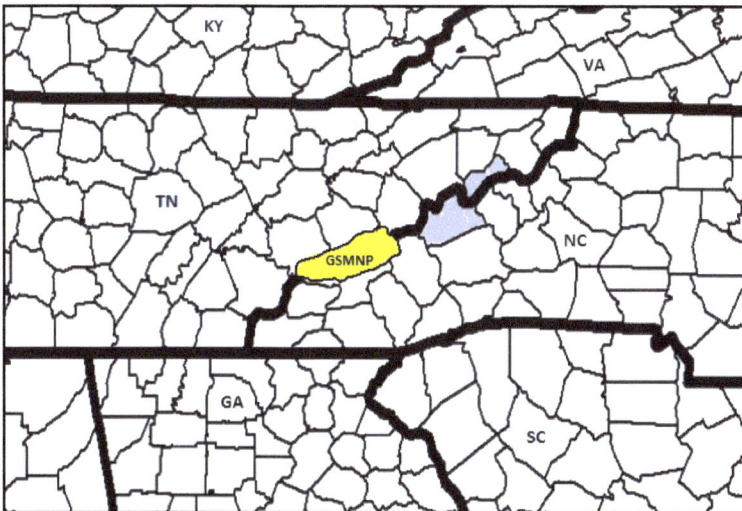

274

Ramp cove supercoil

Paravitrea lacteodens (Pilsbry, 1946)

Diameter: 4.8 mm

Description: Depressed heliciform or heliciform; lip simple; shell with 6.5 tightly coiled whorls; umbilicate; shell glossy, thin and translucent in fresh and live shells; shell sculptured with numerous irregularly spaced; both young and adults with teeth, adults retaining 2-3 rows (sometimes up to four pairs) of radial teeth with 2 in each row (a).

Similar Species: *Paravitrea capsella* is the same size and form of *P. lacteodens* but is generally without paired teeth in juvenile and adult shells; *P. placentula* has paired teeth in only juvenile shells, adults are without.

Habitat: Found among moist leaf litter in mixed hardwood forests, especially rich north facing mountainsides.

Status: G1; Globally Rare; Endemic to Graham County, North Carolina; precious little is known of this apparently range restricted gastropod; not yet recorded in the GSMNP but the species is reported close to park boundaries near Fontana Lake.

Specimen: North Carolina, Graham County, Cheoah Mountains, Yellow Creek, Ramp Cove (FMNH 99150).

Pristilomatidae

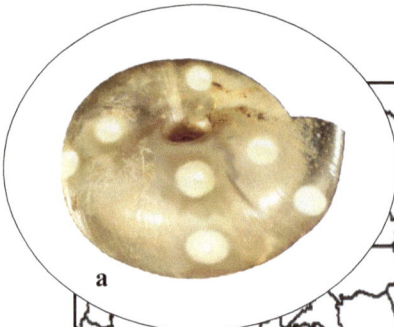

a

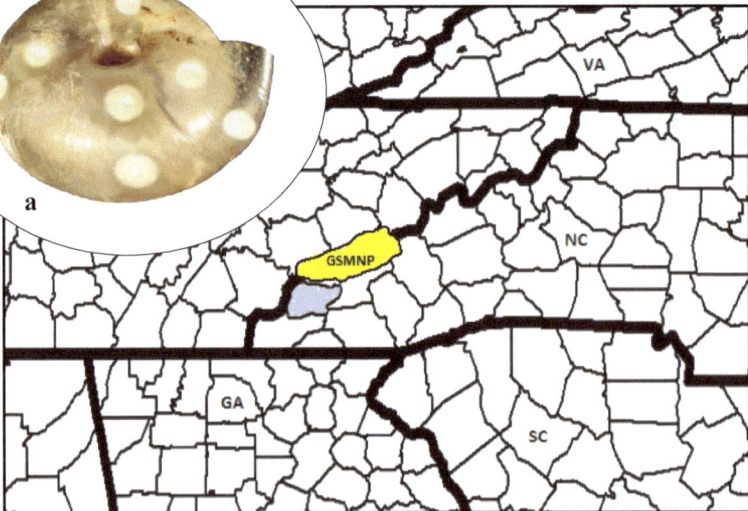

Round supercoil

Pristilomatidae

Paravitrea reesei Morrison, 1937
Common name: round supercoil
Diameter: 3.5-4.7 mm
Description: Depressed heliciform; lip simple; shell with 5-6.5 tightly coiled whorls, the last whorl not expanded; umbilicate, the umbilicus deep and well-like; shell pale-amber; glossy; aperture lunate; indented transverse striae are irregularly spaced; spiral sculpture indistinct above and below; in adult shells 3 rows of rather large teeth of three (b), are regularly spaced and visible through the base of fresh shells (Burch 1962); young shells of 2-2.5 whorls have two conical teeth, the third tooth appearing as shells mature (Pilsbry 1946); teeth are usually without the callus bridge seen in *P. mira*. (a species of northwestern Virginia and eastern Kentucky).
Similar Species: *Paravitrea capsella* is larger and is without lamellae or any teeth at all stages of growth.
Habitat: Found under moist leaf litter, detritus and rocks on wooded hillsides, ravines and river bluffs of mixed hardwood forests.
Status: G3; Rare to Common.
Specimen: Virginia, Pulaski County, opposite Radford (FMNH 249118).

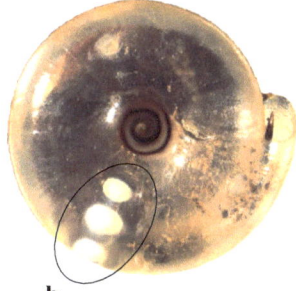

b

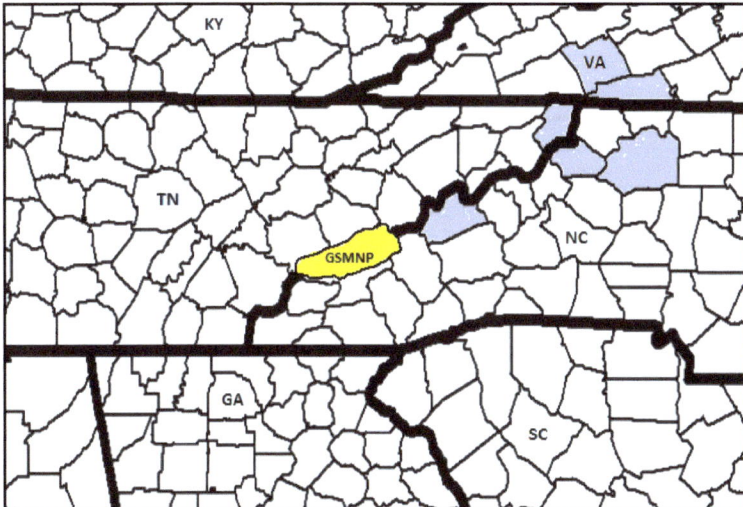

Imperforate glyph

Gastrodontidae

Glyphyalinia solida (H. B. Baker, 1930)

Diameter: 7.5-8 mm

Description: Depressed heliciform; lip simple; shell with 5.25 loosely coiled whorls; imperforate, the umbilicus completely covered in all stages of growth; shell fragile, color corneous to fulvous to almost chestnut, glossy and semi-transparent; no teeth present; indented transverse striae are well-developed, closely and nearly equally spaced, continuing to the base; the spiral striae may be weakly defined but are a constant feature (scope required to view this fine feature).

Similar Species: *Glyphyalinia solida* is typically larger, heavier shelled and has a stronger sculpture than *G. cryptomphala* (page 95); other *Glyphyalinia* species are not imperforate, their umbilicus remaining at least partially or widely open.

Habitat: A species that can be found under moist leaf litter in mixed upland hardwood forest and in talus near rock ledges.

Status: G5; Locally Rare.

Specimen: Tennessee, Marion County, Clear Spring Cave, 2 miles NE of Jasper (author's collection).

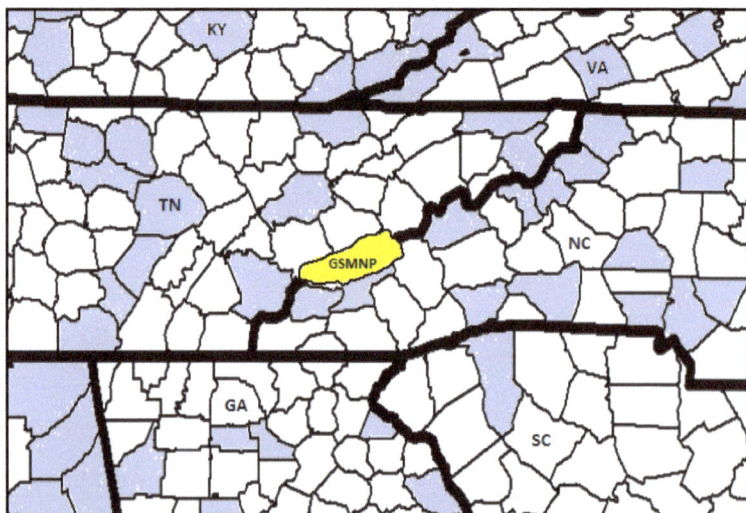

277

Blue-gray glyph

Glyphyalinia ocoae (Hubricht, 1978)

Gastrodontidae

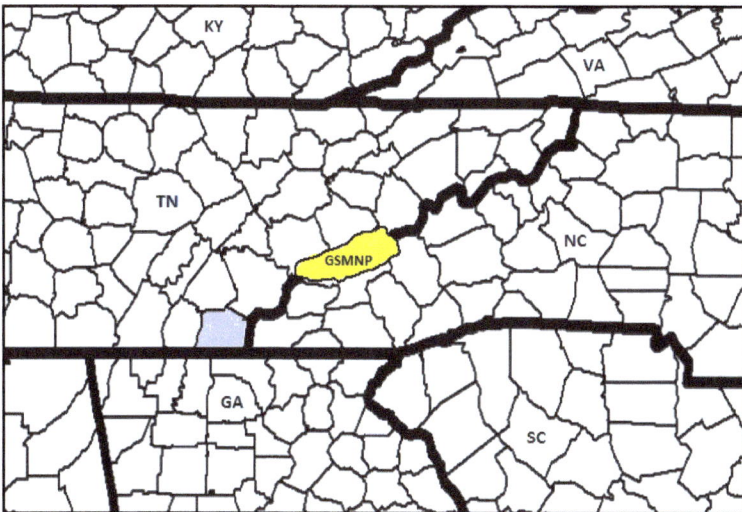

Diameter: 4.8 mm

Description: Depressed heliciform; lip simple; shell with 4.5-5 loosely coiled whorls; perforate, the umbilicus sometimes sealed with a hardened slime cover; no teeth present; shell fragile, shell pale yellow-orange and translucent, glossy; the live animal pale bluish-gray; indented, irregularly spaced transverse striae are well-developed, best developed on top of shell but less conspicuous beneath; although close spiral striae may be weakly defined they remain a constant feature (scope required); deceased and dried animal can be seen through the translucent shell (a), making some portions of the shell darker.

Similar Species: *Glyphyalinia ocoae* has been confused with *G. indentata* but is quite different by its internal anatomy and the shell of *G. ocoae* is smaller and darker in colored.

Habitat: Usually found under moist leaf litter in mixed hardwood forest on hillsides and in ravines; found in drier situations than other *Glyphyalinia* species.

Status: G1; Globally Rare; Endemic to the Ocoee River Gorge, Tennessee.

Specimen: Tennessee, Monroe County, Ocoee Gorge, Cherokee National Forest (author's collection).

Fragile glyph

Glyphyalinia clingmani (Dall, 1890)

Diameter: 6.5 mm

Description: Depressed heliciform; lip simple, the aperture rectangular in shape; shell with 5.5 loosely coiled whorls; umbilicate; shell thin and fragile, greenish-yellow to reddish-horn and translucent, glossy; the live animal dark bluish-black; indented, regularly and closely spaced transverse striae are well-developed, best developed on top of shell much less conspicuous and becoming wider on the base; without notable spiral striae; periphery well rounded.

Similar Species: *Pilsbryna vanattai* is smaller, honey-yellow in color and as not as flat in frontal view; umbilicus and transverse striae more or less the same.

Habitat: Usually found under logs and rocks and moist leaf litter at high elevation in the Black Mountains, specifically Mount Mitchell (Hubricht, 1985).

Status: G1; Globally Rare; Endemic to Mount Mitchell, North Carolina; precious little is known of this beautiful and range restricted species.

Specimen: North Carolina, Mitchell County, Mount Mitchell (Wayne and Amy Van Devender collection).

Gastrodontidae

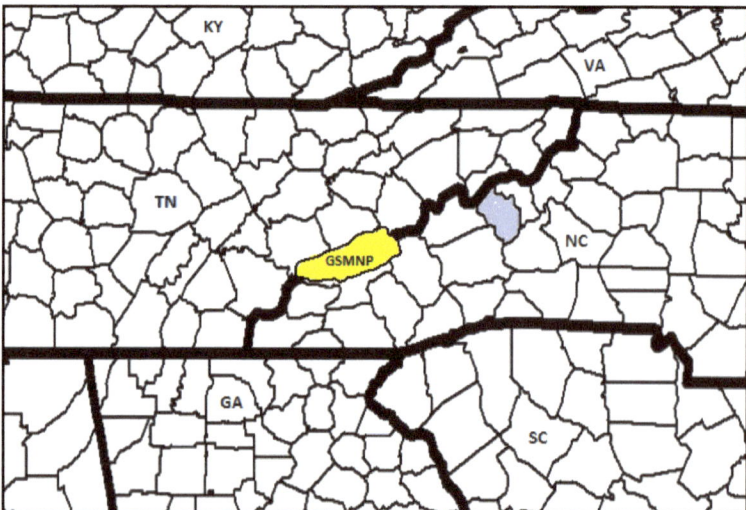

Fragile glyph, *Glyphyalinia clingmani*

Southeastern tigersnail

Discidae

Anguispira strongylodes (Pfeiffer, 1854)

Diameter: 15-20 mm

Description: Depressed heliciform; lip simple; shell with 5 whorls; umbilicate, smaller than other *Anguispira* species found in the GSMNP; color features weak; no teeth present; transverse striae are well-developed and rib-like, having a count of 56 to 62 in the last whorl; chestnut spots generally larger and bolder in other species are reduced in size and intensity; periphery rounded to carinate.

Similar Species: *Anguispira alternata* is larger, has a wider umbilicus and has transverse striae sculpture that is notably finer; *A. mordax* has wider-spaced ribs (a count of 40 to 45 ribs in the last whorl).

Habitat: Limestone regions along forested hillsides adjacent to rivers, usually found under the leaf litter but also among loose rock formations.

Status: G5; Common; not yet recorded in the GSMNP but found in counties that border the park.

Specimen: Tennessee, Smith County, Dixon Springs (FMNH 238349); all images by Jochen Gerber.

.

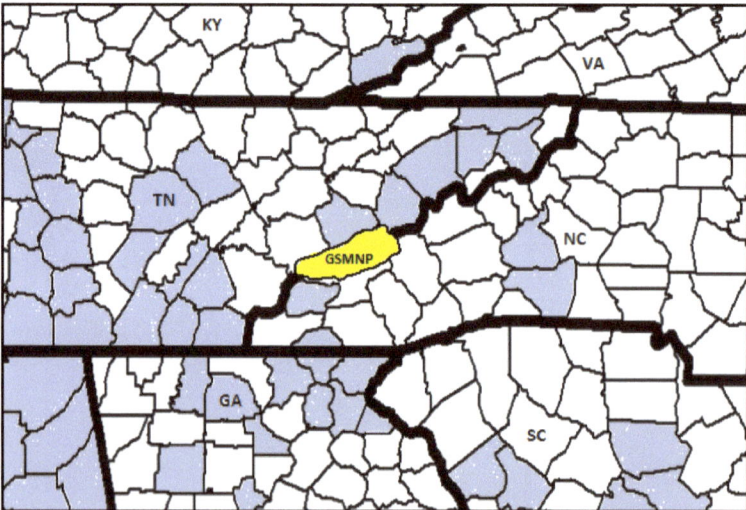

Spiral coil

Helicodiscidae

Helicodiscus bonamicus (Hubricht, 1978)

Diameter: 5 mm

Description: Extremely depressed heliciform; lip simple; shell with 5 whorls; widely umbilicate; in the last whorl there are usually pairs of teeth on the outer and basal walls (a); with spiral striae and conspicuously haired on all whorls (a hand lens of 10X will be useful to view this amazing micro-feature), but these hairs usually lost in older shells.

Similar Species: *Helicodiscus parallelus* and *H. notius* do not have the hairs seen in *H. bonamicus, H. fimbriatus* (figure b) has long flattened fringes.

Habitat: A calciphile found under rocks and leaf litter on river bluffs and in caves in the Nantahala Gorge, North Carolina (Hubricht 1985).

Status: G1; Globally Rare; the species is thought to be Endemic to the Nantahala Gorge, North Carolina.

Specimen: North Carolina, Macon County, Nantahala Gorge, N. of Beechtown (FMNH 329688).

a

b

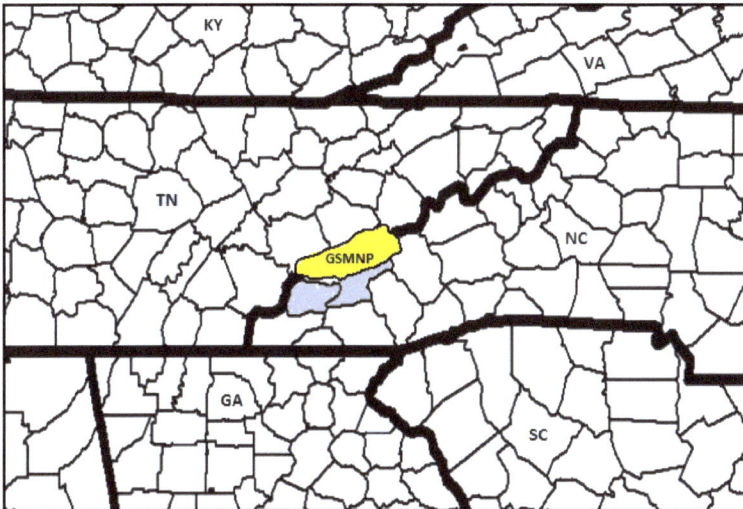

Plain button

Gastrodontidae

Mesomphix inornatus (Say, 1821)

Diameter: 16-21 mm

Description: Depressed heliciform; lip simple; shell with 5 loosely-coiled whorls; perforate, shell thin but not fragile, olive-tan, very glossy; embryonic whorl smooth (a hand lens of 10X can be used to view this feature); no teeth present; a thickening or thin whitish callus that is typically just inside the aperture's bottom of the last whorl; transverse striae (growth wrinkles) and spiral rows of papillae are present; a carnivorous species feeding on other live snails (Dourson pers. obs.).

Similar Species: *Mesomphix perlaevis* has a higher shell profile and the transverse striae are clearly more defined on the embryonic whorl; *M. subplanus* is more compressed and without rows of papillae.

Habitat: A snail of rich, upland mixed hardwood forests of beech and hemlock usually found under and among moist leaf litters and detritus but also a species found thriving along shale banks of road-cuts.

Status: G5; Locally Rare; this is a common land snail of more northern states.

Specimen: Kentucky, Powell County, Red River Gorge (author's collection).

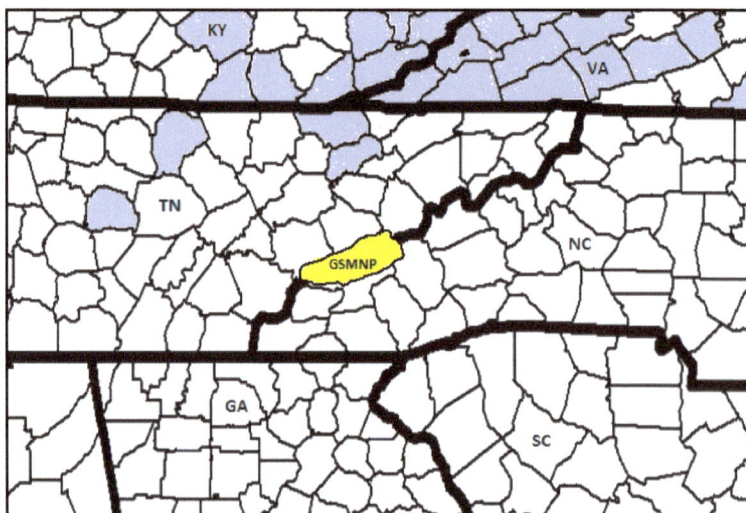

Bidentate dome

Gastrodontidae

Ventridens coelaxis (Pilsbry, 1899)

Diameter: 6.5-6.7 mm

Description: Depressed heliciform; lip simple; shell with 6.5-7 tightly coiled whorls; umbilicate; shell translucent, thin and glossy; transverse striae (growth wrinkles) are relativity distinctive on top but indistinct on the sides and bottom of shell; only traces of spiral striae; within the aperture there are two elongated lamellae or teeth that are present at every stage of growth, live animal pale, not dark as seen in species of the *V. gularis* assemblage; a member of the *V. pilsbryi* group.

Similar Species: *V. lawae* is larger, has a thicker shell and is higher dome shape. *V. lasmodon* has a wider umbilicus.

Habitat: Found in wooded ravines of higher elevation mountainsides under leaf litter and around log structure.

Status: G3; Uncommon; <u>Locally Endemic</u>; not yet reported from the GSMNP.

Specimen: Virginia, from Washington County (author's collection).

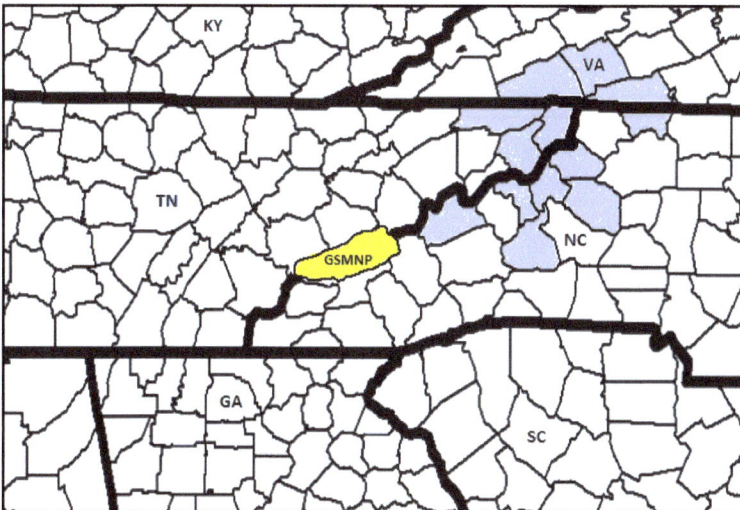

Maze pinecone

Strobilopsidae

Strobilops labyrinthicus (Say, 1817)

Diameter: 2.3-2.5 mm

Description: Dome-shape or bee-hive in form; lip only slightly reflected; shell with 5.5 tightly coiled whorls; perforate; chestnut-brown; elongated teeth or lamellae present in the aperture which can be seen through the bottom of live and fresh dead shells; transverse striae are modified into well-developed ribs on the top and side but are poorly developed on the base (this feature is easily view with a hand lens of 10X); periphery of last whorl of shell is rounded or bluntly angular.

Similar Species: *Strobilops aeneus* (page 192) has a lower shell profile, wider umbilicus and a more angular periphery; *Euconulus* species (page 172) are beehive-shape and have a smooth surface instead of ribs and their lip is simple.

Habitat: Found in mixed hardwood forests, glade-like areas especially around loose soils of limestone substrate.

Status: G5; Common; not yet recorded in the GSMNP.

Specimen: Kentucky, Powell County, Furnace Mountain (author's collection).

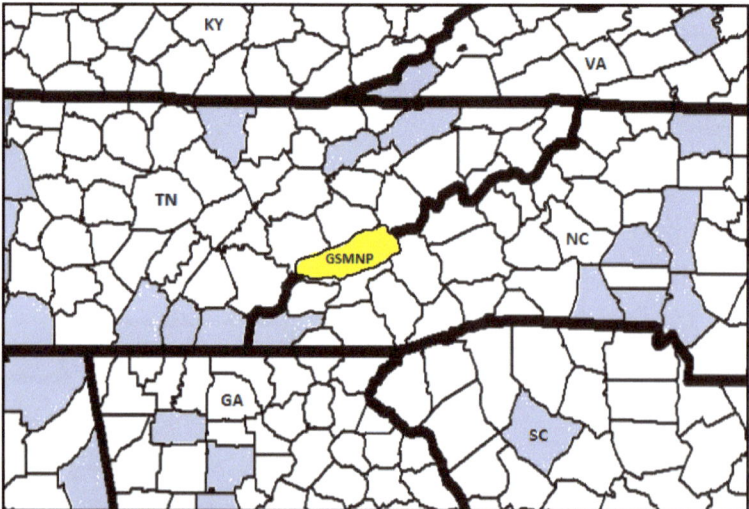

Cohutta slitmouth

Polygyridae

Stenotrema cohuttense (Clapp, 1914)

Diameter: 6-7.3 mm

Description: Pill-shape and compact; shell with 5 tightly coiled whorls; imperforate; reddish horn-colored; covered in dense, short stiff hairs; spiral striae poorly developed on base; basal notch wide; without a distinctive interdenticular sinus; shell periphery well rounded.

Similar Species: *S. hirsutum* (figure a and page 183) is larger, has a boxy profile (not rounded) and has a shallow basal notch; *S. pilula* is smaller and has a well developed interdenticular sinus.

Habitat: A species of leaf litter in mixed hardwood forests on hillsides and in ravines and the base of limestone rock outcroppings.

Status: G2; <u>Locally Endemic;</u> Uncommon with a limited distribution including one county in Tennessee and three counties in Georgia.

Specimen: Tennessee, Polk County, Ocoee Gorge (author's collection).

a

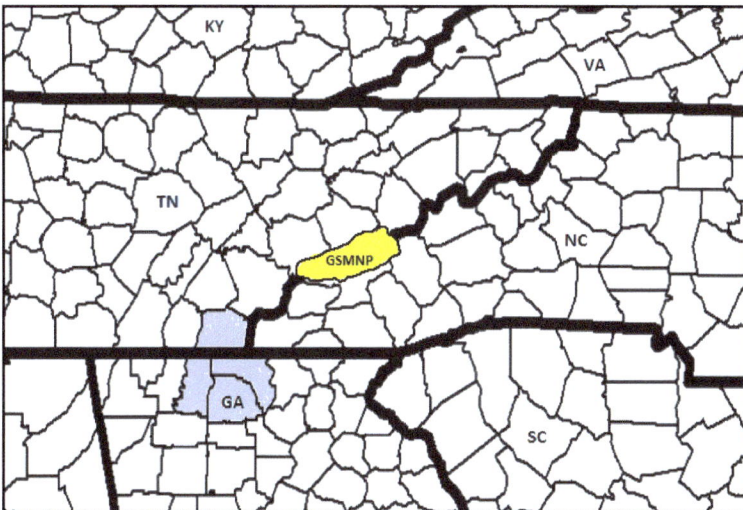

286

Ridge-and-valley slitmouth

Stenotrema edvardsi (Bland, 1856)

Diameter: 7-8 mm

Description: Pill-shape; shell with 5-5.5 whorls; imperforate; tawny-olive to light cinnamon-colored; stiff and well developed hairs, usually present on live and fresh shells but may be lost on older shells (see opposite page of a young animal); transverse striae poorly developed; basal notch and interdenticular sinus are indistinct; fulcrum short and moderately developed; shell periphery distinctly and always angular.

Similar Species: *Stenotrema spinosum* (next page) is much larger and usually found in more calcareous soils or fractures in limestone cliflines, notably more compressed and is without significant hairs; *S. hirsutum* (illustrated in Part I) is larger, has a wider mouth and rounded periphery.

Habitat: Rocky mixed hardwood forests, around and under logs and in leaf litter; very common in hemlock-dominated ravines on hillsides (Hubricht 1985).

Status: G5; Common in the Valley and Ridge regions of Tennessee and the Cumberland Plateau of Kentucky.

Specimen: Kentucky, Powell County, Furnace Mountain (authors collection)

Polygyridae

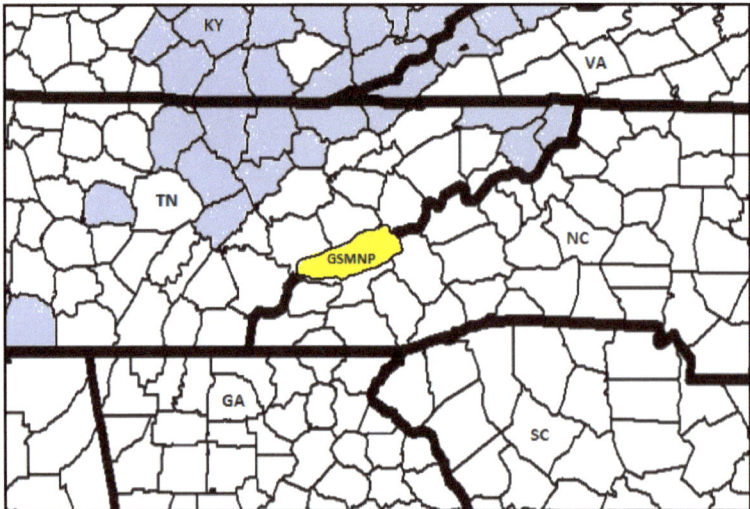

Above image of a juvenile Ridge-and-valley slitmouth, *Stenotrema edvardsi* crawling toward dinner; a colony of maturing sporangium (slime mold) growing from rotting wood. Below, note the numerous hairs that cover the entire shell surface; the longest ones growing from the sutures. These hairs are usually lost as shells mature. Both images from Red River Gorge, Kentucky.

Carinate slitmouth

Polygyridae

Stenotrema spinosum (I. Lea, 1830)

Diameter: 11.8-15 mm

Description: Depressed heliciform, lens-shape; shell with 5.5-6 whorls; imperforate; cinnamon-brown; there is a fringe of long hairs in live specimens, although these hairs are usually lost in older shells; spiral striae poorly developed on base; basal notch and the interdenticular sinus are barely perceptible; the fulcrum (a) is well developed, extending nearly one-third across the cavity; shell periphery sharply and notably angular to carinate.

Similar Species: *Stenotrema spinosum* is one of several unique species in this genus that are flattened and lens-shaped.

Habitat: Usually found on limestone rock faces where it hides in narrow crevices by day, becoming active during moist nights.

Status: G4; Common; not yet recorded in the GSMNP but found in Tennessee counties near the park.

Specimen: Virginia, Scott County; limestone cliffs above the Clinch River and I-24 bridge (author's collection).

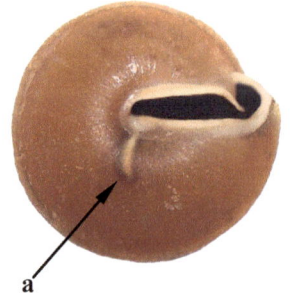

a

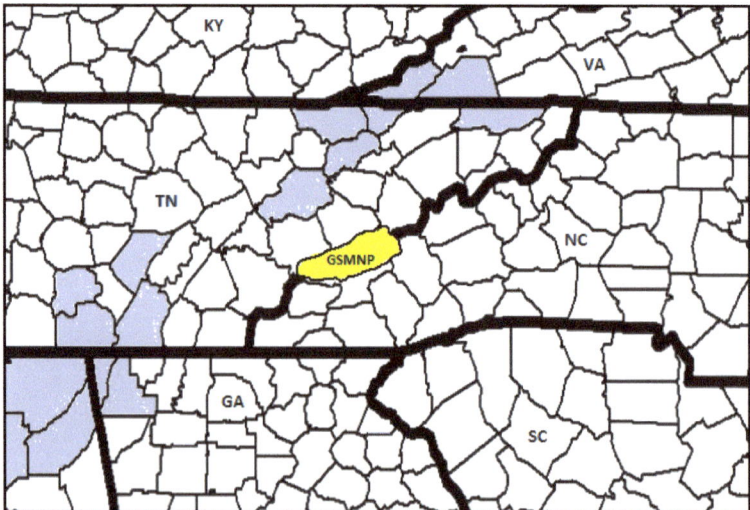

Fringed slitmouth

Polygyridae

Stenotrema barbigerum (Redfield, 1856)

Diameter: 8.7-10 mm

Description: Depressed heliciform, lens-shape; shell with 5 whorls; imperforate; light cinnamon-colored; with long stiff hairs emerging from the sutures; spiral striae poorly developed on base; without a basal notch or interdenticular sinus; shell periphery sharply angular.

Similar Species: *Stenotrema edvardsi* is not nearly as compressed and has shorter, thicker hairs; *S. spinosum* is more compressed, larger and usually without notable hairs.

Habitat: Generally found in fairly dry locations in mixed hardwood forests; often associated with rotting hardwood logs in advanced stages of decay where it can be found in small colonies of up to a dozen or so individuals on the same log.

Status: G4/G5; Uncommon; not yet recorded in the GSMNP but found in North Carolina counties that border the park.

Specimen: Tennessee, Polk County, Ocoee Gorge (author's collection).

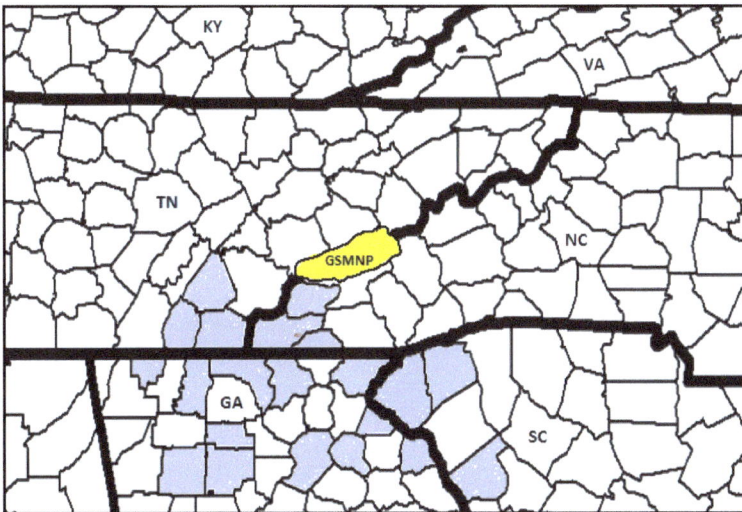

Noonday globe

Polygyridae

Patera clarki nantahala Clench, 1933

Diameter: 15-20 mm

Description: Depressed heliciform; lip reflected; shell with 5.5-6 whorls; imperforate; moderate parietal tooth present; basal tooth smaller but a constant feature; pale buff to tan; no hairs in adult shells; transverse striae well-developed; no spiral striae or if present, seen only as scattered traces on the base; periphery rounded.

Similar Species: *Patera clarki* (illustrated below) has a notably higher shell profile and usually larger parietal tooth.

Habitat: Rich cove hardwood forests under leaf litter, rock talus and around log structure.

Status: G1; Globally Rare; **USFWS Federally Threatened;** the species is Endemic to Nantahala Gorge, North Carolina.

Specimen: North Carolina, Macon County, Nantahala Gorge (author's collection).

Patera clarki clarki

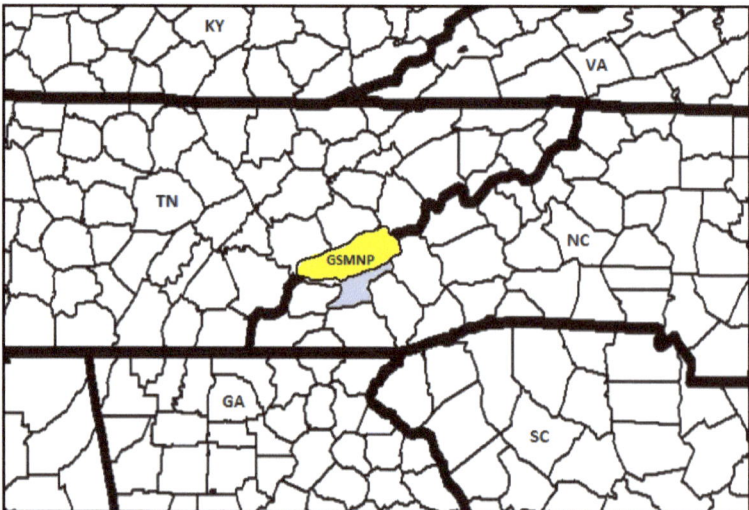

291

Proud globe

Mesodon elevatus (Say, 1821)

Polygyridae

Diameter: 19.8-26.3 mm

Description: Heliciform; lip reflected and thickened; shell with 6-7 whorls; imperforate; shell solid; large parietal tooth usually present; pale yellow to light olive; no hairs; transverse and minute spiral striae a well developed feature and always present; shell periphery well-rounded; occasionally young and adult shells with a light color band.

Similar Species: *Mesodon clausus* is smaller, has a thinner shell and is without a parietal tooth; *M. zaletus* has a notably smaller parietal tooth and has a slightly more compressed shell but this feature varies from site to site.

Habitat: A species of limestone river bluffs and rich wooded slopes in mixed hardwood forests, also found in upland sites under forest litter; can be especially common around the entrances of limestone caves.

Status: G5; Common; not yet recorded in the GSMNP but found in counties in Tennessee and North Carolina that border the park.

Specimen: Kentucky, Knox County, Pine Mountain State Park (author's collection).

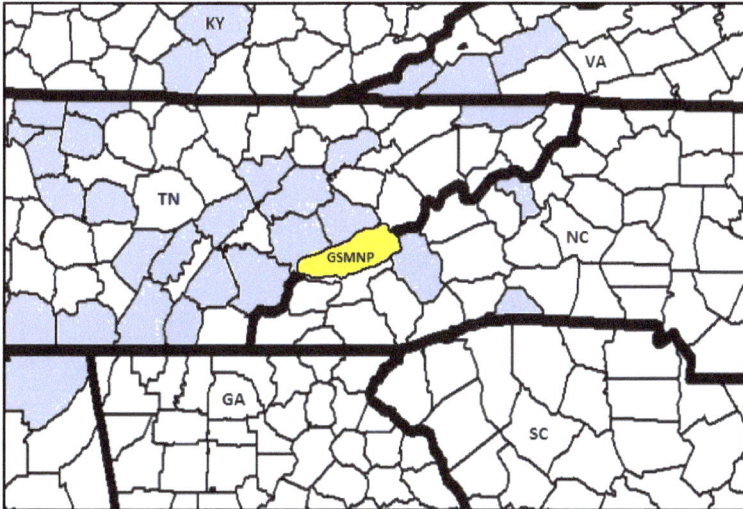

The *Mesodon* Complex (*M. andrewsae*, *M. altivagus* and *M. normalis*)

Mesodon andrewsae

These are the largest and most conspicuous land snails of the southern mountains. In the 1940s, Pilsbry grouped these three species together, describing them as forms related to elevation, ranging across the Southern Appalachians. He also discusses a fourth intermediate form that shares common shell morphology with the three currently recognized species, occurring at elevations between 1100 and 1500 meters. In 1985, Hubricht treated *M. altivagus* and *M. andrewsae* as a single species. It wasn't until 1998 when Emberton conducted his genitalic, allozymic and conchological evolution of the tribe Mesodontini (Pulmonata: Stylommatophora: Polygyridae) that the *Mesodon* complex became more thoroughly understood. Now considered as three distinctive taxa, *Mesodon altivagus* and *M. andrewsae* are restricted to higher elevation peaks (above 1500 meters); *Mesodon altivagus* endemic to the GSMNP, while *M. andrewsae* is found from Pocahontas County, West Virginia to Roan Mountain, including Mt Mitchell, North Carolina. *Mesodon normalis,* the widest ranging of the three species, is a snail of lower elevation mountains, ranging from Rockbridge County, Virginia to Tuscaloosa County, Alabama. The intermediate form discussed by Pilsbry has remained anonymous. While these copious land snails are easily confused, range and elevation limitations should help in their separation. Keep in mind however that like most land snails, there will always be color and shell-form discrepancies, even within the same populations.

M. andrewsae, **Roan Mt, NC**

M. normalis, **Macon Co. NC**

M. altivagus, **GSMNP, NC**

Balsam globe

Polygyridae

Mesodon andrewsae (W.G. Binney, 1879)

Diameter: 20-30 mm

Description: Heliciform; lip reflected; shell with 5.5-6 whorls; imperforate; shell thin for its size, some shells easily dented with finger tips (c); without a parietal tooth; coppery-brown (a) to chocolate-brown (b), pictured here with live animal inside; no hairs; transverse striae not well-developed; spiral striae a strong feature; shell periphery well-rounded; this species has been observed feeding on fresh dead millipedes, hypothetically for the calcium content of the exoskeleton, Roan Mountain, NC (pers. obs.).

Similar Species: *Mesodon thyroidus* has a rimate umbilicus that is slightly open; *M. normalis* is larger with a thicker shell and generally found below 1400 m; *M. zaletus* has a thicker shell and a parietal tooth.

Habitat: A species of high elevation northern hardwood and spruce/fir forests (from 1600 to 2000 meters).

Status: G2/G3; Locally Rare; currently known from around Roan Mountain (North Carolina and Tennessee), north to the West Virginia highlands.

Specimen: All specimens from North Carolina, Mitchell County, Roan Mountain (author's

a

b

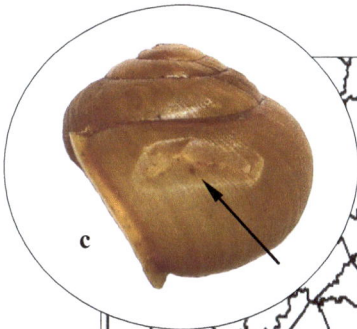

c

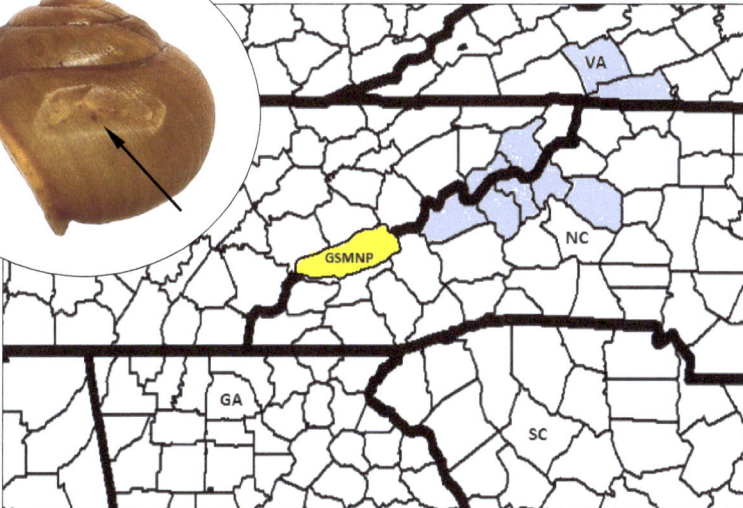

Velvet covert

Polygyridae

Inflectarius subpalliatus (Pilsbry, 1893)

Diameter: 12-16 mm

Description: Depressed heliciform; lip reflected and roundish in bottom view; shell with 4.5-5 whorls; imperforate; large parietal tooth with a moderately wide base; with or without (usually without) a small basal tooth; palatal tooth absent; shell surface dull with fine erect periostracal processes; greenish-olive (figure c).

Similar Species: *Inflectarius verus* has a more squared aperture, a longer basal tooth with a wider base and most importantly (in bottom view) less space between the parietal tooth and lower lip (figures a and b).

Habitat: Above 900 m in northern hardwood and spruce/fir forests; usually occurs under leaf litter but also climbs trees during wet weather.

Status: G2; Endemic to a small region in the states of Tennessee and North Carolina; Uncommon to Locally Rare.

Specimen: North Carolina, Mitchell County, Roan Mountain (author's collection).

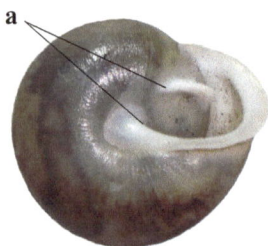

I. subpalliatus, Roan Mt.

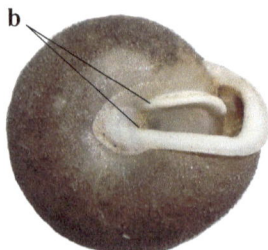

I. verus, GSMNP

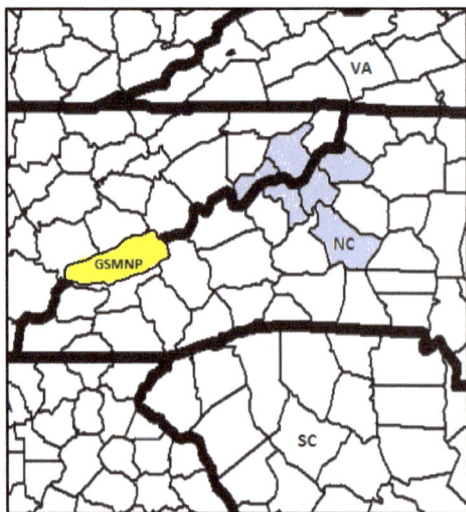

295

Ocoee covert

Polygyridae

Fumonelix archeri Pilsbry, 1940

Diameter: 13-15 mm

Description: Depressed heliciform to heliciform; lip widely reflected; shell with 5-6 whorls; imperforate; large parietal tooth, that crowds the aperture; shell brownish; well-developed transverse striae; without papillae or spiral lines; periphery well rounded.

Similar Species: *Fumonelix wheatleyi* is around the same size but has a notably smaller parietal tooth; *F. jonesiana* has a smaller parietal tooth and is found in high elevation forests of the GSMNP only.

Habitat: Found along rivers and creeks in rich cove hardwood habitats, also found under doghobble and rhododendron thickets.

Status: G1; Endemic to a relatively small area around Tennessee and Georgia; reported from the Ocoee Gorge, Monroe County, Tennessee and along the Conasauga River, Murray County, Georgia and Jacks River, Murray and Fannin Counties Georgia (Biggins 1988, USFWS and Hubricht 1986)

Specimen: Tennessee, Monroe County, Goforth Creek, Ocoee River Gorge (author's collection).

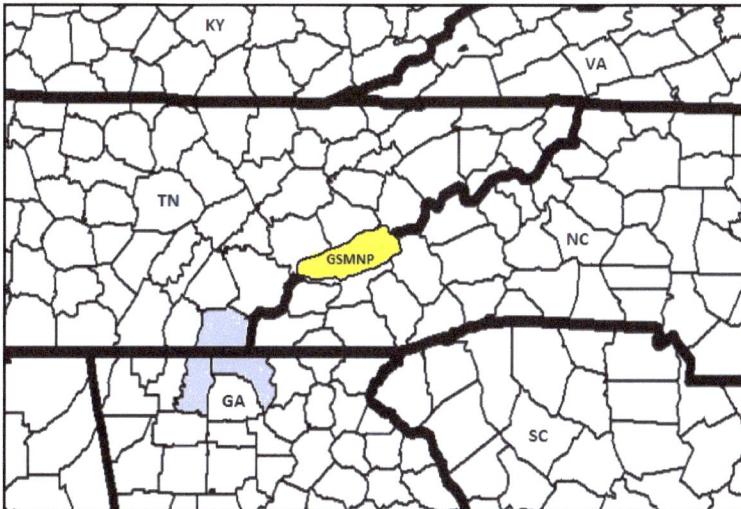

Rock-loving covert

Polygyridae

Fumonelix cherohalaensis Dourson 2012

Diameter: 20-23 mm

Description: Heliciform; lip widely reflected; shell with 5-6 whorls; imperforate; typically toothless but occasionally with a small parietal tooth, shell tannish-brown with tinges of yellow; well-developed transverse striae and, most importantly, spiral striae that appear as closely-spaced, raised wavy lines or beads sometimes seen as low fringes; periphery rounded.

Similar Species: *Fumonelix wheatleyi* is smaller, has a larger parietal tooth and lacks any discernible spiral striae (Pilsbry 1940); *F. orestes* is more depressed and has engraved wavy lines not raised as seen in *F. cherohalaensis*.

Habitat: Found on the undersides of outcropping rock ledges in higher elevation (around 1500 m), northern hardwood forests located along the Cherohala Scenic Drive; type locality is Huckleberry Knob in Graham County, North Carolina.

Status: G1; Globally Rare; Endemic to Graham County, North Carolina and possibly a small area in Monroe County, Tennessee, but this remains to be investigated.

Specimen: North Carolina, Graham County, Huckleberry Knob (author's collection).

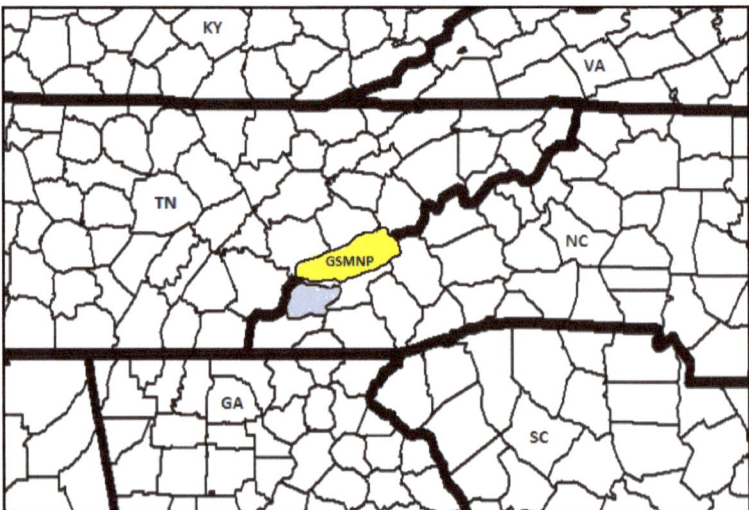

Engraved covert

Polygyridae

Fumonelix orestes (Hubricht, 1975)

Diameter: 14.7-18.3 mm

Description: Heliciform; lip widely reflected; shell with 5 whorls; imperforate; with or without a small parietal tooth; shell thin, pale olive-brown; well-developed transverse and spiral striae that appear as closely-spaced, engraved wavy lines strongest on the base; periphery well-rounded.

Similar Species: *Fumonelix orestes* differs from *F. langdoni* primarily by having pits (not papillae) on the second, third and forth whorls, having more strongly developed spiral striae and when present a smaller parietal tooth.

Habitat: Found on the undersides of rock outcrops in higher elevation (over 1500 m) northern hardwood and spruce/fir forests, Haywood, County, NC.

Status: G1; Globally Rare; Endemic to areas along the Blue Ridge Parkway, North Carolina; although the species has not been recorded from the GSMNP it has been found along portions of the Blue Ridge Parkway that are close to the park.

Specimen: North Carolina, Haywood County, Blue Ridge Parkway (author's collection).

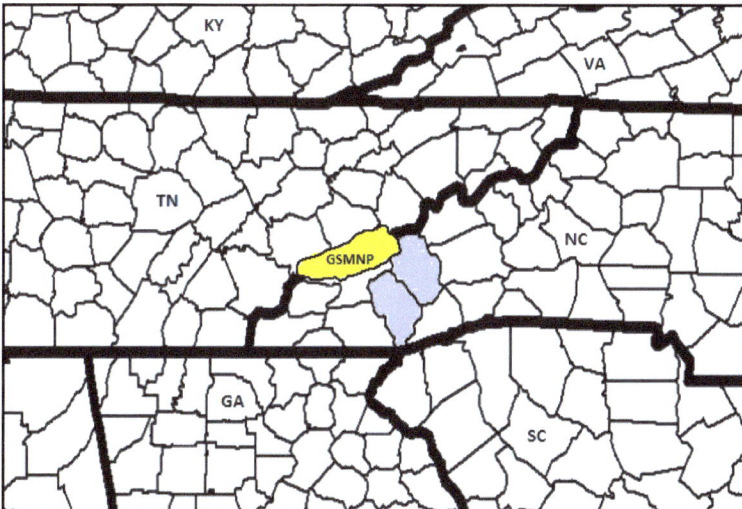

Roan Mountain covert

Polygyridae

Fumonelix roanensis Dourson 2012

Diameter: 15-17 mm

Description: Heliciform; lip widely reflected; shell with 5-6 whorls; imperforate; specimens are with or without (most without) a nearly undetectable parietal tooth and without a basal and palatal tooth; apex usually worn; shell brownish, covered in numerous hairs, especially prevalent in young shells but occasionally on adult shells as well, these hairs typically lost in aging; well-developed transverse striae but without notable spiral striae; periphery rounded; figure (a) was photographed with live animal inside.

Similar Species: *Fumonelix wheatleyi* is more or less the same size, has a higher shell profile and usually has a small parietal tooth.

Habitat: A species of higher elevation (above 1400 m), northern hardwood, beech gap and spruce/fir forests, under leaf litter and around logs.

Status: G1; Globally Rare; this species is Endemic to Roan Mountain, North Carolina and Tennessee.

Specimen: North Carolina, Mitchell County, Roan Mountain around Carver's Gap (author's collection).

a

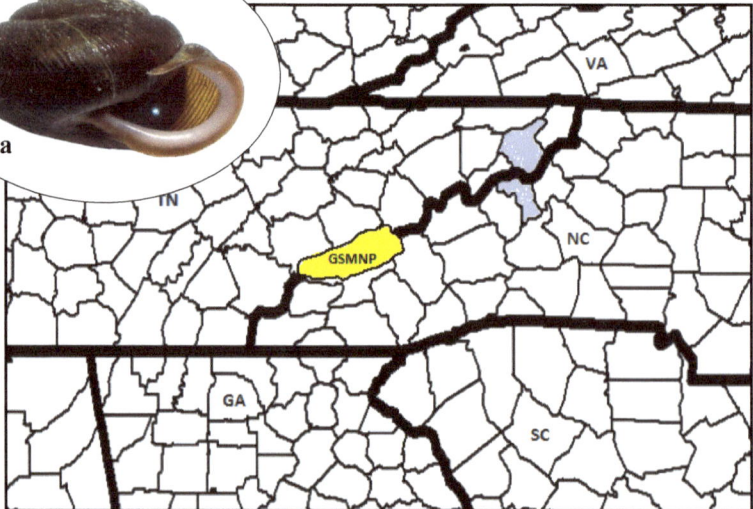

Big-tooth whitelip

Polygyridae

Neohelix dentifera (A. Binney, 1837)

Diameter: 20-30.5 mm

Description: Depressed heliciform; lip reflected; shell with 5-5.5 whorls; imperforate; shell surface dull; parietal tooth low and wide; pale olive; transverse striae well-developed, impressed spiral striae present; no hairs in adult shells; periphery rounded.

Similar Species: *Neohelix albolabris* is larger, with a higher shell profile and no parietal tooth; *Mesodon normalis* has a thinner, higher profile shell and no parietal tooth; *M. zaletus* has a higher shell profile and contains a smaller parietal tooth.

Habitat: Restricted to higher elevation mixed hardwood forests occurring under leaf litter, around rocks and logs on acidic soils.

Status: G5; Locally Rare; not yet recorded in the GSMNP; there is a slight chance that higher elevation forests of the park may harbor this mostly northern species.

Specimen: Kentucky, Letcher County, Pine Mountain (author's collection).

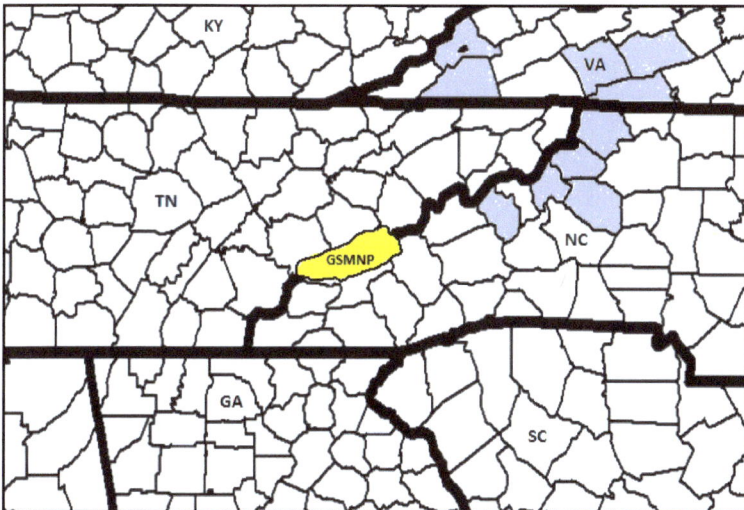

Southeastern whitelip

Polygyridae

Neohelix major (A. Binney, 1837)

Diameter: 35-46 mm

Description: Heliciform; lip reflected, thickened and wide; shell with 5-6 whorls; imperforate; shell solid with a dull surface; without teeth; cream-buff to pale-tan; transverse striae well-developed, impressed wavy spiral striae well-developed; no hairs; periphery broadly rounded; this is the largest native (shelled) land snail in the eastern US.

Similar Species: *Mesodon normalis* is smaller and has a thinner shell; *M. zaletus* is smaller and contains a parietal tooth; *N. albolabris* is smaller and more compressed.

Habitat: A species of mixed hardwood hillsides around rocky limestone outcrops and near cave entrances.

Status: G5; Common; not yet recorded in the GSMNP but reported in counties in Tennessee and North Carolina that border the park.

Specimen: Georgia, Walker County, Pigeon Mountain (author's collection).

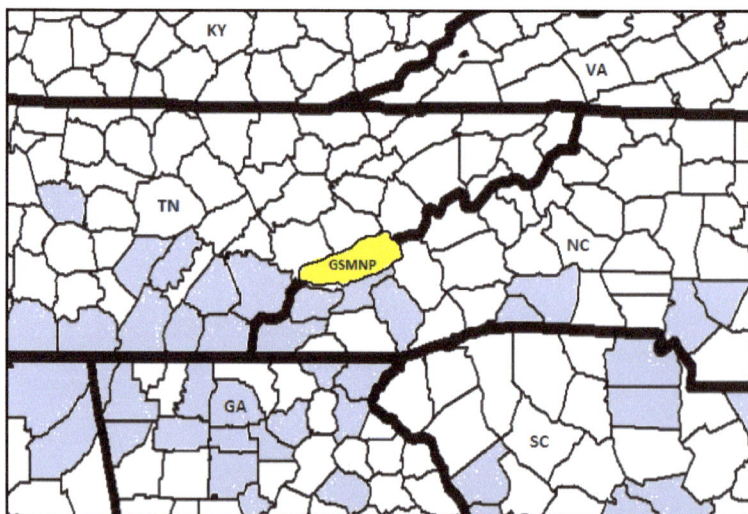

Broad-banded forestsnail

Allogona profunda (Say, 1821)

Diameter: 19-34 mm

Description: Depressed heliciform; lip reflected; shell with 5.5 whorls; umbilicate; basal tooth present but small; shell light tan with a few reddish-brown bands that vary in their width but sometimes shells are found without color bands; shell glossy and thick; transverse striae well-developed, spiral striae are a stronger feature on top but nearly absent on the base; no hairs; periphery rounded.

Similar Species: No other native species in the vicinity of the GSMNP has the multiple color bands of *A. profunda; Anguispira* species have color blotches not bands.

Habitat: Found around limestone and also acidic mountainsides in mixed hardwood.

Status: G5; Uncommon, not yet recorded in the GSMNP but reported in one county in Tennessee that borders the park.

Specimen: Kentucky, Powell County, Furnace Mountain (author's collection).

Polygyridae

plain form, Powell Co, KY

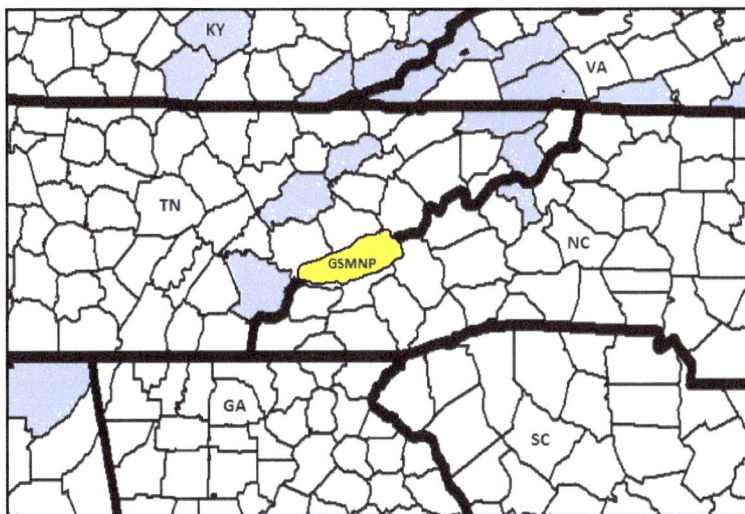

Spike-lip crater

Polygyridae

Appalachina sayana (Pilsbry, 1906)

Diameter: 19.4-27 mm

Description: Depressed heliciform to heliciform; lip reflected, thin and wire-like; shell with 5.5 whorls; umbilicate; shell thin for its large size; small parietal and basal tooth usually present; pale-yellow to pale olive-tan, sometimes with darker streaks; no hairs in adult shells; transverse and minute spiral striae always present; periphery well-rounded.

Similar Species: *Appalachina chilhoweensis* (page 223) has a narrower umbilicus, wider lip and is usually without any notable teeth; *Mesodon* species are generally more globose, are perforate and have heavier shells.

Habitat: A common species of rich upland mixed hardwood forests under leaf litter and other forest debris on both limestone and acidic sites, the species is generally indicative of quality forests.

Status: G5; Uncommon; not yet recorded in the GSMNP but found in North Carolina counties that are close to park boundaries.

Specimen: Kentucky, Powell County, Furnace Mountain (author's collection).

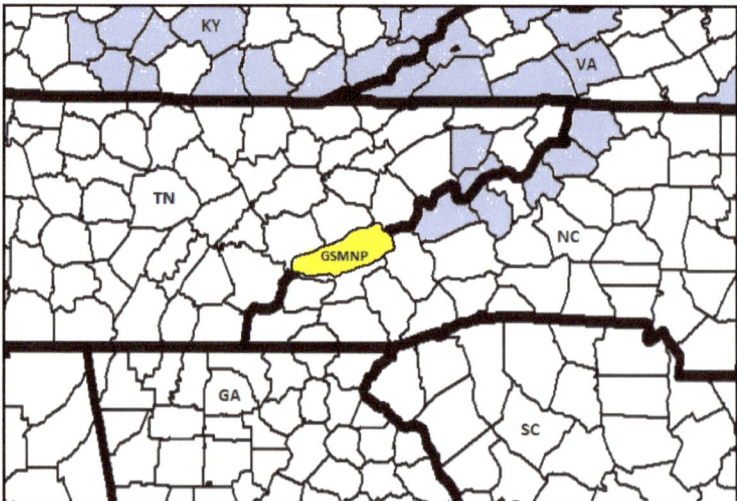

Symbiosis Between Land Snails and Slime Molds?

Above image of a spike-lip crater, *Appalachina sayana* feeding on immature slime mold sporangium in Red River Gorge, Kentucky. There is some speculation that symbiotic relationships exist between land snails and particular species of slime molds. For example, slime molds in the order Physarales precipitate amorphous calcium carbonate on some portion of their mature fruiting bodies (Townsend et al. 2005). *Badhamia utricularis* (figure a) not only coats the outer surface of its fruiting body with calcium, but has a system of calcareous tubes running throughout its core (figure b). This may be coincidence or a clever strategy for attracting land snails who require the mineral for shell building. What if Physarales slime molds sequester and store calcium for the singular purpose of attracting land snails? To ensure the spores are eaten by the visiting snail, the calcium deposits are intermingled with the spores (b). Pushing the hypotheses a bit further. What if slime molds were one of several conduits that help facilitate land snail migration into more acidic environments? These hypotheses await investigation. Figures a & b images from Eumycetozoan Research Project, University of Arkansas.

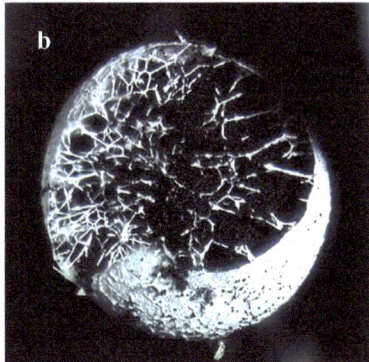

Carter threetooth

Triodopsis anteridon Pilsbry, 1940

Polygyridae

Diameter: 11-14 mm

Description: Depressed heliciform; lip reflected; shell with 5-5.5 whorls; umbilicate; large parietal tooth present, points above the palatal tooth; basal tooth and palatal tooth substantially smaller; palatal tooth situated on a wide buttress (figure a), a widening of the outer lip; shell brown; transverse striae are well developed on the top and sides of the shell continuing well into the umbilicus region; scattered papillae are generally a strong feature especially near sutures and the umbilicus; without hairs in mature shells; periphery is rounded.

Similar Species: *Triodopsis tridentata* is larger, the parietal tooth is smaller and points at or below the palatal tooth.

Habitat: Found in mixed hardwood forests under leaf litter and close to logs; occasionally under logs and rocks.

Status: G3; Locally Rare; an infrequent land snail in the region covered by this book.

Specimen: Tennessee, Carter County, Cherokee National Forest (author's collection).

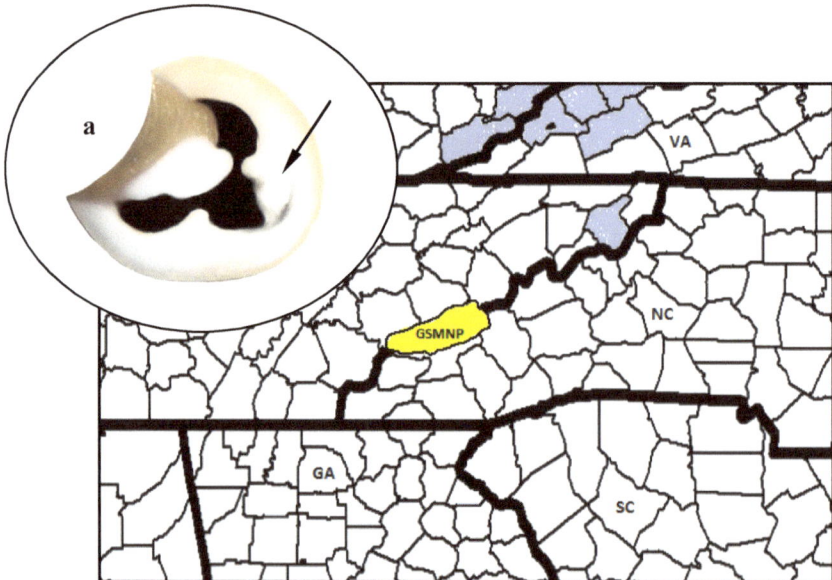

a

Hanging Rock threetooth

Polygyridae

Triodopsis pendula Hubricht, 1952

Diameter: 9-14 mm

Description: Depressed heliciform; lip reflected; shell with 5-5.5 whorls; umbilicate; large parietal tooth which points above the palatal tooth; basal tooth and palatal tooth substantially smaller and of equal size; palatal tooth not situated on a wide buttress (b); shell pale cinnamon-buff; transverse striae well developed on the top and sides of the shell continuing well into the umbilicus region; scattered papillae are a strong feature, especially around the umbilicus and behind the peristome (lip); without hairs in adult shells; periphery rounded.

Similar Species: The palatal tooth of *Triodopsis anteridon* sits on a buttress; *T. tridentata* is larger, the parietal tooth is smaller and points at or below the palatal tooth.

Habitat: Found in upland oak woods under leaf litter and close to logs.

Status: G4; Uncommon; the species is somewhat range restricted in North Carolina and reported from only one county in Virginia.

Specimen: North Carolina, Watauga County, Stone Mountain at Locust Gap (author's collection).

b

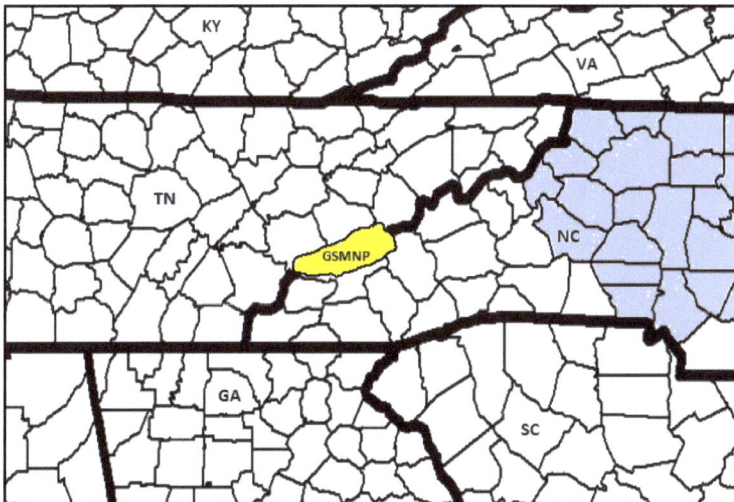

Budded threetooth

Polygyridae

Triodopsis tennesseensis (Walker & Pilsbry, 1902)

Diameter: 19-24 mm

Description: Depressed heliciform; lip reflected and wide; shell with 5-5.5 whorls; umbilicate; parietal tooth moderate, points directly at or below the palatal tooth; basal and palatal tooth small, roughly the same size and do not crowd the aperture in frontal view; on occasion without basal and palatal teeth (Dourson 2010) shell snuff-brown; transverse striae are well-developed on the top, sides and bottom of the shell into the umbilicus region; scattered papillae mostly along the transverse striae and around the umbilicus; without hairs; periphery rounded.

Similar Species: *Triodopsis tridentata* is typically smaller, more compact, has a notably smaller umbilicus and larger basal and palatal teeth.

Habitat: Found under leaf litter in mixed hardwood on hillsides.

Status: G5; Uncommon; not yet recorded in the GSMNP but located in counties that are close to the park.

Specimen: Kentucky, Letcher County, Bad Branch (author's collection).

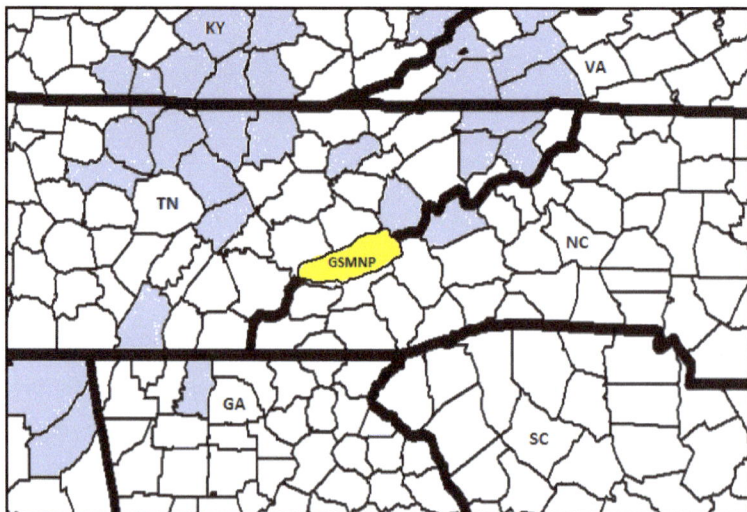

Small mammals are aggressive feeders on a variety of terrestrial snails. Above and below images of a hairy-tail mole gripping a *Triodopsis tridentata,* below the mole bearing its imposing teeth. Carvers Gap, Roan Mountain, Mitchell County, North Carolina.

Virginia mantleslug

Philomycidae

Philomycus virginicus (Hubricht, 1953)

Length: Adults 75-100 mm while crawling

Description: Base color of mantle is a grayish-white. Color pattern consisting of a darker (brownish) broad dorsal band, sometimes bordered by a row of darker spots and a narrow lateral band on each side, connected to the dorsal band by a series of oblique (diagonal) faint, stripes (b, c and d), the whole pattern obscured by a general fine flecking. Young with the pattern brownish-gray, becoming chestnut-brown with age (Hubricht 1953); defense mucus whitish; foot margin off-white; sole of foot gray to whitish.

Similar Species: *Philomycus venustus* (below image) has three darker longitudinal bands often times broken into spots that are also connected by darker oblique bands or spots (a), all of which are notably darker on a usually lighter-colored animal, than is seen in *P. virginicus*.

Habitat: An upland species found in mixed hardwood under the exfoliating bark of large rotting hardwood logs in advanced stages of decay.

Status: G3; Locally Rare; in North Carolina a slug restricted to upper elevations.

Specimen: Figures (b and c) from North Carolina, Avery County, Lutherock (Wayne Van Devender), figure (d) from West Virginia, Coopers Rock State Forest (authors).

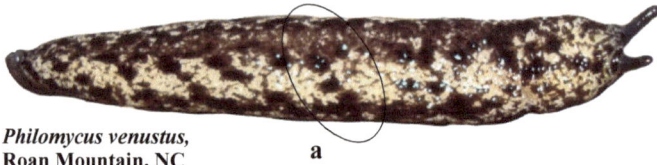

Philomycus venustus,
Roan Mountain, NC   a

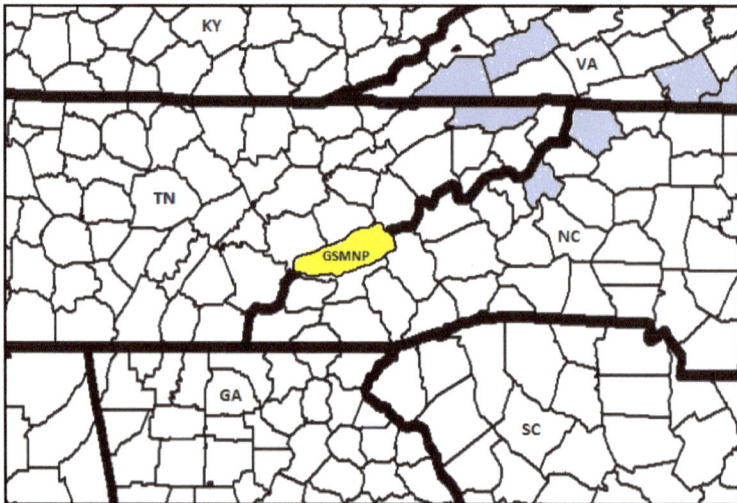

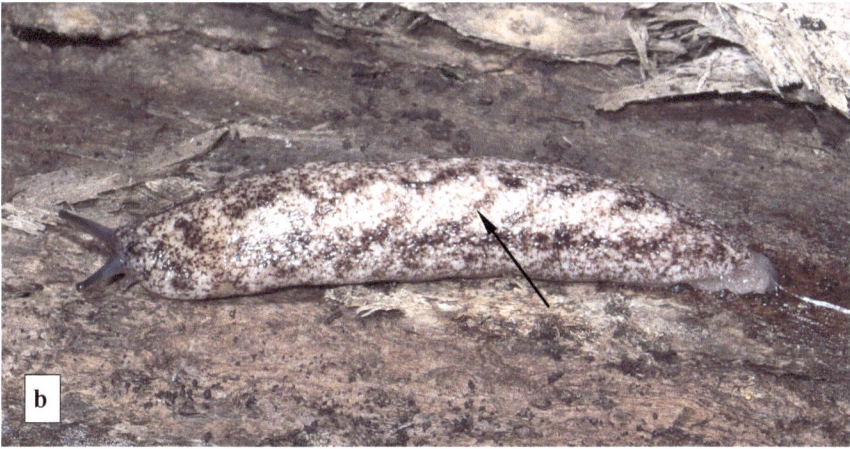

b

c

d

Redfoot mantleslug   Philomycidae

Pallifera ohioensis (Sterki, 1908)

Length: Adults 15-30 mm while crawling

Description: The live animal has a speckled grayish back without an inter-rupted black line down the center of the mantle; defense mucus whitish; sole of foot with two blood-red bands running lengthwise along the edge (figure a showing underside of slug) seen in this overturned slug in the picture below.

Similar Species: *Pallifera dorsalis* (page 247) is around the same size and lean in form but is without the red bands and has, on occasion, an interrupted set of black spots running down the center of its back.

Habitat: A snail of humid forests, usually found under bark, deep leaf litter under decaying logs or in rock talus; in West Virginia, it occurs in upper eleva-tion red spruce forests around sandstone.

Status: G3; Locally Rare; this species was only recently discovered (2013) on Roan Mountain, NC at around 1500 meters in a disappearing beech-gap forest; also reported from Ohio, Indiana and West Virginia.

Specimen: Blackwater Falls, Tucker County, West Virginia (author's photo collection).

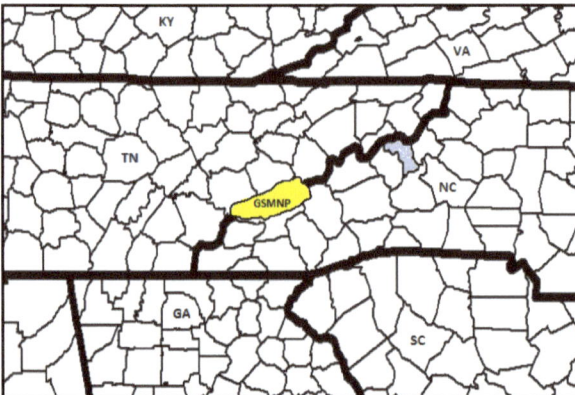

Size of slug crawling

Exotic slugs that have crossed country borders are included in this section. Most have been accidently introduced into North America by way of plants, construction lumber, potting soils and tree nurseries. A full color plate of additional (exotic) slugs that may eventually find their way to degraded lands of the GSMNP are included (page 316).

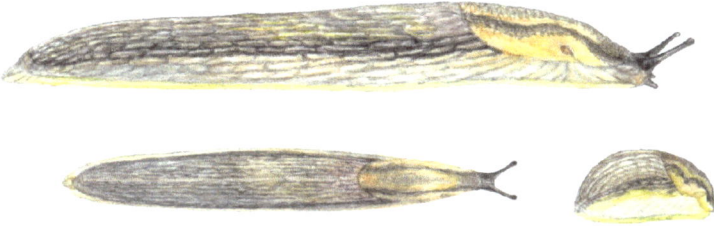

I was here first!
Take your appetite
somewhere else!

Ze'll turn moi
into Escargo!

Dusky Arion
Arion subfuscus (Draparnaud, 1805)

Length: Adults 50-60 mm while crawling

Description: Orange and brown with lateral light-brownish bands; orange saddle-like mantle which covers only the anterior (front) part of the dorsal region; breathing pore is in the front half of the mantle in all *Arion* (as opposed to *Deroceras* and *Limax*); reproductive pore located behind the breathing pore; defense posture of slug (a); defense mucus bright yellow or orange; foot pale-yellow.

Similar Species: *A. fasciatus* is smaller and has a colorless defense mucus; *L. maximus* is much larger, has a different body color and clear defense slime.

Habitat: Found in mostly degraded habitats of cities and urban areas, usually living among or under stones, old wood piles and window wells; this pest species is especially problematic in gardens and can reach substantial numbers.

Status: Locally Rare, but can be quite common in disturbed areas (e.g. old homesteads located just north of Fontana Lake).

Specimen: Illustrations from Grimm et al. (2009)

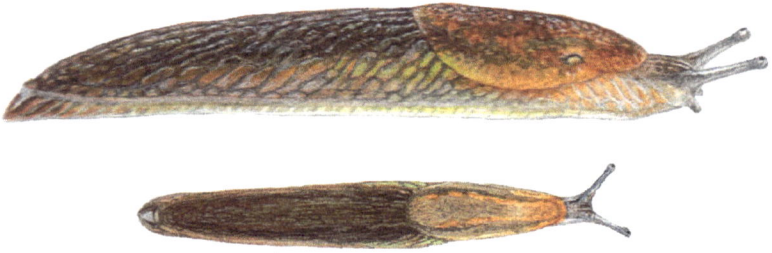

© Aleta Karstad

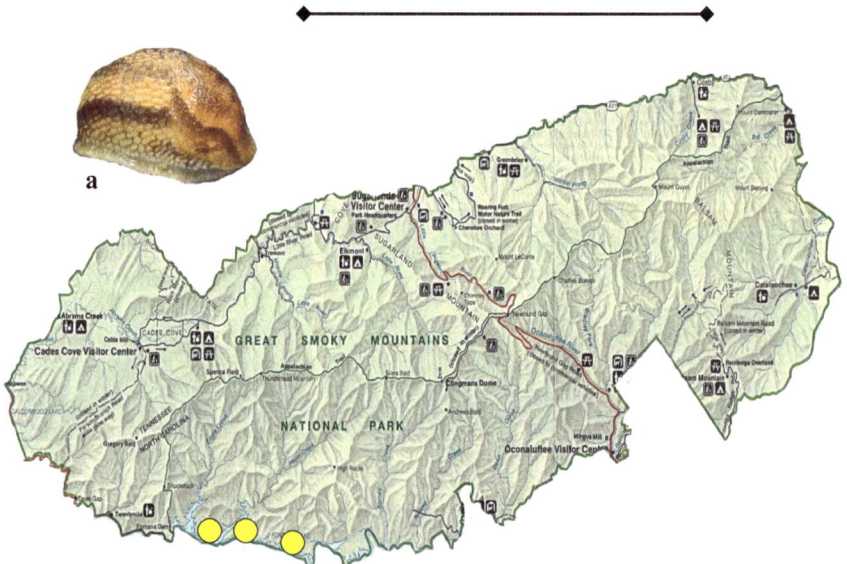

a

Orange-banded arion

Arion fasciatus (Nilsson, 1823)

Length: Adults 30-40 mm while crawling

Description: Gray overall, gray above, fading to gray-white on the sides below dark lateral bands; gray to yellowish saddle-like mantle covers only the anterior (front) part of the slugs dorsal region (back); breathing pore is in the front half of the mantle in all *Arion* (as opposed to *Deroceras* and *Limax*); reproductive pore in front of breathing pore; defense mucus clear; margins of the foot grayish-white.

Similar Species: Native slugs in the GSMNP except for *Deroceras laeve* have mantles that nearly cover the entire body; the mantle of *D. laeve* is longer, has no lateral bands; *A. subfuscus* is larger, the reproductive pore is located behind the breathing pore and the defense slime is yellow or orange.

Habitat: Found in mostly degraded habitats of cities and urban areas, usually living among or under stones, old wood piles and window wells; this pest species is especially problematic in gardens and can reach large numbers.

Status: Locally Rare; only common in and around urban areas.

Specimen: Illustrations from Grimm et al. (2009).

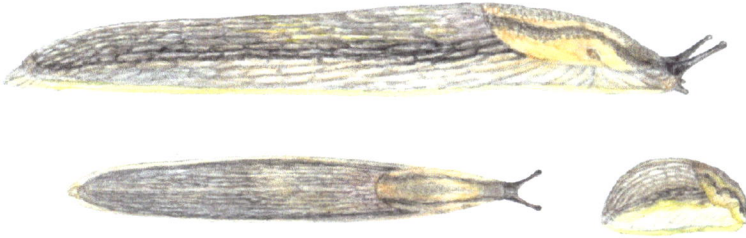

© Aleta Karstad

Tiger slug

Limax maximus Linnaeus 1758

Length: Adults 100-200 mm while crawling

Description: Mantle and body black spotted or mottled rather than with continuous bands; body color yellowish gray or light brownish gray; tentacles reddish brown; posterior end of slug keeled but not up to the mantle; defense mucus clear; margins of the foot and sole whitish.

Similar Species: This sizeable slug looks like no other native or exotic slug in the GSMNP.

Habitat: Very common in degraded habitats of cities and urban areas, under railroad ties used for landscaping; the species feeds mostly on dead vegetation and fungi; other foods include roots, fruit, leafy crops and carrion (Grimm et al. 2009); also reported to crawl onto porches to consume pet food (MacGregor pers. comm. 2010); especially problematic in gardens and can reach large numbers.

Status: Locally Rare; known from around old homesteads in Cades Cove.

Specimen: Illustrations from Grimm et al. (2009)

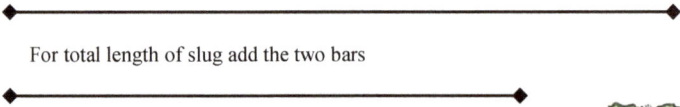

For total length of slug add the two bars

© Aleta Karstad

Exotic Slugs that may Eventually be Found in Degraded Habitats of the GSMNP (Slugs are illustrated proportionate)

Arion ater (100-200 mm)

Limax maximus (100-200 mm)

Shell of slug

Testacella haliotidea (100 mm)

Limacus flavus (75-100 mm)

Arion circumscriptus (30-40 mm)

Arion subfuscus (50-70 mm)

Arion fasciatus (30-40 mm)

Lehmannia valentiana (70 mm)

Deroceras panormitanum (30-40 mm)

Milax gagates (50-60 mm)

Arion intermedius (15-20 mm)

Deroceras reticulatum (35-50 mm)

Boettgerilla pallens (30-40 mm)

© Aleta Karstad

Native Aquatic Snails of the GSMNP Region

All specimens from the North Carolina State Museum of Natural Sciences.
Shells are not propionate, scale bars represent true size of each species.

Phanorbella trivolvis
NCSM 58010

Micromenetus dilatatus
NCSM 31706

Helisoma anceps
NCSM 31762

Gyraulus parvus
NCSM 23521

Ferrissia rivularis
NCSM 31516

Ferrissia fragilis
NCSM 47660

Elimia clavaeformis
NCSM 5922

Elimia proxima
NCSM 5869

**Pleurocera canaliculatum
undulate,** NCSM 6598

Campeloma decisum
NCSM 5827

Galba humilis
NCSM 31550

Pseudosuccinea columella
NCSM 31567

Leptoxis praerosa
NCSM 31384

Physella gyrina
NCSM 9601

Physella heterostropha
NCSM 51385

Pesticide Free Alternatives for Invasive Land Snail Control

Commercial slug baits and repellents contain a molluscicide that kills gastropods. Although they are widely used, they may be toxic to pets, fish, wildlife and humans. Less toxic methods have been employed by organic gardeners for years and are effective in controlling persistent slugs and shelled snails. Included here are several suggestions.

Gather individuals and place in boiling water or drench them in salt water.

Maintain a compost heap near the garden (which attracts the slugs).

Place boards, moist newspaper or cardboard, old scraps of carpet in the garden, which provide cover for the slugs during daytime hours; the slugs can than be easily harvested and euthanized.

Placing cups in the garden filled with beer or grape juice will attract and drown the slugs.

Placing wet dog food in piles close to infestations will attract slugs, check in the evening and simply dispose of the food and slugs.

Inverted orange or grapefruit rinds scattered among the vegetables will collect slugs which can than be collected, mashed and composted.

Releasing ducks (ducks may also eat young plants) and chickens (Rhode Island Reds are best) loose in the garden will keep the plot clear of slugs and snails. Other birds reported to eat pesky gastropods include blackbirds, crows, jays, owls, robins, seagulls, starlings, and thrushes.

Lava rock, the spiny fruits of sweet gum trees, wood ash, oak leaves, pine needles, coffee grounds, cedar chips, Epsom salts, builders sand, nut shells, oat bran, lime, and talcum power are reported to act as barriers or repellents to slugs or are toxic to them.

Watering the soil with liquid seaweed extract has just enough of an alkaline effect that slugs detest, however not enough to significantly change soil pH.

Human hair, pet fur, and horsehair will entangle slugs. Ground beetles (particularly *Carabidae* beetles), box turtles, toads, frogs, lizards, salamanders, lighting bug larvae, and brown snakes are reported to eat slugs and shelled snails, so providing cover for these organisms in and around your garden will help in controlling unwanted gastropods.

Springtime is the best time of year to deal with these uninvited guests. Spring cultivation of the soil will help kill hibernating slugs and their eggs.

Removing slug slime from your hands can be accomplished with white vinegar and warm water or consider using chopsticks for picking up those slimy gastropods.

GLOSSARY

Abcission: when a plant drops one or more of its parts like leaves, fruit or seeds.

Amorphous: shapeless.

Algific: It contains an unusual type of ecosystem characterized by algific ("cold-producing") talus, a loose-rock slope affected by the movement of cold air.

Angular periphery: shell having an angular rather than a round contour (also referred to as carinate).

Angular lamella: the tooth on the parietal wall of the aperture to the right of the parietal lamella in dextral shells.

Aperture: opening or mouth of the snail shell.

Apex: the tip of a gastropod shell farthest away from its aperture.

Axis: The imaginary line around which the whorls of a coiled shell are formed.

Basal: pertaining to, situated at or forming, the base; the part of the shell furthest from the apex; the ventral part of the aperture lip.

Basal lamella: tooth on the bottom left side of the aperture below the columellar lamella.

Basal tooth: calcareous deposit on the basal apertural lip.

Body Whorl: the last whorl of a shell, measured from the outer lip back to a point immediately above the outer lip.

Breathing pore: opening in mantle or mantle edge for passage of air into the air sac or lung cavity.

Calcareous: composed of carbonate of lime (calcium carbonate).

Callus: a deposit of lime or shell material, often as a thickening near the umbilicus.

Carinate: the periphery of the shell appears compressed, creating a sharper angular appearance with a ridge-like rim.

Cations: minerals contained in the soil like magnesium, aluminum and calcium.

Caudal: situated in or near the tail or posterior end.

Columella: the internal column around which the whorls revolve; the axis of a spiral shell.

Columellar lamella: tooth on the columellar wall of the aperture.

Columellar wall: the left side of the aperture in dextral shells.

Compressed shell: a shell that is heliciform in shape.

Corneous: horn-like in color.

Crypsis: the ability of an organism to avoid observation or detection by other organisms.

Depressed heliciform: shell with flattened spiral.

Dextral: spiraled to the right when the shell is held so that the apex is up and the aperture is facing the observer.

Dioecious: having both sexes.

Dessication: drying out.

Embryonic whorl: the earliest whorls that are formed in the egg.

Epiphragm: a hardened mucous covering sealing the aperture that prevents desiccation during dry weather.

Endemic: native or restricted to a certain country

Foot: the locomotory organ of mollusks; it is often modified for digging, grasping prey; the long broad ventral surface of the animal.

Globose: globular, formed like a globe, spherical.

Growth line: a line on the surface of a shell indicating a rest period during growth.

Heliciform: shell that has an elevated spire or globose form.

Hermaphrodite: possessing both male and female reproductive organs.

Hirsute: covered with hairs.

Holarctic: a region which includes the northern parts of North America, Greenland, Europe and Asia.

Imperforate: lacking an opening on the ventral or anterior end of the shell.

Impressed: marked by a furrow.

Infra-parietal lamella: the tooth on the parietal wall to the left of the parietal lamella in dextral shells.

Lamella: a fold, "tooth", or raised callus in the aperture of a shell.

Lip: the edge of the aperture; also called the peristome.

Loosely coiled: having few, widely expanding whorls.

Lower palatal lamella: lowermost of the two major teeth often found on the palatal wall.

Lunate: shaped like a half-moon.

Mantle: a membranous flap or outer covering of the softer parts of the mollusk; it secretes the shell.

Mouth: the opening or aperture of a gastropod shell.

Mucus: a viscid, slippery secretion, slime.

Operculate: bearing an operculum or cover to close the aperture.

Operculum: a horny or calcareous plate that serves the purpose of closing the aperture when the snail withdraws into its shell.

Outer lip: the outer edge of the aperture.

Palatal depression: indentation of the shell surface at the location of the palatal lamellae.

Palatal tooth: refers to the tooth located on the outer lip.

Palatal wall: the right side of the aperture in dextral shells.

Papillae: small calcium deposits that appear as minute bumps on the surface of the shell.

Parietal lamella: major tooth in the middle of the parietal wall of the aperture.

Parietal tooth: refers to the tooth located on inner wall of the aperture or body of the snail shell.

Penultimate whorl: the next to the last whorl in an adult shell.

Perforate: a minute opening in the umbilicus.

Periostracum: the chitinous external layer covering most mollusk shells.

Periphery: the part of the whorl farthest from the central axis.

Radula: a ribbon-like organ with many fine teeth used in rasping food.

Reflected: turned outward, e.g., a portion of the apertural lip of some snails' shells.

Ribs: prominent protrusions on the shell surface.

Rib-striae: transverse undulation of the shell surface which are more prominent than striae but less prominent than ribs.

Rimate: the lip of the aperture slightly covers the umbilicus.

Rounded periphery: an evenly curved periphcry; not angular or carinate.

Slug: a common designation for a snail without an external shell. The shell is either rudimentary and enclosed or is wanting entirely.

Senescence: biological aging.

Spiral: winding, coiling, or circling around a central axis; the form of the shell of most snails.

Spiral striae: surface features that are indented or raised on the shell surface running parallel with the whorls.

Spire: all the whorls above the aperture.

Striae: surface features that are either indented or raised in the shell surface.

Succiniform: Shell shape: shell that is higher than wide with a very large aperture (mouth). The spire is generally brief and the body whorl very expanded.

Suture: indentation of the shell surface where two whorls meet.

Taiga: also known as boreal forest, is a biome characterized by coniferous forests consisting mostly of pines, spruces and larches

Tentacles: elongated, flexible organs on the head of snails used for feeling or sensing light. Geophile snails have two pairs of tentacles, with an eye at the tip of each tentacle of the larger, upper pair. Snails in the genus Carychidae and the freshwater snails have one pair of tentacles with an eye at the base of each tentacle.

Tightly coiled: having many, narrowly expanding whorls.

Tooth: A hard, calcareous nodule or projection in or around the aperture of the shell of many snail species; usually of diagnostic value in snail taxonomy.

Transverse striae: surface features indented or raised in the shell surface running perpendicular with the whorls.

Truncate: cut off at the end; terminating abruptly; ending in a transverse line.

Umbilicate: an opening or cavity at the base of a gastropod shell, with the opening more than a narrow perforation.

Umbilicus: an opening in the center of the columella or axis of the shell.

Upper palatal lamella: uppermost of the two major teeth often found on the palatal wall.

Whorl: one complete turn of a gastropod shell.

Xeric: containing very little moisture; dry.

BIBLIOGRAPHY

ABBOTT, R.T. 1989. Compendium of Landshells. American Malacologists, Inc. Melbourne, Florida. 240 pp.

ANDERSON, R.C. & A.K. PRESTWOOD. 1981. Lungworms. Pp. 266-317 in W. R. Davidson, F. A. Hayes, V. F. Nettles, and F.E. Kellogg, eds., Diseases and parasites of white-tailed deer. Miscellaneous Publication. No. 7. Tall Timbers Research Station. Tallahassee, FL. 458 pp.

ARTHUR , M. A., L. M. TRITTON, & T. J. FAHEY. 1993. Dead bole mass and nutrients remaining 23 years after clear-cutting of a northern hardwood forest. Canadian Journal of Forest Reserves 23:1298-1305.

ATKINSON, J. W. 1998. Food manipulation and transport by a carnivorous land snail, *Haplotrema concavum.* Invertebrate Biology 117(2):109-113.

BONATO, V., K. G. FACURE, & W. UIEDA. 2004. Food habits of bats of subfamily Vampyrinae in Brazil. Journal of Mammology 708-713.

BOYCOTT, A. E. 1934. The habitats of land mollusca in Britain. Journal of Ecology 22:1-38.

BRANSON, B. A. 1968. Two new slugs from Kentucky and Virginia. The Nautilus 81:127-133.

BRANSON, B. A. 1973. Kentucky land mollusca: checklist, distribution, and keys for identification. Kentucky Department of Fish and Wildlife Resources. Frankfort, KY. 67 pp.

BRANSON, B. A. & D. L. BATCH. 1988. Distribution of Kentucky land snails (Mollusca: Gastropoda). Transactions of the Kentucky Academy of Science 49:101-116.

BRAUN, E. L. 1940. An ecological transect of Black Mountain, Kentucky. Ecological Monographs 10:193-241.

BURCH, J. B. 1955. Some ecological factors of the soil affecting the distribution and abundance of land snails in eastern Virginia. Nautilus 69:26-29.

BURCH, J. B. 1962. How to know the eastern land snails. Dubuque: William C. Brown Company.

BURCH, J. B. 1969. Land mollusks of the southern Appalachians. In: the distributional history of the biota of the southern Appalachians. Part I: Invertebrates. D. C. Holt, ed. VPI State University. Blacksburg, Virginia. Pgs. 247-264.

BURCH, J. B. & T. A. PEARCE. 1990. Terrestrial Gastropoda. Pp. 201-309 in D. L. Dindall (ed.), Soil Biology Guide. Wiley and Sons, Inc., New Jersey

CAMERON, R. A. D. 1970. Differences in the distributions of three species Helicid snails in the limestone district of Derbyshire. Proceedings of the Royal Society of London 176:130-159.

CHASE, R. & K. C. BLANCHARD. (2006) The snail's love-dart delivers mucus to increase paternity. Proceedings of the Royal Society Biology. 273:1471-1475.

CONEY, C. C., W. A. TARPLEY, J. C. WARDEN, & J. W. NAGEL. 1982. Ecological studies of land snails in the Hiwassee River basin of Tennessee, U.S.A. Malacological Review 15:69-106.

DALLINGER, R. 1993. Strategies of metal detoxification in terrestrial invertebrates. *in*: Dallinger, R. and Rainbow, P. S. (eds.). Ecotoxicology of Metals in Invertebrates. Lewis Publishers. Boca Raton, Florida. Pp. 245-289.

DALLINGER, R., & A. W. WIESER. 1984. Patterns of accumulation, distribution and liberation of Zn, Cu, Cd and Pb in different organs of the land snail Helix pomatia L. Comparative Biochemistry and Physiology-Part C: Toxicity and Pharmacology. 79 (1): 117-124.

DALLINGER, R. & A. W. WEISER. 1984a. Molecular fractionation of Zn, Cu, Cd, and Pb in the midgut gland of *Helix pomatia* L. Comparative Biochemistry and Physiology. 79:125-129.

DALLINGER, R., B. BERGER, C. GRUBER, P. HUNZIKER & S. STUZENBAUM. 2000. Metallothioneins in terrestrial invertebrates: structural aspects, biological significance, and implications for their use as biomarkers. Cellular and Molecular Biology 46:331-346.

DOUGLAS, D. A. 2011. Land snail species diversity and composition among different disturbance regimes in central and eastern Kentucky forests. (MS thesis). Eastern Kentucky University. Richmond, KY.

DOUGLAS, D. A, D. DOURSON, & R. C. CALDWELL. 2013. The land snails of White Oak Sinks, Great Smoky Mountains National Park. Southeastern Naturalist.

DOURSON, D. 2007. A selected land snail compilation of the Central Knobstone Escarpment on Furnace Mountain in Powell County Kentucky, USA. Journal of the Kentucky Academy of Science 68(2):119-131.

DOURSON, D. 2007. Survey Protocol for Cheat Threetooth (*Triodopsis platysayoides*) (Revised). West Virginia Department of Natural Resources, United States Fish and Wildlife Service.

DOURSON, D. 2008. The feeding behavior and diet of an endemic West Virginia land snail, *Triodopsis platysayoides*. American Malacological Bulletin. 26:153-159

DOURSON, D. 2009. A natural history of the Bladen Nature Reserve and its gastropods. 148p. Goatslug Publications, Bakersville, North Carolina.

DOURSON, D. 2010. Kentucky's land Snails and their ecological communities. 298 pp. Goatslug Publications, Bakersville, NC.

DOURSON, D. 2012. Four new land snail species from the southern Appalachian Mountains. Journal of the North Carolina Academy of Science 128(1):1-10.

DOURSON, D. & J. BEVERLY. 2009. Diversity, substrata divisions and biogeographical affinities of land snails at Bad Branch State Nature Preserve, Letcher County Kentucky, USA. Journal of the Kentucky Academy of Science. 68(2):119-131.

DOURSON, D. & M. GUMBERT. 2004. A survey of terrestrial mollusca in selected areas of the Great Smoky Mountains National Park, North Shore Road Project. Report submitted to Arcadis G & M of North Carolina, Inc. 76 pp.

DOURSON, D. & K. LANGDON. 2012. Land snails of selected rare high elevation forests and heath balds of the Great Smoky Mountains National Park. Journal of North Carolina Academy of Science: Summer 2012, Vol. 128, No. 2, pp. 27-32.

EMBERTON, K. C. 1991. The genitalic, allozymic and conchological evolution of the tribe Mesodontini (Pulmonata: Stylommatophora: Polygyridae). Malacologia 33:71-178.

EMBERTON, K. C. 1994. Polygyrid land snail phylogeny: external sperm exchange, early North American biogeography, iterative shell evolution. Biological Journal of the Linnean Society 52:241-271.

FAIRBANKS, H. L. 1998. Clarification of the taxonomic status and reproductive anatomy of *Philomycus batchi* Branson, 1968. The Nautilus 112(1):1-5.

FOOTE, B. A. 1959. Biology and life history of the snail-killing flies belonging to the genus Sciomyza. Annals of the Entomological Society of America. 52:31-32.

FOURNIE, J. & M. CHETAIL. 1984. Calcium dynamics in land gastropods. American Zoologist 24:857-870.

GEIGER, R. 1965. The climate near the ground. Harvard University Press. Cambridge.

GETZ, L. L. 1974. Species diversity of terrestrial snails in the Great Smoky Mountains. The Nautilus 88:6-9.

GOSZ, J. R., G.E. LIKENS, & F. H. BORMANN. 1973. Nutrient release from decomposing leaf and branch litter in the Hubbard Brook Forest, New Hampshire. Ecological Monographs 43:173-191.

GRAVELAND, J. R. 1996. Avian eggshell formation in calcium-rich and calcium-poor habitats: importance of snail shells and anthropogenic calcium sources. Canadian Journal of Zoology 74:1035-1044.

GRAVELAND, J. R., & R. VAN DER WAL, J. H. 1994. Poor reproduction in forest passerines from decline of snail abundance on acidified soils. Nature 368:446-448.

GRIMM, F. W., R. G. FORSYTH, F. W. SCHUELER, & A. KARSTAD. 2009. Identifying land snails and slugs in Canada: introduced species and native genera. Canadian Food Inspection Agency. Ontario, Canada. 168 pp.

GUHA, M. M. & R. L. MITCHELL. 1966. The trace and major element composition of some deciduous trees: II. Seasonal changes. Plant and Soil 24:90-112.

HAMES, R. S., K. V. ROSENBERG, J. D. LOWE, & S. E. DHONDT. 2002. Adverse affects of acid rain on the (Wareborn, 1992) distribution of the woodthrush, *Hylocichla mustelina*, in North America. Proceedings of the National Academy of Science. 99 (17): 11235-11240

HICKMAN, C. P., L. S. ROBERTS & A. LARSON. 2003. Animal diversity. 3rd ed. McGraw-Hill, NY.

HUBRICIIT, L. 1964. Land snails from the caves of Kentucky, Tennessee, and Alabama. Bulletin of the National Speleological Society. 26 (1):33-36.

HUBRICHT, L. 1985. The distributions of the Native Land Mollusks of the Eastern United States. Fieldiana Zoology New Series, No. 24, Publication 1359: Field Museum of Natural History, Chicago, Illinois. 191 pp.

HUBRICHT L., R. S. CALDWELL, & J. G. PETRANKA. 1983. *Vitrinizonites latissimus* (Pulmonata: Zonitidae) and *Vertigo clappi* (Pupillidae) from eastern Kentucky. The Nautilus 97:20-22.

JACOT, A. P. 1935. Molluscan populations of old growth forests and re-wooded fields in the Asheville Basin of North Carolina. Ecology 16:603-605.

JENKINS, M. A., S. JOSE, & P. S. WHITE. 2007. Impacts of an exotic disease and vegetation change on foliar calcium cycling in Appalachian forests. Ecological Applications 17(3):869-881.

KALISZ, P. J. & J. E. POWELL, 2003. Effect of calcareous road dust on land snails and millipedes in acid forest soils of the Daniel Boone National Forest. Forest Ecology and Management 186:177-183.

KARLIN, E. J. 1961. Ecological relationships between vegetation in the distribution of land snails in Montana, Colorado, and New Mexico. American Midland Naturalist 65:60-66.

KELLER, H. & K. SNELL 2002. Feeding activities of slugs on Myxomycetes and macrofungi. Mycologia, 94:757-760.

KOENE J. M, T. S. LIEW, K. MONTAGNE-WAJER, M. SCHILTHUIZEN 2013. A syringe-like love dart injects male accessory gland products in a tropical hermaphrodite. PLoS ONE 8(7): e69968. doi:10.1371/journal.pone.0069968

LEE, J. C. 1994. The amphibians and reptiles of the Yucatán Peninsula. Cornell University Press, Ithaca /. London.

LOUV, R. 2005. Last child in the woods: Saving our children from nature-deficit disorder. Algonquin Books of Chapel Hill, North Carolina.

LUTZ, H. J., & R. F. CHANDLER. 1946. Forest soils. Wiley and Sons, Inc. N. Y. 514 p.

MACHENSTED, U. & K. MARKEL. 2001. Chapter 4: Radular structure and function. Pp. 213-236. In: The biology of terrestrial mollusks. G. M Barker, ed. CABI. New York, New York.

MARTIN, A. C., H. S. ZIM, & A. L. NELSON. 1951. American wildlife and plants: A guide to wildlife food habits. McGraw-Hill, Inc., New York.

MCHARGUE, J. S. & W. R. ROY. 1932. Mineral and nitrogen content of the leaves of some forest trees at different times in the growing season. Botanical Gazette 94:381-393.

MORITZ, C., K. S. RICHARDSON, S. FERRIER, J. STANISIC, S. E. WIL-LIAMS, & T. WHIFFIN. 2001. Biogeographic concordance and efficiency of taxon indicators for establishing conservation priority in a tropical rainforest biota. Proceedings of the Royal Society 268:1875 -1881.

NATION, R. 2005. The influence of soil calcium on land snail diversity in the Blue Ridge Escarpment of South Carolina. Dissertation presented to Clemson University. Clemson, South Carolina.

NATURESERVE. 2019. NatureServe Explorer: An online encyclopedia of life [web application]. Version 7.0. NatureServe, Arlington, VA. U.S.A. Available http://explorer.natureserve.org. (Accessed: 8/28/19)

NEKOLA, J. C. 1999. Terrestrial gastropod richness of carbonate cliff and associated habitats in the Great Lakes region of North America. Malacologia 41(1):231-252.

NEKOLA, J. C. 2003. Large-scale terrestrial gastropod community composition patterns in the Great Lakes region of North America. Diversity and Distribution. 9:55-71.

NEKOLA, J. C. & B. F. COLES. 2010. Pupillid land snails of eastern North America. American Malacological Bulletin. 28:29-57.

NICKLAS N.L. & R. J. HOFFMAN. 1981. Apomictic parthenogenesis in a hermaphroditic terrestrial slug, Deroceras laeve (Müller). Biological Bulletin. 160:123-135.

PATTERSON, C. M. & J. B. BURCH. 1966. The chromosome cycle in the land snail *Catinella vermeta* (Stylommatophora: Succineidae) Malacologia. 3:309.

PEARCE, T. A. 2008. When a snail dies in the forest, how long will the shell persist? Effect of dissolution and micro-bioerosion. American Malacological Bulletin 26:111-117.

PEARCE, T. A. & A. GAERTNER. 1996. Optimal foraging and mucus-trail following in the snail-eating snail *Haplotrema concavum* (Gastropoda: Pulmonata). *Malacological Review* 29:85-99.

PETRANKA, J. G. 1982. The distribution and diversity of land snails on Big Black Mountain, Kentucky. (A thesis). University of Kentucky. Lexington, KY.

PETRANKA, J. W. 1998. Salamanders of the United States and Canada. Washington, D. C., USA: Smithsonian Institution Press.

PILSBRY, H. A. 1940. Land mollusca of North America (north of Mexico), Vol. I. Part 2. The Academy of Natural Sciences of Philadephia Monographs.

PILSBRY, H. A. 1946. Land mollusca of North America (north of Mexico), Vol. II. Part 1. The Academy of Natural Sciences of Philadephia Monographs.

PILSBRY, H. A. 1948. Land mollusca of North America (north of Mexico), Vol. I. Part 2. The Academy of Natural Sciences of Philadephia Monographs.

POLLARD, E. 1975. Aspects of the ecology of *Helix pomatia* L. Journal of Animal Ecology 44:305-329.

POTTER, C. S., H. L. RAGSDALE, & C. W. BERISH. 1987. Resporption of foliar nutrients in a regenerating southern Appalachian forest. Oecologia 73:268-271.

RAWLS, H. & R. YATES. 1971. Fluorescence in Endodontid Snails. The Nautilus 85:17-20.

REID, F. A. 2006. Mammals of North America. Peterson Field Guide. New York: Houghton-Mifflin.

RICHTER, K. O. 1980. Evolutionary aspects of mycophagy in Ariolimax columbianus and other slugs. In: D. L. Dindal, ed., Soil Ecology as Related to Land Use Practices. Proceedings of the VII International Colloquium of Soil Biology, Washington D. C.

RICKLEFS, R. E. & K. K. MATTHEWS. 1982. Chemical characteristics of the foliage of some deciduous trees in southeastern Ontario. Canadian Journal of Botany 60:2037-2045.

RIMMER, C. C., K. P. MCFARLAND, D. C. EVERS, E. K. MILLER, Y. AUBRY, D. BUSBY & R. J. TAYLOR. 2005. Mercury concentrations in Bicknell's Thrush and other insectivorous passerines in montane forests of northeastern North America. Ecotoxicology 14:223-240.

ROODY, W. C. 2003. Mushrooms of West Virginia and the Central Appalachians. The University of Kentucky Press. Lexington, Kentucky 519 p.

SATHEESHKUMAR, P., A. B. KHAN, & D. SENTHILKUMAR. 2010. Marine organisms as potential supply for drug finding-a review study. Middle-East Journal of Scientific Research 5 (6): 514-519.

SCHEIFLER, R., C. SWARTZ, G. ECHEVARRIA, A. DE VAUFLEURY, P. BADOT, & J. MOREL. 2003. "Nonavailable" soil cadmium is bioavailable to snails: evidence from isotopic dilution experiments. Environmental Science Technology 37(1):81-86.

SHEARER, A. & J. W. ATKINSON. 2001. Comparative analysis of food-finding behavior of an herbivorous and a carnivorous land snail. Invertebrate Biology 120:199-205.

SLAPCINSKY, J. & B. COLES. 2004. Revision of the genus *Pilsbryna* Gastropoda: Pulmonata: Gastrodonatidae) and comments on the taxonomic status of *Paravitrea tridens* Morrison, 1935. The Nautilus 118(2):55-70.

SYMONDSON, W. O. C. 2004. Chapter 2: Coleoptera (Carabidae, Staphylinidae, Lampyridae, Drilidae, and Silphidae) as predators of terrestrial gastropods. pp. 37-84 in *Natural Enemies of Terrestrial Mollusks*. CABI Publishing. Oxford, UK.

THABAH, A. & G. LI, Y. WANG, B. LIANG, L. HU, S. ZHANG, & G. JONES. 2007. Diet, echolocation calls, and phylogenetic affinities of the great evening bat (IA IO; Vespertilionidae): another carnivorous bat. Journal of Mammology. 88(3):728-735.

THOMPSON, F. 2011. An annotated checklist and bibliography of the land and freshwater snails of Mexico and Central America. 903 p. Available: http://www.flmnh.ufl.edu/malacology/mexico-central_america_snail_checklist/ [Online: 16 June 2008.]

TOWNSEND, J. S., H. C. ALDRICH, L. D. WILSON, J. R. MCCRANIE. 2007. First report of sporangia of a myxomycete (*Physarum pusillum*) on the body of a living animal, the lizard *Corytophanes cristatus'*. Mycologia. 97(2):246-248.

VAN DEVENDER, A. S. 1985. Status report on the Miry Ridge Supercoil Paravitrea clappi" Contract # 40181-0536 US Fish & Wildlife Service Endangered Species Office, Asheville, NC 28801, November.

VAN DEVENDER, A. S. & R. W. VAN DEVENDER. 2003. Surveying the land snails of the southern Appalachians. Program and Abstracts of American Malacological Society: 61.

VESTERDAL. I., & K. RAULUND-RASMUSSEN 1998. Forest floor under seven tree species along a soil fertility gradient. Canadian Journal of Forest Reserves 28:1636-1647.

WALLACE, M.S., R. RAUCK, R. S. FISHER, G. CHARAPATA, D. ELLIS, & S. DISSANAYAKE. 2008. Ziconotide 98-022 Study Group. Intrathecal ziconotide for severe chronic pain: safety and tolerability results of an open-label, long-term trial. Anesthesia and Analgesia 106: 628-637.

WAREBORN, I. 1970. Environmental factors influencing the distribution of land mollusks of an oligotrophic area in southern Sweden. Oikos 2:285-291.

WORLD RESOURCES INSTITUTE (WRI). 1992. Global Biodiversity: A Policy Maker's Guide. World Research Institute. International Union for the Conservation of Nature and United Environmental Protection, Washington, D. C.

Index of Scientific Names

A

acerra, Ventridens 143
aeneus, Strobilops 192
albolabris, Neohelix 224
Allogona 302
alternata, Anguispira 161
altispira, Stenotrema 185
altivagus, Mesodon 222
andrewsae, Mesodon 294
andrewsae, Mesomphix 154
andrewsae, Paravitrea 114
Anguispira 160, 259
anteridon, Triodopsis 305
Appalachina 218, 260
appressa, Patera 209
arboreum, Carychium 57
arboreus, Zonitoides 90
arcellus, Ventridens 144
archeri, Fumonelix 296
Arion 313
armifera, Gastrocopta 66
ater, Arion 316
aurea, Pilsbryna 268

B

barbigerum, Stenotrema 290
blandianum, Punctum 80
blarina, Paravitrea 271
bollesiana, Vertigo 74
bonamicus, Helicodiscus 282
bryanti, Discus 123

C

capnodes, Mesomphix 153
capsella, Paravitrea 109
carolinianus, Philomycus 235
caroliniensis, Glyphyalinia 100
Carychium 57
Catinella 259
cherohalaensis, Fumonelix 297
chersinus, Euconulus 173
chilhoweensis, Appalachina 223
christyi, Fumonelix 217
circumscriptus, Arion 316
clappi, Carychium 58
clappi, Paravitrea 111
clarki, Patera 207
clarki nantahala, Patera 291

clausus, Mesodon 200
clingmani, Glyphyalinia 279
clingmanica, Fumonelix 216
Cochlicopa 53
coelaxis, Ventridens 284
cohuttense, Stenotrema 286
collisella, Ventridens 137
Columella 55
concavum, Haplotrema 169
concordialis, Succinea 263
contracta, Gastrocopta 65
corticaria, Gastrocopta 68
cryptomphala, Glyphyalinia 95
cupreus, Mesomphix 155
cumberlandiana, Glyphyalinia 104

D

decussatus, Ventridens 136
demissus, Ventridens 141
denotatum, Xolotrema 230
dentatus, Euconulus 174
dentifera, Neohelix 300
depilatum, Stenotrema 186
Deroceras 253
Discus 120
dorsalis, Pallifera 247

E

edvardsi, Stenotrema 287
elevatus, Mesodon 292
elliotti, Zonitoides 89
Euchemotrema 195
Euconulus 172
excentrica, Vallonia 194
exigua, Striatura 266
exiguum, Carychium 60
exile, Carychium 59

F

fallax, Triodopsis 228
fasciatum, Euchemotrema 198
fasciatus, Arion 314
ferrea, Striatura 84
ferrissi, Inflectarius 206
fimbriatus, Helicodiscus 129
flavus, Limacus 316
flexuolaris, Philomycus 237
fosteri, Pallifera 251

fraternum, *Euchemotrema* 197
fulvus, *Euconulus* 176
Fumonelix 211, 260

G

gagates, *Milax* 316
Gastrocopta 63
Gastrodonta 88
Glyphyalinia 93, 259
gouldii, *Vertigo* 75
gularis, *Ventridens* 133
Guppya 87

H

haliotidea, *Testacella* 316
Haplotrema 168
Hawaiia 85
Helicodiscus 120, 259
hemphilli, *Pallifera* 249
Hendersonia 172
hexodon, *Helicodiscus* 124
hirsutum, *Stenotrema* 183
hopetonensis, *Triodopsis* 227

I

indentata, *Glyphyalinia* 101
Inflectarius 195, 202, 260
inflectus, *Inflectarius* 203
inornatus, *Mesomphix* 283
intermedius, *Arion* 316
interna, *Gastrodonta* 92
intertextus, *Ventridens* 142

J

jessica, *Anguispira* 166
jonesiana, *Fumonelix* 213
junaluskana, *Glyphyalinia* 97

K

kendeighi, *Haplotrema* 170
knoxensis, *Anguispira* 163

L

labyrinthicus, *Strobilops* 285
lacteodens, *Paravitrea* 275
laeve, *Deroceras* 253
lamellidens, *Paravitrea* 115
langdoni, *Fumonelix* 212

lapidaria, *Pomatiopsis* 52
lasmodon, *Ventridens* 140
latior, *Mesomphix* 151
latior form monticola, *Mesomphix* 152
latissimus, *Vitrinizonites* 171
lawae, *Praticolella* 196
lawae, *Ventridens* 139
Lehmannia 316
ligera, *Ventridens* 145
Limacus 316
Limax 314
lubricella, *Cochlicopa* 54
Lucilla 77, 259

M

magnifumosum, *Stenotrema* 184
major, *Neohelix* 301
maximus, *Limax* 315
Megapallifera 233
meridionalis, *Striatura* 83
Mesodon 195, 218, 260
Mesomphix 147, 259
Milax 316
milium, *Vertigo* 70
Millerelix 191
minuscula, *Hawaiia* 85
minutissimum, *Punctum* 78
mordax, *Anguispira* 164
morseana, *Cochlicopa* 53
mutabilis, *Megapallifera* 243
multidens, *Helicodiscus* 125
multidentata, *Paravitrea* 116

N

nannodes, *Carychium* 61
Neohelix 218, 260
nigrimontanus, *Discus* 122
nodopalma, *Pilsbryna* 269
normalis, *Mesodon* 221
notius, *Helicodiscus* 127
Novisuccinea 51

O

occulta, *Hendersonia* 177
ocoae, *Glyphyalinia* 278
ohioensis, *Pallifera* 311
oklahomarum, *Mediappendix* 261
orestes, *Fumonelix* 298

oscariana, Vertigo 71
ovalis, Novisuccinea 51
ovata, Vertigo 69

P

pallens, Boettgerilla 316
Pallifera 233, 260
panormitanum, Deroceras 316
parallelus, Helicodiscus 128
Paravitrea 107, 259
parvula, Vertigo 73
Patera 207, 260
patuloides, Zonitoides 91
patulus, Discus 121
pendula, Triodopsis 306
pentadelphia, Glyphyalinia 105
pentodon, Gastrocopta 67
perigrapta, Patera 210
perlaevis, Mesomphix 149
petrophila, Paravitrea 110
Philomycus 233, 260
pilsbryi, Ventridens 135
pilula, Stenotrema 187
Pilsbryna 259
placentula, Paravitrea 112
plicata, Millerelix 193
Pomatiopsis 52
praecox, Glyphyalinia 96
Praticolella 195
profunda, Allogona 302
Punctum 77

Q

quadrilamellata, Pilsbryna 270

R

reesei, Paravitrea 274
reticulatum, Deroceras 316
rhoadsi, Glyphyalinia 102
roanensis, Fumonelix 299
rugeli, Inflectarius 204
rugeli, Mesomphix 157

S

sayana, Appalachina 303
scintilla, Lucilla 86
sculptilis, Glyphyalinia 99
secreta, Pallifera 245

simplex, Columella 55
singleyana, Lucilla 265
smithi, Punctum 81
solida, Glyphyalinia 277
spinosum, Stenotrema 289
Stenotrema 179, 260
stenotrema, Stenotrema 181
sterkii, Guppya 87
Striatura 77, 259
Strobilops 191, 260
strongylodes, Anguispira 281
subfuscus, Arion 313
subpalliatus, Inflectarius 295
subplanus, Mesomphix 158
Succinea 259
suppressus magnidens, Ventridens 134

T

tennesseensis, Triodopsis 307
ternaria, Paravitrea 274
theloides, Ventridens 138
thyroidus, Mesodon 201
togatus, Philomycus 239
tridens, Paravitrea 273
tridentata, Triodopsis 229
tridentata, Vertigo 72
Triodopsis 225, 260
trochulus, Euconulus 175

U

umbilicaris, Paravitrea 117
umbilicaris dentatus, Paravitrea 118

V

valentiana, Lehmannia 316
Vallonia 191
vanattai, Pilsbryna 267
variabilis, Paravitrea 113
varidens, Paravitrea 272
Ventridens 131, 259
venustus, Philomycus 241
vermeta, Catinella 262
Vertigo 63
verus, Inflectarius 205
virginicus, Philomycus 309
Vitrinzonites 168

vulgata, Triodopsis 226
vulgatus, Mesomphix 150

W

wetherbyi, Fumonelix 214
wheatleyi, Fumonelix 215
wheatleyi, Glyphyalinia 103

X

Xolotrema 225

Z

zaletus, Mesodon 219
Zonitoides 88

Index of Common Names

A

Appalachian gloss 91
Appalachian pillar 53
Appalachian scrubsnail 196
Appalachian slitmouth 184
Appalachian thorn 58
Appalachian tigersnail 165

B

Balsam globe 294
Bark snaggletooth 68
Blade vertigo 70
Black mantleslug 249
Black Mountain disc 122
Black striae 84
Blue-foot lancetooth 169
Blue-gray glyph 278
Bidentate dome 284
Big-tooth covert 213
Big-tooth whitelip 300
Bottleneck snaggletooth 65
Bright glyph 103
Brilliant glyph 96
Brilliant granule 87
Broad-banded forestsnail 302
Broad button 151
Brown-banded arion 315
Bronze pinecone 192
Brown bellytooth 92
Brown hive 176
Brown spot 50
Brown-spotted mantleslug 241
Budded threetooth 307

C

Capital vertigo 71
Carinate slitmouth 289

Carolina mantleslug 235
Carter threetooth 305
Carved glyph 101
Changeable mantleslug 243
Cherokee supercoil 110
Cherrystone drop 177
Cinnamon covert 215
Clifty covert 214
Cohutta slitmouth 286
Comb snaggletooth 67
Common button 150
Compound coil 128
Copper button 155
Copper dome 138
Corneous dome 134
Cove slitmouth 182
Crossed dome 136
Cumberland liptooth 193

D

Deep-tooth shagreen 204
Delicate vertigo 74
Dentate supercoil 116
Detritus ambersnail 261
Dimple supercoil 109
Dished threetooth 226
Domed disc 121
Dusky arion 313
Dusky button 153
Dwarf globelet 199
Dwarf proud globe 208

E

Engraved bladetooth 209
Engraved covert 298

F

File thorn 61
Flamed tigersnail 161
Flat bladetooth 209
Flat button 158
Foster mantleslug 251
Four blade bud 270
Fragile glyph 279
Fringed coil 129
Fringed slitmouth 290
Fuzzy covert 205

G

Glass spot 79
Glassy grapeskin 170
Globose broad button 152
Globose dome 145
Glossy covert 217
Glossy dome 143
Glossy supercoil 112
Golden dome 144
Grand globe 221
Gray-foot lancetooth 167
Great Smoky slitmouth 186
Green gloss 89

H

Hairy slitmouth 183
Hanging Rock threetooth 306
Highland slitmouth 185
High mountain supercoil 114
Hill glyph 104
Hollow dome 140
Honey bud 267
Honey vertigo 72

I

Ice thorn 59
Inland slitmouth 181
Imperforate glyph 277
Iroquois vallonia 194

L

Lamellate spot 81
Lamellate supercoil 115
Light glyph 97

M

Magnolia threetooth 227
Maze pinecone 285
Meadow slug 253
Median striate 83
Mimic threetooth 224
Minute gem 85
Mirey Ridge supercoil 111
Mountain button 154
Mountain pillsnail 198
Mountain tigersnail 164

N

Northern threetooth 229
Noonday globe 291

O

Oartooth bud 269
Obese thorn 60
Ocoee covert 296
Oldfield coil 86
Open supercoil 117
Orange-banded arion 315
Ornate bud 268
Oval ambersnail 51
Ovate vertigo 69

P

Pale mantleslug 247
Perforate dome 141
Pink glyph 105
Plain button 283
Proud globe 292
Pygmy slitmouth 187
Pyramid dome 142

Q

Queen crater 223
Quick gloss 90

R

Ramp cove supercoil 275
Redfoot mantleslug 311
Ribbed striate 265
Ridge-and-valley slitmouth 287
Roan Mountain supercoil 299
Roan supercoil 272

Rock-loving covert 297
Round supercoil 276
Rounded dome 139
Rustic tigersnail 162

S

Sawtooth disc 123
Sculpted glyph 102
Sculpted supercoil 274
Sculptured dome 137
Severed mantleslug 245
Shagreen 203
Shrew supercoil 271
Silk hive 175
Slender walker 52
Smallmouth vertigo 73
Small spot 78
Smoky Mountain covert 204
Smooth button 149
Smooth coil 265
Southeastern tigersnail 281
Southeastern whitelip 301
Spike-lip crater 303
Spiral coil 282
Spiral mountain glyph 100
Spotted ambersnail 263
Suborb glyph 99
Suborb ambersnail 262
Summit covert 216

T

Talus covert 212
Thin glyph 95
Thin pillar 54
Throaty dome
Tiger slug 314
Toothed supercoil 118
Toothless hive 174
Toothless pupa 55
Tree thorn 57
Toothed globe 219
Toothy coil 124
Twilight coil 125

U

Upland pillsnail 197

V

Variable mantleslug 239
Variable supercoil 113
Variable vertigo 75
Varnish button 156
Velvet covert 295
Velvet wedge 230
Virginia mantleslug 309

W

Wandering globe 222
White foot supercoil 273
Whitelip 224
White-lip globe 201
Wild hive
Winding mantleslug 237
Wrinkled button 157

Y

Yellow dome 135
Yellow globelet 200

ABOUT THE AUTHOR

Dan Dourson is a biologist/naturalist/illustrator who has spent most of his adult life dedicated to the preservation, conservation, and understanding of the planet's more obscure species. An employee of the United States Forest Service for nearly 20 years, during his tenure he worked to manage and conserve snails, mussels, bats, reptiles, and amphibians. In addition, Dan has conducted biological inventories for multiple agencies and has worked for more than a decade studying land snails in the Great Smoky Mountains National Park as part of the All Taxa Biodiversity Inventory through the Discover Life in America initiative. He served as staff biologist for Belize Foundation for Research and Environmental Education in Belize, Central America for seven years. Dan has authored and illustrated nearly a dozen books including the recent *Biodiversity of the Maya Mountains, Belize, Central America*, *Kentucky's Land Snails and their Ecological Communities, Land Snails of West Virginia, Land Snails of Belize, Reptiles and Amphibians of the Red River Gorge and the Greater Red River Basin, Kentucky,* and *Wildflowers and Ferns of the Red River Gorge and the Greater Red River Basin, Kentucky.* Dan remains committed to conservation work protecting the earth's most amazing and underappreciated organisms. His allegiance with the natural world is clearly reflected through his writing and simple lifestyle. He divides his time between the USA and Belize with his wife, Judy.

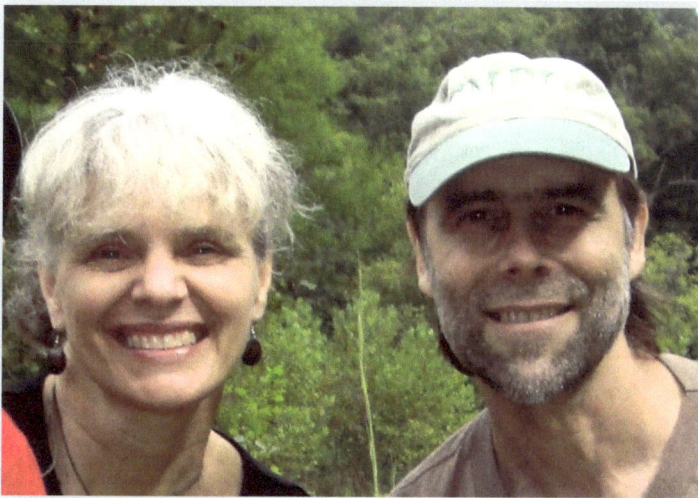

www.ingramcontent.com/pod-product-compliance
Lightning Source LLC
Chambersburg PA
CBHW040928030426
42334CB00002B/2